Long-wave Optics

Long-wave Optics

The Science and Technology of Infrared and Near-millimetre Waves

GEORGE W. CHANTRY

National Physical Laboratory, Teddington, Middlesex, United Kingdom

Volume 2: Applications

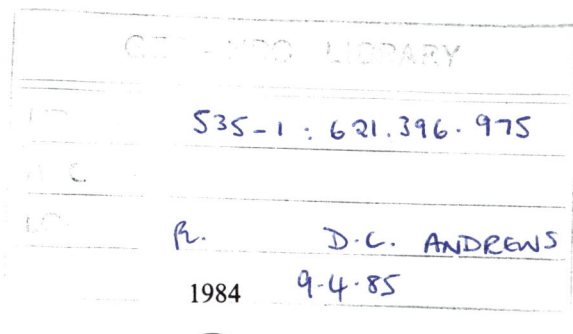

ACADEMIC PRESS

(Harcourt Brace Jovanovich, Publishers)

London Orlando San Diego San Francisco New York
Toronto Montreal Sydney Tokyo São Paulo

ACADEMIC PRESS INC. (LONDON) LTD
24/28 Oval Road,
London NW1

United States Edition published by
ACADEMIC PRESS INC.
(Harcourt Brace Jovanovich, Inc.)
Orlando, Florida 32887

Copyright © 1984 by
ACADEMIC PRESS INC. (LONDON) LTD

British Library Cataloguing in Publication Data
Chantry, G. W.
  Long-wave optics.
  Vol. 2: Applications
  1. Infra-red rays
  I. Title
  535′.012   QC547

  ISBN 0-12-168102-5

Typeset by Macmillan India Ltd., Bangalore
and printed in Great Britain by
Thomson Litho Ltd., East Kilbride, Scotland

To my wife Diana and my children Richard, Catherine and Paul whose vigorous protests at being deprived of my time and company did much to expedite the completion of this book.

Preface

Infrared science and technology is far from being a new field of endeavour, indeed it has a venerable history stretching back almost to the middle of the last century. What is new is the integration of the infrared into the general area of electromagnetic physics and engineering and the deployment of infrared devices into a bewildering and rapidly increasing number of technical areas. As some token of this, one could say with fair accuracy that nearly all the significant work in the field has been done in the last twenty years or so and that most of the books devoted to the art, written more than ten years or so ago, are quite out of date. This state of affairs is true, of course, of technology as a whole and some areas, such as semi-conductor device fabrication, have had even shorter periods before obsolescence set in, but the rate of change in infrared technology is faster than most.

This rapid rate of advance has been fired undoubtedly by the evident fact that infrared devices provide the best solutions for a huge range of technically important problems, but this "market need" would have gone unrequited were it not for the "technology push" provided by the development of some radically new components. Principal amongst these were infrared lasers and infrared detectors with sensitivities approaching the fundamental limits. The lasers gave the infrared engineer coherent radiation and opened up to him all the possibilities which had been enjoyed for so long by his microwave colleagues. The detectors made possible new applications of incoherent radiation and the interaction between these two areas, by a "knock-on" effect, led to the opening up of quite novel areas, for example ultra-high resolution spectroscopy and high-definition infrared imaging. Infrared devices are now used almost across the board: examples would include, analysis, pollution monitoring, imaging, movement sensing, espionage, medicine, machining, materials fabrication, thermonuclear research, telecommunications and warfare. In this latter, heat-seeking missiles have repeatedly proven that they are the most deadly and effective air-to-air weapons available.

The subject is now clearly far too large to be covered in adequate depth in a single volume and such a treatment would anyway be beyond the reach of a single author no matter how diligent. What is attempted here is an overview which hopefully covers all the important topics in sufficient depth to enable the interested reader to continue his or her reading in the learned literature. To this end the book includes a general bibliography and also a particular bibliography which includes over 1500 references.

In writing the book I have received help and information from a huge number of people spread all over the world. To list them all would be an impossible task. All I can do is to mention those who have contributed individual points. I therefore thank Mohammed Afsar, John Baker, Andre Bellemans, Tom Blaney, David Buckingham, Ken Button, James Calderwood, the late John Chamberlain, Alan Costley, Mike Cudby, Myron Evans, John Fleming, Werner Frank, Alistair Gebbie, Ludwig Genzel, Bob Gott, Jose Goulon, Armand Hadni, John Harries, Dean Hodges, Harry Jones, Udo Kaatze, Fritz Kneubuhl, David Knight, Derek Martin, Jim McConnell, Elisabeth Nicol, Takeshi Oka, Terry Parker, Roger Partridge, Carl Pidgeon, Reinhard Pottel, Ernest Putley, Jean Louis Rivail, Joseph Sattler, George Simonis, Des Smith, the late Robin Smith, Tony Stradling, Wally Stone, Mary Tobin, Brian Walker, Harry Willis and David Whiffen. Especially though I would like to thank Dr James R. Birch of the National Physical Laboratory and Professor George R. Wilkinson of King's College London who read the manuscript critically in its early drafts and made countless suggestions for improvements and for the removal of obscurities.

December 1983   G. W. CHANTRY

Contents

Contents of Volume 1
Principles

Chapter 5
Spectroscopy at Optical and Longer Wavelengths

5.1 Atomic spectroscopy

5.1.1 *Introduction*

Atomic spectroscopy began with the classic work of Bunsen and Kirchoff who, in the 1860s, proved that the brilliant colours, characteristically produced when many inorganic salts are vapourised into hydrocarbon flames, result from the sharp-line emission spectra of the corresponding atoms or ions. They also showed that when the continuous radiation from a hot thermal source was passed through a vapour containing the atoms in question, some, but not all, of the lines could be observed in absorption. Emission spectroscopy quickly established itself as an essential tool for elemental analysis and it soon had some spectacular successes, such as Lockyer's discovery of helium in the sun, but atomic absorption spectroscopy had to wait until the second half of this century when, following the pioneering work of Walsh [446] in Australia, it too established itself as an essential weapon in the analyst's armoury.

The emission spectrum of an atom is complex because the highly excited atom (often in fact an ion) which is produced in the hot flame or in a discharge, will, on relaxing back to its ground state, pass, by radiative cascades, through very many excited states and thus generate a very large number of lines. In absorption, on the other hand, only the lines which terminate on the ground-state can be observed with measurable intensity since, at ambient temperatures, the population of the electronically excited states will be negligible because of the very small Boltzmann factors. The strongest of the few lines which are observed in absorption are usually called *resonance lines* because the ground-state atoms (by far the largest part of the total population) are in resonance with the radiation field at these frequencies. The yellow "D" lines of sodium at 588·995 and 589·592 nm are well known examples. Resonance radiation [139] propagates very slowly through the atomic vapour because the chance of photon emission or absorption is very high—approaching unity. A photon which is emitted by one atom is therefore promptly reabsorbed by the next and this, in its turn, quickly re-emits the photon and so on. The photons therefore diffuse through the vapour in a random, "drunkards walk" or Monte Carlo trajectory with a mean free path roughly equal to the average interatomic separation in the vapour. The net effect of this very strong coupling between the electromagnetic field and the vapour is that this latter develops a macroscopic polarisation which is oscillating at the resonance frequency.

The emitted resonance line is usually broader than the absorption line because the source will usually be hotter than the absorbing vapour and the Doppler broadening will be greater. When resonance radiation is therefore incident on cool vapour one can get the very interesting phenomen of *self-reversal*, that is the radiation leaving the specimen shows a spectrum in which the resonance line appears as a close doublet—its centre having been reduced in intensity very much more than its wings. Michelson did some elegant work on the shapes of self-reversed lines [64]. His interferometer was almost the ideal instrument for this type of measurement at that time, though nowadays long path length, high-finesse Fabry–Perot etalons would be preferred. Atomic absorption is a very sensitive technique and is nowadays perhaps even more important than the emission technique, but of necessity one must be careful to avoid self-reversed sources since otherwise there will be little absorption in the specimen and the sensitivity will be lost. Ways of confining an arc discharge to be almost coincident with the exit window of the lamp [447] so that there will be the very minimum of cold absorbing path length have been developed and now it is possible to buy sources for virtually every element of analytical significance. The substance being analysed is usually vapourised in a flame to provide the free atoms to do the absorption.

The development of the laser has seen a renewed interest in atomic emission spectroscopy as a subject in its own right. This is especially true of the long-wave infrared transitions which, lying as they do beyond the reach of the photographic plate, were previously rather poorly characterised. Many of the most important laser lines arise in atoms and since they frequently involve transitions between excited states they tend to be at longer wavelengths. The reason for this is that the total range of energies for neutral moderately heavy atoms may be only of the order of 10 eV or less and that the first, or resonance, transition may well use up most of this. All the remaining levels may thus be squeezed into a band only 2 eV or so across and all the transitions within this band will, of necessity, lie in the red or infrared region of the spectrum. Second ionization energies are usually larger than first so the laser lines from ionised atoms can lie at shorter wavelengths. Thus the best blue and green laser lines come from argon and krypton ions, A^+ and Kr^+ respectively. Even so, there are relatively few strong laser lines in the blue and ultraviolet whereas there are very many infrared laser lines. Lines from neon are known out to beyond 100 μm and lines from helium have been found at 95·8 and 216·3 μm. The neodymium line at 1·063 μm is one of the most intense laser lines known and it, together with the line at 1·315 μm from atomic iodine, are currently the best candidates for the driver of a laser fusion machine [448].

Welcome though this resurgence of interest in atomic spectroscopy is, there still remains the old problem that we have no way of calculating the energy levels of complex atoms and, in the more complicated cases, no way of even describing them properly. For the majority all one can do is to assemble a diagram, sometimes called a Grotrian diagram, in which the transitions are merely indicated. Some progress has been made using numerical methods in high-speed computers and from these, as expected, one finds that the infrared transitions, arising as they do from fairly excited states, offer the best chance of description and assignment. The magnitude of the task, however, can be gauged from the fact that of the huge number of lines revealed by the Connes [449] in the high-resolution infrared spectra of heavy atoms such as holmium, the majority have yet to be assigned. With the lighter atoms, however, much more progress can be made and a brief account of this will be given in the following sections.

5.1.2 Hydrogen-like spectra

The tantalising patterns of lines in atomic spectra intrigued scientists throughout the last century but only qualitative progress, for example grouping the lines into various series and progressions, was made until Balmer, in 1896, discovered that the strong lines of atomic hydrogen could be represented by the simple formula

$$\tilde{v} = R\left[\frac{1}{2^2} - \frac{1}{n^2}\right] \quad n = 3, 4, 5 \ldots \tag{5.1.1}$$

where R is a constant (equal to $109\,677\cdot6$ cm^{-1}). Balmer's work was extended and checked by Rydberg (after whom the constant is named) and by Ritz who subsequently showed that in *all* atomic spectroscopy the wave number of a line was given by the difference of two terms. This discovery, which has a natural explanation in quantum mechanics, is still known as the *Ritz Combination Principle*. Subsequent to Balmer's work further series representable by similar formulae were found. In fact *all* the discrete lines in the hydrogen atomic spectrum can be represented by the common formula,

$$\tilde{v} = R\left[\frac{1}{m^2} - \frac{1}{n^2}\right] \quad m < n. \tag{5.1.2}$$

The series with $m = 1$ is called the Lyman series and occurs in the far ultraviolet; the series with $m = 3$ is called the Paschen series and occurs in the near infrared; those with $m = 4$ (Brackett) and $m = 5$ (Pfund) occur still further into the infrared. Other series at still longer wavelengths must exist but they will be weak and difficult to observe. Within a series it is conventional to refer to the line at $n = m + 1$ as α, $n = m + 2$ as β etc. Lyman α at 121·5 nm is one of the most important energy transfer vectors in astrophysics. The sun is intensely bright at this wavelength. Some quasars (thought to be the nuclei of extremely violent galaxies lying near the edge of the observable universe) have red-shifts so large that Lyman α can be observed in the visible and this contributes to the abnormally large apparent luminosities of these remote objects.

The first successful explanation of equation (5.1.2) came from Bohr within the framework of the old quantum theory. He imagined that the electron was circling round the proton, like a planet round the sun, but constrained to move only in orbits for which the angular momentum was quantised—that is was an integral multiple of $h/2\pi$. In such an orbit the velocity will be given by

$$v = \frac{e^2 Z}{2\varepsilon_0 n h},$$ (5.1.3)

where, for generality, we have introduced an atomic number Z not necessarily one as it must be for hydrogen itself, and where n is an integer—the so-called principal quantum number. It follows at once that

$$\mathscr{E}_n = -\tfrac{1}{2} m v^2 = -\frac{m e^4 Z^2}{8 \varepsilon_0^2 n^2 h^2}.$$ (5.1.4)

The term values in wavenumber units then become

$$T_n = \frac{\mathscr{E}_n}{hc} = -\left[\frac{m e^4 Z^2}{8 \varepsilon_0^2 h^3 c}\right] \frac{1}{n^2}.$$ (5.1.5)

When the fundamental constants (from Table 1.2) are inserted into the quantity in brackets on the RHS of (5.1.5), one finds, with $Z = 1$,

$$R_\infty = 109\,737\cdot318 \text{ cm}^{-1},$$ (5.1.6)

in good agreement with experiment. The agreement is made still better by noting that the proton, though heavy, is not infinitely massive when compared with the electron (in fact $m_p = 1836\cdot683 m_e$). If the participation of the proton in the orbital motion is taken into account, one finds

$$R_H = \frac{R_\infty}{(1 + m_e/m_p)},$$ (5.1.7)

which gives $R_H = 109\,678\cdot603 \text{ cm}^{-1}$, in complete agreement with experiment. It will be noted that because the nuclear mass enters into the expression for the appropriate Rydberg constant, the lines of deuterium will not be exactly coincident with those of normal hydrogen. That is, there are isotope shifts. In the Balmer series these are of the order 4 cm^{-1} but in the Lyman series they are as much as 26 cm^{-1}.

The first indications that the simple Balmer formula did not give the whole truth came from Michelson [64], who, using his celebrated interferometer, and the method of visibility of fringes, was able to show that the red Balmer line (H_α) had fine structure. Michelson was able to show that there were two strong components separated by 0·329 cm^{-1} and he could see some indication of a weaker third component lying between the other two. The main difficulty which bedevilled these measurements was the large Doppler width of hydrogen lines. Studies on deuterium in a discharge tube cooled to liquid hydrogen temperature improved matters and allowed the triplet structure to be clearly seen. As usual, experimental work of this calibre stimulated the theorists into intense activity and eventually Sommerfeld came up with a plausible explanation. He postulated the existence of elliptical orbits in addition to the circular ones of Bohr. This, of itself, would not produce any fine structure since the energy still depended only on n but when the varying mass of the electron along its orbit—a relativistic consequence of the varying velocity—was taken into account splittings resulted. Sommerfeld introduced another quantum number k to specify the ellipticity of the orbit. The actual splitting was determined by the dimensionless fine structure constant.

$$\alpha = e^2/4\pi\varepsilon_0 hc = \frac{1}{137\cdot035965\ldots} \tag{5.1.8}$$

Sommerfeld's theory gave excellent numerical agreement with the observed splittings but it could not give a reasonable explanation for the appearance of the third component which should be strictly forbidden.

The next big advance in atomic spectroscopy was the development of quantum (or wave) mechanics. The hydrogen atom was, in fact, one of the few cases which could be solved exactly and Bohr's quantisation condition—which had originally appeared to be totally arbitrary—now was shown to be a perfectly natural consequence of physically necessary boundary conditions. The wave-equation for a particle bound to a massive point by a Coulomb potential is [450]

$$\nabla^2\psi + \frac{8\pi^2 m}{h^2}\left[\mathscr{E} + \frac{Ze^2}{4\pi\varepsilon_0 r}\right]\psi = 0. \tag{5.1.9}$$

Transforming this to the more appropriate spherical polar coordinates r, θ, ϕ, gives

$$\frac{\partial^2\psi}{\partial r^2} + \frac{2}{r}\frac{\partial\psi}{\partial r} + \frac{1}{r^2 \sin\theta}\frac{\partial}{\partial\theta}\left(\sin\theta\frac{\partial\psi}{\partial\theta}\right) + \frac{1}{r^2\sin^2\theta}\frac{\partial^2\psi}{\partial\phi^2}$$
$$+ \frac{8\pi^2 m}{h^2}\left(\mathscr{E} + \frac{Ze^2}{4\pi\varepsilon_0 r^2}\right)\psi = 0. \tag{5.1.10}$$

This equation can be separated in all three variables and from this separation three quantum numbers emerge as constants of separation. We have n as before but in addition there is a subsidiary or azimuthal quantum number l which is integral and which is specified by the relation

$$|l| \leqslant n. \tag{5.1.11}$$

There is also a magnetic quantum number m which represents the number of possible integral projections of l, regarded as a vector, on a fixed direction in space. By tradition wave functions with $l = 0$ are called s orbitals, those with $l = 1$ are called p orbitals, those with $l = 2$ are called d orbitals and those with $l = 3$ are called f orbitals. This practice stems from an old division of alkali atom spectra (see later) into so-called sharp, principal, diffuse and fine series and the subsequent discovery that the optical electron involved in these series had $l = 0, 1, 2, 3$ respectively.† Orbitals with $l = 0$, i.e. s states, are spherically symmetrical, orbitals with $l = 1$ involve a factor of $\cos \theta$ and are dumb-bell shaped whilst those with $l = 2$ and 3 are even more complicated in shape. The lowest two s-type wave functions are

$$\psi_{1s} = 2 \left(\frac{1}{a_0} \right)^{3/2} \exp(-\rho/2), \tag{5.1.12a}$$

$$\psi_{2s} = \frac{2 - \rho}{2\sqrt{2}} \left(\frac{1}{a_0} \right)^{3/2} \exp(-\rho/2), \tag{5.1.12b}$$

where a_0 is the so-called Bohr radius $4\pi\varepsilon_0 h^2/me^2$ and ρ is a dimensionless radial parameter

$$\rho = 2rZ/na_0 \tag{5.1.13}$$

The lowest p-type orbital, pointing say along the x axis is

$$\psi_{2p} = \frac{1}{4\sqrt{2\pi}} \left(\frac{1}{a_0} \right)^3 \rho \cos \theta \exp(-\rho/2). \tag{5.1.12c}$$

with equivalent orbitals pointing along y and z.

The higher p, d, f orbitals are more complicated but their angular dependence is given uniquely by the associated Legendre functions or else by the set of spherical harmonics. They are listed in most standard textbooks [451, 452]. The overall state of the atom when the electron is in an orbital with azimuthal quantum number l is usually denoted by using the equivalent capital letter. Thus one has S, P, D, F etc. electronic states. To specify which is which, the principal quantum number is given first: thus the ground state of hydrogen is 1S. Electromagnetic radiation being a vector field has negative parity so electric dipole allowed transitions must be between states of opposite parity in

---

† This letter code to cover higher l values is continued alphabetically, thus for $l = 4, 5, 6$, etc. one uses g, h, i, etc.

order that the matrix elements be non-zero. In the atomic case, since the parity changes alternately with l, the selection rule demands that l change by ± 1. In the state notation, Lyman α would therefore be specified as $1S \rightarrow 2P$.

The wave functions (5.1.12) are not themselves physically observable quantities—a unique feature of quantum mechanics—but their squares and some of their products are. Thus the chance of finding the electron in a shell thickness dr at a distance r is given by taking the square of the radial wave function (or its square modulus if a complex solution of 5.1.10 is adopted) and multiplying by $4\pi r^2 \mathrm{d}r$. As an example for the ground state

$$P(r)\mathrm{d}r = 4\pi r^2 \frac{1}{\pi}\left(\frac{1}{a_0}\right)^3 \exp(-\rho)\mathrm{d}r. \qquad (5.1.14)$$

This function has its maximum value at $r = a_0$—the quantum mechanical justification for Bohr's "orbits". The integral of (5.1.14) over all r is unity, as it must be since the chance of finding the electron in the totality of space must be a certainty. It is most important to note, however, in this context, that although (5.1.14) is zero at $r = 0$, the wave function itself is not zero at the nucleus. In fact *all* s orbitals have a finite value at the origin. The energies, corresponding to the various stationary states given by the wave functions (5.1.12), are found, just as in the Bohr theory, to depend only on n: there is complete degeneracy of l and m states. However the experimental discovery that the electron has a spin, and hence a magnetic moment, showed that an extra term should be included in the wave equation to cover the possibility of magnetic interactions between the orbital and self-spinning motions. With this extra term included, the l degeneracy is lifted, splittings can arise and one has an explanation of the observed fine structure. For atoms such as hydrogen, one combines l and s (the conventional labelling for the electron spin quantum number, equal to $\pm\frac{1}{2}$, not, of course, to be confused with the label for orbitals with $l = 0$), into a total angular momentum quantum number $j = l \pm s$. States with the same n and the same j are still found to be degenerate but those with the same n but different j will now differ in energy. This theory gives an explanation for the H_α splitting (Fig. 5.1) which agrees very well with observation when the selection rules $\Delta j = 0, \pm 1$ and $\Delta l = \pm 1$ are applied.

The near exact agreement between two theories so different as those of Sommerfeld and Schrodinger was quite a puzzle. It was resolved by the monumental work of Dirac who produced a wave-mechanical equation, for the electron, which was also Lorentz invariant and therefore compatible with the postulates of relativity [454]. Dirac's theory involved negative energy states which were assumed to be completely filled in the normal vacuum state and hence unobservable (compare the analogous situation of filled energy bands in semiconductors in section 5.4). This did give a rather severe difficulty that the total energy of the vacuum would be infinite and that the charge and mass of the electron would be likewise infinite but it was eventually found that the theory could be *renormalised* (tantamount to subtracting away the awkward infinities) and the whole renormalised theory could be re-scaled so that the charge and mass of the electron had their observed values [455]. In this form Dirac's theory is usually called Quantum Electrodynamics (or QED for short). The theory is not expressible analytically and much of its more esoteric branches are formidably daunting but one can usually express a solution as a power series in α. Since α is very small (at least for electrons) these series rapidly converge and good numerical answers result. In fact some theorists have even gone so far as to say that QED is the most precise physical theory that has yet been produced! Certainly in all cases where it has been applied the agreement with experiment has proved exact. Within the framework of QED, the point charge electron will be surrounded by a virtual cloud of electron-hole (i.e. positron) pairs that it has caused to spring into ephemeral existence. This phenomenon is sometimes called polarisation of the vacuum. The electron will then carry out orbital motion about these transitory

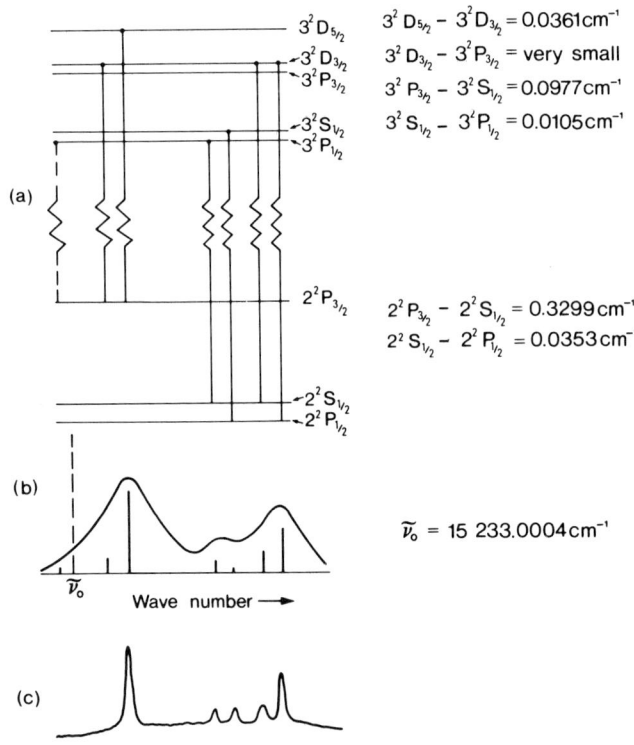

FIG 5.1. Fine-structure of H_α. In (a) is shown the energy level configuration after Kuhn and Series [453]. In (b) is shown the 300 K Doppler-broadened profile together with the positions of the components and an indication, $\bar{\nu}_0$, of the position of the single line of the simple Bohr theory. In (c) is shown the saturation, or Lamb-dip, spectrum recorded by Hansch and his colleagues [457].

partners and because of the rotatory component of this motion it will acquire a magnetic moment and hence a spin. This demonstration by Dirac that Lorentz invariance automatically leads to an intrinsic spin was one of the triumphs of the early development of quantum theory. Since the imposition of relativity automatically leads to magnetic interactions one has a satisfactory account of the puzzle mentioned earlier. When the Dirac theory is applied to the hydrogenic atom it leads to energy levels for states with quantum numbers n and j given by [452].

$$\mathscr{E}_{n,j} = -\frac{RZ^2 ch}{n^2}\left[1 + \frac{\alpha^2 Z^2}{n}\left(\frac{1}{j+\frac{1}{2}} - \frac{3}{4n}\right)\right], \qquad (5.1.15)$$

where R is the appropriate Rydberg constant for the atom in question. This equation is identical to that given by Sommerfeld if $j + \frac{1}{2}$ is set equal to k. Equation (5.1.15) gives an excellent account of the hydrogen ($Z = 1$) atom spectrum fine structure. Thus for H_α, the two strong components have shifts

$$\left.\begin{aligned}(n = 3, j = 3/2) \to (n = 2, j = 1/2), \ \Delta\tilde{\nu} = 0.402\,\text{cm}^{-1}\\(n = 3, j = 5/2) \to (n = 2, j = 3/2), \ \Delta\tilde{\nu} = 0.073\,\text{cm}^{-1}\end{aligned}\right\} \qquad (5.1.16)$$

with respect to the simple Bohr theory. The separation of these two components is then $0.329\,\text{cm}^{-1}$ in splendid agreement with observation.

The Dirac theory does, however, like all the previous accounts give identical energies for states with the same n and j. Thus for example the pair $2S_{\frac{1}{2}}$, $2P_{\frac{1}{2}}$ are degenerate. However as the optical spectrum of hydrogen was studied at higher and higher resolution, evidence began to emerge that this presumed degeneracy was not quite true. The experiments were very difficult because of the large Doppler width but they all seemed to show that the nS_j states were displaced upwards by small amounts compared to the nP_j states. The eventual proof that this was indeed so came in a famous radio frequency experiment carried out in 1947 by Lamb and Retherford [456] in which the stimulated emission due to the transition $2S_{\frac{1}{2}} \to 2P_{\frac{1}{2}}$ at $1.0577\,\text{GHz}$ (i.e. $0.03528\,\text{cm}^{-1}$) was detected. The success of this experiment depended on the $2S_{\frac{1}{2}}$ state being metastable in the sense that the only state to which it can decay is $2P_{\frac{1}{2}}$ and, this being so close in frequency, the spontaneous decay rate is very slow. Considerable concentrations of $2S_{\frac{1}{2}}$ atoms therefore build up in discharges through hydrogen—very much higher than the concentrations which might be expected on thermal grounds. Indeed had Lamb and Retherford had to rely on thermal population they would not have observed any signal. The upper pairs, $3S_{\frac{1}{2}} - 3P_{\frac{1}{2}}$ etc. are therefore unobservably weak and the ground state $1S_{\frac{1}{2}}$, which incidentally would have the largest shift, cannot be studied by radio frequency methods because there is no nearby $P_{\frac{1}{2}}$ state. In acknowledgement of the skill and perspicacity of the experimenters and of the fundamental significance of their work, these shifts of nS states in hydrogenic atoms are universally known as *Lamb shifts*. The Lamb shifts are bigger the smaller is n; thus for $n = 1$ it is $8.049\,\text{GHz}$, but as hinted above it is difficult to establish this by straightforward spectroscopy because the Doppler broadening of Lyman α is so large. Recently, however, the introduction of the technique of double-photon, Doppler-free spectroscopy with a tunable dye laser as the primary source has enabled the accurate determination [457] of the frequency of the transition $1S_{\frac{1}{2}} \rightarrow 2S_{\frac{1}{2}}$ (an even parity transition allowed in a *double* photon absorption) and from this the Lambshift of $1S_{\frac{1}{2}}$ has been found to be 8·053 GHz.

The explanation of the Lamb shift can be found, as shown by Bethe [458] within the same basic formalism, i.e. QED, introduced by Dirac. The mathematical development is rather difficult but the basic physics can be presented in a simple form due to Welton [459]. This starts from the idea (see section 4.2.5) that the lowest state of the quantised electro-magnetic field, i.e. the vacuum state, has zero-point energy. There will consequently be fluctuations of this zero-point energy which will manifest themselves as a randomly fluctuating electromagnetic field. The presence of this stochastic field will cause the electron to move about in a rapid irregular fashion and thus appear to be "spread-out". This spreading reduces the value of ψ^2 in regions where it is high and increases it in regions where it is low. For hydrogenic atoms, as remarked above, S states have finite values of ψ^2 at the nucleus and the QED effects therefore reduce the electrostatic interaction between the proton and the electron. The binding energy is thus reduced and these S states are displaced upwards. The effect will obviously be the less the further out on average the electron is, so the shifts will decrease as n increases, in agreement with experiment. We have already seen that the $1S_{\frac{1}{2}}$ state is displaced by 0.2684 cm^{-1} and the $2S_{\frac{1}{2}}$ state by 0.0353 cm^{-1}, the $3S_{\frac{1}{2}}$ state is calculated to be displaced by only 0.0105 cm^{-1}.

To round off this discussion of fine structure in the hydrogen atom spectrum one should mention that the proton, having half integral spin, can interact with the electron magnetically leading to hyperfine structure. Thus the ground state $1S_{\frac{1}{2}}$ of the hydrogen atom is split by 1420·405 752 MHz. Transitions between these split states are not allowed by electric dipole selection rules but they are allowed by magnetic dipole selection rules. Signals from cosmic hydrogen, at this frequency, provide the radio astronomer with one of his most effective ways of mapping the distribution of hydrogen within the galaxy. At this wavelength (~ 21 cm) scatter by galactic dust is virtually non-existent and although the signal is intrinsically weak because of the magnetic dipole selection rules and because of the microscopic population difference there is so much hydrogen in the spiral arms of the galaxy that these are easily detected.

5.1.3 *Atoms with more than one electron*

As soon as the atom contains more than one electron there is no possibility of solving the wave equation analytically and only approximate or numerical methods may be applied. The first approximation, and one so widely used that most people are not even aware that it is an approximation, is to assume that the electrons can still be labelled by the four quantum numbers n, l, m, s. As is well known, this approach gives a very natural explanation of the periodic classification of the elements, when the further assumption is made that the energy still depends primarily on n. In general, however, one is reduced to a purely descriptive account of atomic spectra, having to rely entirely on experiment to give the term values [460]. The exceptions where some sort of numerical headway can be made are the rare gases, which have a closed outer shell and the alkali metals which have a single "optical" electron lying outside a closed shell.

The simplest rare gas, helium, has just two electrons: these can be in any of the orbitals and with their spins either parallel or opposed except that if they are in the *same* orbital the spins *must* be opposed. This is a manifestation of the Pauli Exclusion Principle which in atomic terms prohibits any two electrons from having the same values of n, l, m and s. The interaction between the two electrons is both electrostatic and magnetic so one needs some rule for arriving at the externally observable properties of the atom in its various excited states. One can think of two extreme cases: (1) where the two spins combine to give a resultant $S = s_1 + s_2$ and the two orbital components combine to give a resultant $L = l_1 + l_2$, this is called *Russell–Saunders* coupling, and (2) where each electron combines its parameters to give a resultant $j_i = l_i + s_i$ and then the j_i combine to give an externally observable total angular momentum j. This is called *j–j coupling*. Helium provides a near perfect example of Russell–Saunders coupling, some heavy atoms provide good examples of j–j coupling but unfortunately most heavy atoms represent intermediate cases which are not describable in elementary terms. For helium, however, S and L are well defined (sometimes said to be "good" quantum numbers) and the electronic states therefore divide into two systems, those for which $S = 0$ and those for which $S = 1$. As is usual in quantum mechanics, the degeneracy of a state with an angular momentum quantum number S is $2S + 1$, so the first series is composed of singlets and the second of triplets. As is also usual the degeneracy is slightly removed and under high resolution lines of the second set can actually be resolved into triplets. The value of S is normally given as a left upper superscript on the term symbol so the ground state of helium would be described as 1^1S_0. Transitions between the singlet and the triplet system are forbidden by the electric dipole selection rules and in fact such transitions are hardly ever observed spectroscopically. The gas behaves as though it were made up of two non-combining forms once called parhelium ($S = 0$) and orthohelium ($S = 1$). Transitions between the two can, however, take place on collision with the walls of the containing vessel. The true resonance line of helium $1^1S_0 \rightarrow 2^1P_1$ lies in the vacuum ultraviolet at 58·44 nm and one has the remarkable feature that the first excitation in helium is about 4/5 of the ionisation energy. The lowest state of the triplet system 2^3S_1 lies at 19·72 eV above the ground state but is highly metastable since any transition to the ground state is totally forbidden by the electric dipole selection rules. In a discharge through helium, high concentrations of this metastable form build up and the transition $2^3S_1 \rightarrow 2^3P_2$ at 1·0829 μm behaves like a resonance line. The role that this metastable, and the only slightly less metastable 2^1S_0, play in providing the inversion mechanism of the helium/neon laser has been discussed in section 2.4.3. Laser action has also been observed in the far infrared in pure helium gas. The lines at 95·8 μm $(3^1P \rightarrow 3^1D)$ and 216·3 μm $(4^1P \rightarrow 4^1D)$ have found some technical use, but since their frequencies can now be determined to high precision their study is also of some theoretical interest. Empirically it has been found that modified forms of the Balmer formula can be applied to helium and other two electron atoms, e.g. Li^+. One of the best forms is the Rydberg–Ritz formula

$$\mathscr{E}_n = \frac{-(Z-1)^2 R_H}{(n - \delta(l))^2}, \tag{5.1.17}$$

where the quantity in the denominator is an effective quantum number sometimes called n^*. The quantum defect $\delta(l)$ then represents the shielding effect of the second electron on the optical one: $\delta(l)$ depends on l but not on n so equation (5.1.17) is useful in practice because once one has one term in a "ladder" one can calculate all the rest. The screening interpretation of $\delta(l)$ is confirmed by the observation that it is largest for S terms, where the optical electron does the most penetration, and that it falls for P, D, F, terms etc. For helium average values are $\delta(0) = 0·1471, \delta(1) = -0·0065, \delta(2) = 0·0076$. These give a fairly reasonable account of the term values with the exception of the lowest state which not surprisingly is rather out of line—calculated, 150 854 cm^{-1}; observed, 198 311 cm^{-1}. The formula is, however, not of that much use in the infrared where one calculates a frequency by subtracting two terms quite similar in magnitude. Errors therefore play a more significant role than would be the case in the calculation of an ultraviolet line. As an example, using (5.1.17) and the δ values given above, one finds the $(3^1P \rightarrow 3^1D)$ line to be at 87·2 μm instead of the observed 95·8 μm. To do any better though requires very complex variational approaches to the numerical solution of the wave equation. Using a series expansion with 203 parameters, Pekeris [461] found the ground state energy to be

$$1^1S_0 = -198\,312·01 \text{ cm}^{-1} \pm 0·1 \text{ cm}^{-1}.$$

Herzberg [462], by means of an ingenious system of measurements (including amongst them the weak intercombination line $1^1S_0 \rightarrow 2^3P_1$), found the experimental value to be

$$1^1S_0 = -198\,310·82 \text{ cm}^{-1} \pm 0·15 \text{ cm}^{-1}.$$

The difference between these two agrees well with the expected Lamb shift of 1·3 cm^{-1}.

The spectra of the higher noble gases, neon, argon, krypton and xenon, are much more complex. This is basically because the magnetic and electrostatic interactions within the atom are of the same order of magnitude. It is not easy therefore to devise simple term designations which describe the situation accurately. Most spectroscopists continue to use the old Paschen notation in which the ground state is 0s, the excited states 1s, 2s, 2p etc. (see Fig. 2.33). The levels do tend to group themselves in this way and the splittings can be accommodated by introducing a subscript. Thus within the complex of neon lines in the red region making up $3s \rightarrow 2p$ there is the component $3s_2 \rightarrow 2p_4$ which can be had in stimulated emission giving the well-known 0·6328 μm helium/neon laser line. This notation is a little confusing since the numbers which appear are not the true n quantum numbers and it is also not very informative. However any other system, for example the Racah [463], is bound to be rather cumbersome. The neon multiplets get closer and closer as n increases—a common feature of so-called Rydberg atoms—and at very high values the atom becomes essentially hydrogenic. Many laser lines are known involving these multiplets [464]. So far the series $4p \rightarrow 3d$ (3·77 − 10·98 μm), $5p \rightarrow 4d$ (12·82 − 22·8 μm), $6p \rightarrow 5d$ (20·48 → 41·74 μm), $7p \rightarrow 6d$ (35·60 → 68·33 μm) and $8p \rightarrow 7d$ (72·15 → 198·6 μm) have been observed and assigned. Some further lines probably of the $9p \rightarrow 8d$ and $10p \rightarrow 9d$ series have been observed but not assigned.

The alkali atom spectra bring us back to a relatively simple situation again and it is not surprising that these spectra were analysed and understood long before the more complex spectra of other elements were attempted. Basically a spectrum is complex if there is more than one electron, each in an incomplete shell, which has $l > 0$. When this is not so, as it is for alkali atoms where there is just the single optical electron, one gets a simple spectrum characterised by the presence of simple series. The various series in the spectrum of say sodium have been known for a long time as mentioned earlier. The ionisation energies of alkali atoms are very low (Li, 5·363 eV; Na, 5·12 eV; K, 4·318 eV; Rb, 4·159 eV; (Cs, 3·89 eV) so the whole of the spectrum is confined to the long-wave optical region. For caesium, as an example, nearly all of the spectral lines occur in the infrared. A term (Grotrian) diagram for caesium is shown in Fig. 5.2 and the actual term values (in cm^{-1}) in Table 5.1. All the states can be represented by the Ritz–Rydberg formula with $(Z - 1)^2$ simply replaced by 1 since the optical electron sees only a single surplus positive charge. As with helium, the representation works better for states with $l > 0$. One also notices that the splitting of states with the same n but different j gets smaller and smaller as l increases. For the F states of caesium these splittings are only of the order 0·2 cm^{-1}.

It is interesting to note that using the Ritz–Rydberg formula, it is possible to deduce the ionisation energy of caesium using solely infrared data. Thus from the lines

$$6P_{1/2} \rightarrow 5D_{3/2} = 3321 \text{ cm}^{-1}$$

and

$$6P_{1/2} \rightarrow 6D_{3/2} = 11411 \text{ cm}^{-1}$$

one has that $5D_{3/2} \to 6D_{3/2} = 8090 \text{ cm}^{-1}$. Assuming that $\delta(2)$ is constant, i.e. that it is independent of n, one then deduces its value to be 2·4405. The ionisation energy of the lowest D state, i.e. $5D_{3/2}$ is then $16\,751 \text{ cm}^{-1}$. Adding the transition energy for $5D_{3/2} \to 6P_{1/2}$ (i.e. 3321 cm^{-1}) gives the ionisation

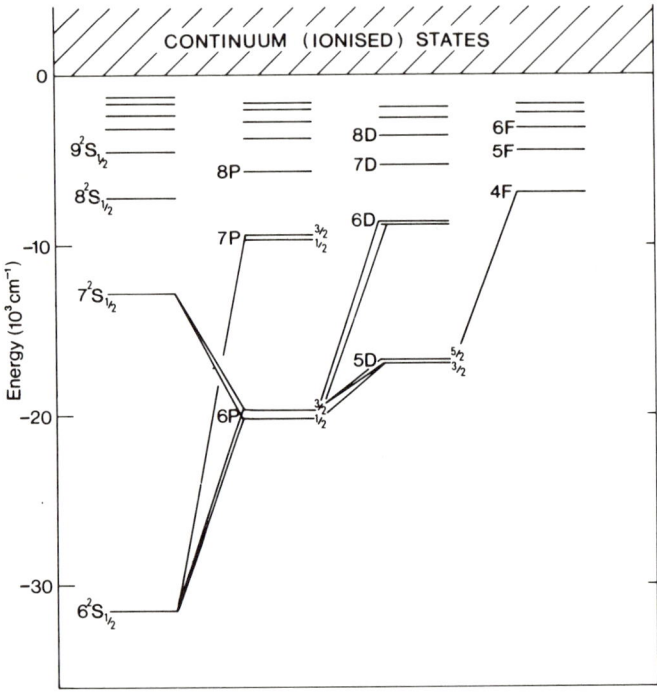

FIG. 5.2. Energy levels of the caesium atom corresponding to the promotion of the outermost "optical" electron. Some observed transitions are also indicated.

TABLE 5.1
Energy levels for the caesium atom

S States			P States			D States			F States		
n	$\mathcal{E}$	$\delta(0)$	n_j	$\mathcal{E}$	$\delta(1)$	n_j	$\mathcal{E}$	$\delta(2)$	n	$\mathcal{E}$	$\delta(3)$
6	31 406·7	4·131	$6\frac{1}{2}$	20 228	3·671	$5\frac{3}{2}$	16 907	2·452	4	6936	0·023
			$6\frac{3}{2}$	19 674	3·638	$5\frac{5}{2}$	16 809	2·445			
7	12 870·9	4·080	$7\frac{1}{2}$	9 641	3·626	$6\frac{3}{2}$	8 817	2·472	5	4437	0·025
			$7\frac{3}{2}$	9 460	3·594	$6\frac{5}{2}$	8 775	2·463			
						$7\frac{3}{2}$	5 358	2·474	6	3079	0·030
						$7\frac{5}{2}$	5 337	2·466			

energy of $6P_{1/2}$ as $20072\ cm^{-1}$ and, finally, adding the $6P_{1/2} \rightarrow 6S_{1/2}$ resonance line at $11\,718\ cm^{-1}$ gives $31\,250\ cm^{-1}$, in quite good agreement with the observed value of $31\,407\ cm^{-1}$.

5.2 Infrared spectroscopy of molecules

5.2.1 *Structure of molecules*

A molecule consists of two or more atoms held together by strong bonding forces, occasionally ionic, as in NaCl, but more usually covalent and thus owing the attraction to the interplay of exchange forces. These arise when an atomic orbital on one of the atoms overlaps with an atomic orbital of the other. It is simplest to discuss the homonuclear case but the same arguments would apply equally to heteronuclear molecules. Considering the simplest case, i.e. H_2, then at large separations of the two atoms the total energy would be doubly degenerate and equal to twice that of an individual atom but on close approach of one hydrogen atom to the other, the 1s orbitals will start to overlap, the degeneracy will be lifted and two possible states will result. These have wave functions

$$\psi_+^{ab} = \frac{1}{\sqrt{2}}\,(\psi_{1s}^a + \psi_{1s}^b), \qquad (5.2.1a)$$

and

$$\psi_-^{ab} = \frac{1}{\sqrt{2}}\,(\psi_{1s}^a - \psi_{1s}^b). \qquad (5.2.1b)$$

The corresponding potential energies are respectively lower and higher than that of the separated atoms. Mathematical analysis shows [465] that the increase in the kinetic energy of (5.2.1a), which arises because each electron, now circumnavigating two protons, has to travel a longer distance, is less than the decrease in the potential energy so one has a bonding situation. Contrariwise in (5.2.1b) one has an antibonding situation.

The functions of (5.2.1), by analogy with the atomic case, are called *molecular orbitals*. The electron density of (5.2.1a) shows a peak at the mid-bond position whereas that for (5.2.1b) shows a node in this position. One can thus derive an electrostatic interpretation of chemical bonding since the attraction which each nucleus feels for the electron density peak lying between them is more than the force of repulsion which they feel for each other. One can extend this analysis to other pairs of atoms and indeed to polyatomic molecules but in most cases a major modification is encountered in that instead of getting overlap of the simple s, p, d, etc. orbitals, one gets overlap of *hybrid* orbitals formed from suitable mixtures of these primitive types. The carbon atom provides the best known example of this. By combining an s and a p orbital, one gets two combinations which point away from one another at

180°; by combining an s with two p's one gets three equal orbitals in a plane inclined at 120° to one another; finally by combining the s with all three p orbitals, one gets four equal hybrids directed towards the vertices of a tetrahedron which has the carbon atom at its centre. The concept of hybridisation is enormously important since it explains the shape of molecules. Thus BF_3 is a plane trigonal molecule and CF_4 is a tetrahedral molecule.

Chemical bonds of either kind can be divided into two main types, σ bonds in which the electronic cloud around the two nuclei has full cylindrical symmetry and π bonds in which there is merely a plane of symmetry. One can think of the σ bonds as arising from "head-on" overlap of atomic orbitals and π bonds from sideways overlap (see Fig. 5.3). A multiple bond usually has one σ plus one or two π components.

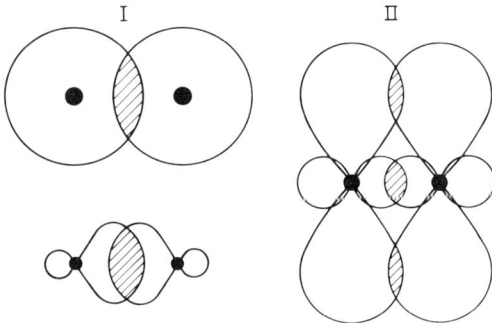

FIG. 5.3. The formation of σ and π bonds in diatomic molecules. In I only single bonds are formed either by overlap of simple or of hybrid orbitals. In II a double bond made up of a σ plus a π bond is formed. The shaded areas give a schematic indication of the regions of enhanced electron density. For II the paper forms a plane of symmetry.

Chemical bonds are very resistant to stretching and in general to angular deformation with respect to other bonds in the molecule but rotation about pure σ bonds is an exception since the cylindrical symmetry ensures that this will always be a low energy process. When there is a π bond present, however, rotation about the internuclear axis becomes a much higher energy process. This is the explanation of *cis/trans* stereoisomerism of the maleic/fumaric acid type, viz.

cis   trans

Maleic acid   Fumaric acid

The maximum bond multiplicity is three since this corresponds to σ bonding, involving possible p_z hybridisation, combined with π bonding involving the p_x and p_y orbitals. In the case of triple bonds, there is again cylindrical symmetry because of the degeneracy of the p_x and p_y pair. For most molecules this treatment in terms of pair-wise bonding is perfectly adequate but for aromatic molecules and conjugated molecules in general, e.g.

one has to consider the electrons as moving around more than two nuclei, that is they become delocalised. In this case a slight deformation of one part of the molecule may make its effects apparent in a remote part of the molecule. In saturated molecules, on the other hand, it is a good approximation to consider only very local effects.

When one comes to consider the energy levels of molecules and the allowed transitions between them, one should in principle include the complete set of all the electrons and all the nuclei, form a complete Hamiltonian and insert this into the time-dependent wave equation to derive the eigenvalues, that is the stationary states. This is, of course, an impossible programme: even the classical three body problem is insoluble and the quantum mechanical equivalent is even more recondite. No exact *ab initio* solution can be expected but approximate solutions have been derived and these have, whenever a comparison with experiment is possible (for example the H_2^+ and H_2 molecules), been found to give excellent agreement. One assumes therefore that quantum mechanics is correct and its description of the energy levels of molecules valid; but the determination of the actual energies of the stationary states is to be regarded as an experimental problem. Fortunately it turns out to be an excellent approximation to assume first-order separability of rotational vibrational and electronic motion. Thus in considering rotation one assumes the molecule to be a *rigid-rotor*, when considering vibration one assumes the molecule to be a *harmonic-oscillator* and when considering electronic excitations one assumes that one may use molecular orbital theory and the Born–Oppenheimer approximation.[†] The separate problems are much more tractable and lead to analytical expressions for the energy levels. These give good accounts of the observed spectra and if small corrections are included to take account of the interactions between the three fundamental types, an even

---

† That is the assumption that electronic motions are so fast in comparison with nuclear motions that the electronic cloud can be assumed to follow the nuclear motion instantaneously.

better account is obtained. Electronic excitations are rare in the infrared so the bulk of what follows will be concerned with rotational and vibrational excitation.

The nuclear spin has usually only a very small effect in molecular spectroscopy merely producing hyperfine splittings but in the case of molecules having a centre of symmetry there can be a much more obvious effect. This is because if one writes the overall wave function

$$\psi = \psi_{el}\psi_{vib}\psi_{rot}\psi_{nuc},$$

where the subscripts refer to the electronic, vibrational, rotational and nuclear parts, the Pauli exclusion principle demands that ψ be antisymmetric, that is change sign upon interchange of symmetrically connected particles. Now in a normal (that is gerade) electronic ground state without vibrational excitation and with zero spin nuclei, the Pauli principle would demand that ψ_{rot} be antisymmetrical. This can only be possible if alternate rotational levels are missing. This is found to be the case for O_2 and CO_2 for example, but in confirmation of the theory *all* the rotational levels are observed for the isotopic forms $^{16}O\ ^{18}O$ and $^{16}O\ C^{18}O$. It follows that on excitation of an antisymmetric vibrational mode the other set of levels alone will be permitted. Hyperfine splittings due to nuclear spin are of the order of kHz for spin–spin interactions and of the order of tens of MHz for quadrupole interations. These can be detected by coherent microwave spectroscopy but since they are seldom of any consequence in infrared spectroscopy where the resolution is commonly of the order of GHz they will not be discussed further here.

5.2.2 Rotational spectroscopy of molecules

In classical as well as quantum mechanics, rotors are divided into four classes depending on the relationships amongst their moments of inertia; thus we have

(a)	Linear rotors	$I_B = I_C,\ I_A = 0$	(5.2.1a)
(b)	Spherical rotors	$I_B = I_C = I_A$	(5.2.1b)
(c)	Symmetric tops	$I_B = I_C \neq I_A$	(5.2.1c)
(d)	Asymmetric rotors	$I_B \neq I_C \neq I_A$	(5.2.1d)

The linear and spherical rotors can clearly be thought of as special cases of the symmetric top. A rotor can only be a symmetric top by symmetry if it has a three-fold or higher rotational axis of symmetry and this axis is then the *figure axis*. In quantum mechanics, it turns out that the important quantity is not so much the moment of inertia as its reciprocal and we therefore define three new parameters:

$$A = h^2/8\pi^2 I_A, \quad B = h^2/8\pi^2 I_B \quad \text{and} \quad C = h^2/8\pi^2 I_C. \quad (5.2.2)$$

In terms of these, the energy of the symmetric rotor, to second order, can be written [466],

$$\mathscr{E}(J, K) = BJ(J+1) + (A-C)K^2 - D_J J^2(J+1)^2 - D_{JK} J(J+1)K^2 - D_K K^4$$
(5.2.3)

where J is the rotational quantum number, a measure of the overall quantised angular momentum, and K is a second quantum number which gives the projection of J on the figure axis. It therefore can take the $(2J+1)$ values $-J, \ldots 0 \ldots +J$. The first two terms in (5.2.3) give the rigid rotor contribution to the rotational energy whereas the last three give the contributions due to the non-rigid or centrifugal distortion effects. These latter arise because the bonds are not infinitely stiff and the molecule gets bigger when it rotates faster. It follows that the centrifugal distortion coefficients are related to the Hooke's law constants, that is the force constants of the bonds and they therefore represent a form of rotation/vibration interaction. However the Ds are much smaller than the As and Bs and the last three terms only become significant for large J or K. To complete this description it should be mentioned that each J, K level has a degeneracy of $(2J+1)$ arising from the fact that the projection of J onto an arbitrary direction in space is also quantized and described by the magnetic quantum number M. Normally any direction in space is as good as any other and the degeneracy is complete and has no spectral consequences. If, however, a magnetic or an electric field is applied to the molecule, this will "label" space by specifying a unique direction and then the $(2J+1)$ components will have different energies and the spectral lines will break up into two or more components. This is known as the *Zeeman Effect* for magnetic fields and as the *Stark Effect* for electric fields. Neither effect is unique to molecules and in fact both were known for atomic spectra before they were observed for molecular spectra; however they are nowadays much more used in molecular spectroscopy and in fact the Stark effect is routinely used as a means of modulating a microwave spectrometer.

For a rotor with $I_B = I_C$ by symmetry, the dipole moment necessarily lies along the figure axis and it follows that a change in K, i.e. a rotation about the figure axis, cannot alter the dipole moment as seen from an external point in space. Translated into quantum language, this gives the selection rules for infrared absorption or emission.

$$\Delta J = \pm 1 \quad \Delta K = 0.$$
(5.2.4)

The lines in the spectrum are therefore at energies which follow from applying these rules to (5.2.3). However in practice one nearly always calculates the frequency or wave number and redefining B one writes for example

$$\tilde{\nu}(J'') = 2B(J''+1) - 4D_J(J''+1)^3 - 2D_{JK}(J+1)K^2.$$
(5.2.5)

In the literature B values are usually quoted in cm^{-1}. The double prime suffix on J in (5.2.5) is a universal convention to indicate the lower state of a transition—a single prime is used for the upper state. From (5.2.5) it will be seen that to first order the spectrum consists of a set of regularly spaced "lines" but to second order the line positions will depart from regularity and each line will break up into a close multiplet due to the D_{JK} term. In microwave spectroscopy where line widths are very small, due to the low pressures used, and the instrumental resolution is very high because of the availability of coherent tunable sources, it is common to resolve these multiplets but this is very unusual in infrared spectroscopy using conventional instruments. Symmetric rotors are usually divided into two classes:

$$\text{Prolate symmetric rotors} \quad I_B = I_C > I_A, \text{ i.e. } B = C < A, \quad (5.2.6a)$$

$$\text{Oblate symmetric rotors} \quad I_B = I_C < I_A, \text{ i.e. } B = C > A. \quad (5.2.6b)$$

The second term in equation (5.2.3) changes sign for the two cases and this causes a different distribution of the overall intensity amongst the lines in the K sub-structure making up a given rotational "line". This comes about because of the Boltzmann factor which gives the population of a given state relative to the ground state, thus

$$N(J, K) = N_0 \exp[-\mathscr{E}(J, K)/kT]. \quad (5.2.7)$$

For a prolate top, the exponential term diminishes if J or K increases and the intensity is mostly in the $K = 0$ component. For an oblate top on the other hand the exponential factor *increases* and the maximum intensity may well be in a high K line well removed from the $K = 0$ component. This can lead to serious errors if line positions—taken to be the peaks of the unresolved profiles—are used via a plot of $\tilde{v}/(J'' + 1)$ versus $(J'' + 1)$ to determine the rotational constants [467]. The line intensities are found by approximating the rotational partition function by using an integration to replace the summation (since F_{rot}/kT will be small). The result is

$$I(J, K) = C v_{JK} \left[\frac{(J'' + 1)^2 - K^2}{(J'' + 1)(2J'' + 1)} \right] g_{JK} N(J, K), \quad (5.2.8)$$

where C is a constant, v_{JK} the line frequency and g_{JK} the statistical weight of the lower level. The intensities will therefore increase at first as J'' increases but will eventually decrease as the effects of the exponential term start to be manifest. There will therefore be a line of maximum intensity and this will not be the $J'' = 0$ line. The result of this is to make several fairly heavy molecules have strong readily observable rotational spectra in the far infrared even though their first few lines are in the microwave region. Thus for the oblate symmetric top NF_3 ($B = 0.3563 \text{ cm}^{-1}$, $A = 0.1948 \text{ cm}^{-1}$, $D_J = 4.75 \times 10^{-7} \text{ cm}^{-1}$ and $D_{JK} = -8.57 \times 10^{-7} \text{ cm}^{-1}$) whose first line is at 0.7126 cm^{-1}, that is 21.36 GHz, there is a strong far-infrared [467] absorption spectrum which has been observed as far as $J'' = 64$ at 45.965 cm^{-1}. Another example is shown in

Fig. 5.4, the pure rotational spectrum of CH_3Cl. This is interesting because the B values for the two isotopic species $CH_3{}^{35}Cl$ and $CH_3{}^{37}Cl$ are different and the spectrum is made up of two sets of lines, one for each species.

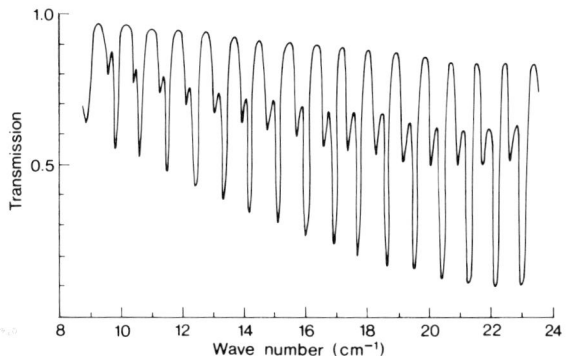

FIG. 5.4. Far-infrared spectrum of methyl chloride, pressure 2 torr, path length 100 mm. The stronger lines are due to $CH_3{}^{35}Cl$ and the weaker lines to $CH_3{}^{37}Cl$.

The spherical rotor is, as mentioned earlier, a special case of the symmetric rotor with $A = B = C$. K is no longer uniquely defined since all directions in space are momently equivalent and the line position formula reduces to

$$\tilde{v}_{J^{\prime}}\,(\text{cm}^{-1}) = 2B(J'' + 1) - 4D_J\,(J'' + 1)^3. \tag{5.2.9}$$

Each line is now $(2J + 1)^2$ degenerate since the K degeneracy is additional to the M degeneracy. Spherical rotors are only of marginal interest to infrared spectroscopists measuring rotational spectra since they are deprived by their symmetry not only of a dipole but of a quadrupole as well. The lowest order multipole moment which they can possess is an octopole (T_d molecules such as CH_4) or a hexadecapole (O_h molecules such as SF_6). Pure rotation absorption in these spherical molecules would obviously be exceedingly weak and it might be expected that it could only be detected by the most sophisticated and sensitive methods. However it is found that the centrifugal distortion effects in XY_4 molecules [468] induce a measurable dipole moment ($\mu \sim 10^{-5}$ D) and it is possible to observe, by fairly normal methods, a weak spectrum made up of $J \to (J + 1)$ lines in the far infrared [469] and Q-branch (i.e. $\Delta J = 0$) transitions in the microwave region [470].

Linear molecules are another special case of the symmetric rotor but this time with $B = C$ and $A = \infty$. The net effect of having an infinite A is that only those states with $K = 0$ are populated or in other words exist. Each level is therefore only $(2J + 1)$ fold degenerate (due to M) but otherwise things are very similar to the spherical rotor case and the line positions are given by equation (5.2.9) as it stands. An example of a linear rotor spectrum is shown in Fig. 5.5 which comes from the careful work of Fleming [471] at the National

Physical Laboratory; it is of CO and the signal-to-noise ratio is so high (see Chapter 4) that the weak lines due to ^{13}CO and $^{12}C^{18}O$ at natural abundance can be readily discerned. Applying equation (5.2.8) to the linear rotor case, it

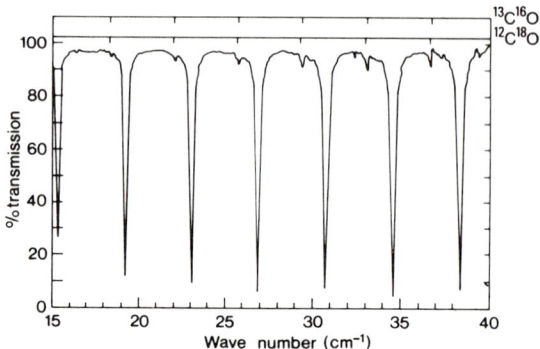

Fɪɢ. 5.5. Far-infrared spectrum of carbon monoxide, pressure 422 torr, path length 903 mm, from Fleming and Chamberlain [471].

follows that to the rigid rotor approximation the intensity of an absorption line will be given by

$$I(J'') = C[2B(J'' + 1)^2 \exp[-BhcJ''(J'' + 1)/kT]\} \qquad (5.2.10)$$

where C is a constant for a given molecule. The most intense line therefore occurs for that value of J'' which is a root of

$$(2J'' + 1)(J'' + 1) = 2kT/hcB$$
$$= 1\cdot39(T/B) \qquad (5.2.11)$$

when B is expressed in cm^{-1}. Thus for CO ($B_0 = 1\cdot922\,529$, $D_0 = 6\cdot134 \times 10^{-6}$ cm^{-1}), the most intense line at room temperature occurs for $J'' = 9$ at $38\cdot43$ cm^{-1}. It is interesting to observe that the most intense line does not arise from the most heavily populated lower level. The line population, bearing in mind the $(2J + 1)$ fold degeneracy, will be given by

$$N(J) = (2J + 1)\exp[-BhcJ(J + 1)/kT] \qquad (5.2.12)$$

and this has its maximum when

$$J = \sqrt{\left(\frac{kT}{2Bhc}\right)} - \tfrac{1}{2}. \qquad (5.2.13)$$

For CO again, the most populated level therefore occurs for $J'' = 7$.

Rotational spectroscopy has one of its most important applications in the determination of the shapes and dimensions of gas-phase molecules. Thus for a linear molecule a plot of line centre wave number divided by $(J'' + 1)$ against $(J'' + 1)^2$ gives a straight line whose intercept is $2B$ and whose slope is $-4D$. If the molecule is just diatomic, the internuclear separation follows immediately from the B value, thus

$$B = \frac{0 \cdot 168\,576}{r^2} \left(\frac{m_1 + m_2}{m_1 m_2} \right) \mathrm{cm}^{-1}, \qquad (5.2.14)$$

where r is in nanometres and the unit for m is the relative atomic mass (^{12}C $= 12 \cdot 000$). If the linear rotor has more than two atoms (NNÓ for example) then the spectra of isotopic forms are necessary if all the individual bond lengths are to be deduced. One very important point here is that the molecular potential energy curve $V(r)$ (i.e. a plot of energy versus internuclear separation) will not in general be symmetric so the B value thus deduced will not equal the value appropriate to the equilibrium position at the bottom of the curve (see Fig. 5.11). This equilibrium separation (in fact unattainable for real molecules because of zero point motion) is usually called r_e and the corresponding B value is labelled B_e. The vibrational ground state B value which is what you get from rotational spectroscopy is then called B_0. B_0 and B_e are very close but detectably different and it is usual to apply a correction to B_0 to get B_e. This can be done analytically if a functional form is assumed for $V(r)$ or purely experimentally by determining values of B for the various vibrationally excited states B_1, B_2 etc. A plot of B_n against $(n + \frac{1}{2})$ has an ordinate intercept of B_e. Linear molecules with a centre of symmetry have no dipole moment so no first-order rotational spectrum is observed. They may, however, have a magnetic moment (e.g. O_2) and then a pure rotational spectrum may be observed but it will be much weaker than that arising from an electric dipole moment because μ_0 is so much less than ε_0 (see section 1.2.1). Nevertheless, because of the high concentration of O_2 in the atmosphere and because of the long path lengths involved, pure rotation lines of O_2 tend to be very prominent in the far-infrared spectrum of the atmosphere. The O_2 spectrum is complex because the molecule is a diradical with a triplet (in fact $^3\Sigma_g$) ground state and the overall mechanical angular momentum has to be combined with the electronic angular momentum to give the quantum number J. The mechanical angular momentum is defined by the quantum number N, but strictly speaking this is not a "good" quantum number. Nevertheless, apart from the $N = 1$ (ground) state, one can describe all the higher states in terms of a triplet made up from an $(N, J = N)$ level accompanied about $2\,\mathrm{cm}^{-1}$ below by a close doublet consisting of $(N, J = N+1)$ and $(N, J = N-1)$. The selection rules are $\Delta J = 0, \pm 1$ and $\Delta N = 0, \pm 2$. The $\Delta N = \pm 2$ rule reflecting in classical terms the geometrical fact that because a rotation through $180°$ about a perpendicular axis gives an identical configuration the magnetic dipole is rotating twice as fast as the molecule. In quantum mechanical terms because the nuclear spin of the oxygen atoms is zero the states with N odd do not exist. There will, as a consequence of the selection rules, be a complex microwave absorption near 60 GHz—essentially arising from spin-flip ($\Delta N = 0$, $\Delta J = \pm 1$) transitions—but at higher frequencies the spectrum is relatively simple

consisting of the $N \rightarrow N + 2$ triplets made up from a strong central component
$(J = N + 1 \rightarrow J = N + 1)$ accompanied at lower and higher frequencies respectively by the weaker components $(J = N \rightarrow J = N + 1)$ and $(J = N + 1$
$\rightarrow J = N + 2)$ [472]. The triplets (12·28, 14·16 and 16·24 cm^{-1}) and (23·85,
25·80 and 27·81 cm^{-1}) are well known to stratospheric spectroscopists. The
central line wave numbers are given quite well by the simple linear rotor
formula and the selection rule $\Delta N = +2$ thus

$$\tilde{v}(\text{cm}^{-1}) = 2·875\,36[2N + 3]. \qquad (5.2.15)$$

From the B_0 value (1·437 68 cm^{-1}) it follows that r_0 is 1·21 Å (i.e. 1·21
$\times 10^{-10}$ m).

The analysis of the rotational spectra of asymmetric rotors is very difficult
and in fact no closed form expression can be given for the energy levels as a
function of J. In fact even labelling the $(2J + 1)$ sub-levels of a given J presents
problems since K is not defined. The modern system is derived from imagining
the molecule distorted in two ways: firstly forced into being an accidental
prolate top, and secondly into being an accidental oblate top. In the prolate
limit the molecule would have a defined K value, say K_a, and in the oblate limit
it would have another value, say K_c. A given level can therefore be labelled
$J_{K_a K_c}$. This way of looking at things gives considerable insight into the energy
levels of an asymmetric rotor and is illustrated for the case of the H_2O
molecule in Fig. 5.6. The H_2O molecule is a very asymmetric rotor with

$$A_0 = 27·877\,\text{cm}^{-1}, \quad B_0 = 14·512\,\text{cm}^{-1} \quad \text{and} \quad C_0 = 9·285\,\text{cm}^{-1}.$$

The asymmetry is measured by the *asymmetry parameter* $\kappa = (2B - A - C)/$
$(A - C)$ and in terms of this parameter the energy levels of an asymmetric rotor
may be written

$$\mathscr{E}\,(\text{cm}^{-1}) = \tfrac{1}{2}(A + C)J\,(J + 1) + \tfrac{1}{2}(A - C)E\,(J_{K_a K_c}), \qquad (5.2.16)$$

where $E\,(J_{K_a K_c})$ is a dimensionless function of κ. $E\,(J_{K_a K_c})$ cannot of course be
written in closed form for arbitrary $J_{K_a K_c}$ but it has been listed for very
many values; thus for water $E(0_{00}) = 0$, $E(1_{01}) = \kappa - 1$, $E(1_{11}) = 0$, $E(1_{10})$
$= \kappa + 1$, etc. The calculated energy levels and parities of H_2O ($\kappa = -0·437\,72$)
are displayed in the centre of Fig. 6.4 and some observed transitions are
indicated. These obey the selection rule

$$\Delta J = 0, \pm 1$$

and the parity selection rules

$$(+ +) \longleftrightarrow (- -) \quad \text{and} \quad (+ -) \longleftrightarrow (- +).$$

These selection rules, interestingly, divide the H_2O levels into two non-
communicating sets, rather reminiscently of *ortho* and *para* hydrogen or of
singlet and triplet helium. The prolate and oblate limits, derived by first letting
C_0 become equal to B_0 and second by letting A_0 become equal to B_0, are also

shown in the Figure. One can imagine that as one approaches the actual condition from either extreme the $K \neq 0$ levels split into two levels and these diverge more and more as the asymmetry increases. This gives a good idea of the origin of the complexity of asymmetric rotor spectra but the symbolism cannot always be used to derive the selection rules. This is because we have

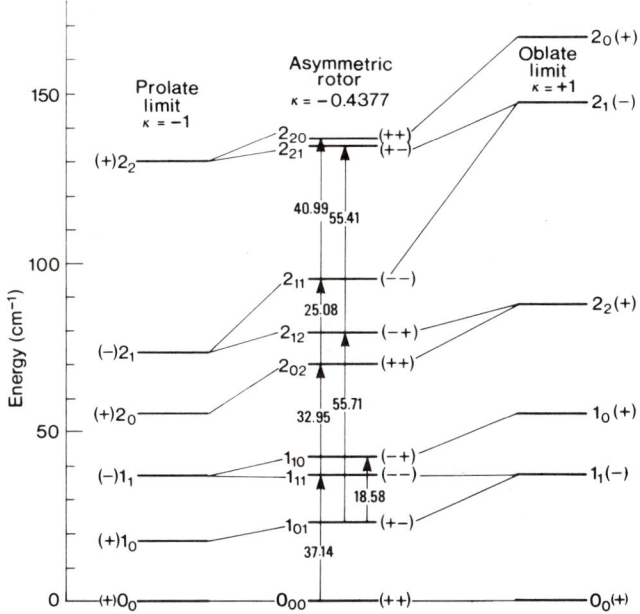

FIG. 5.6. Some low-lying energy levels and the corresponding transitions for H_2O. The energy levels for the equivalent prolate and oblate symmetric rotors are shown to the right and left respectively.

manipulated the molecule to make it merely an *accidental symmetric top* and the dipole moment will not necessarily lie in the correct direction. Thus for water, a "b-type" asymmetric rotor, the dipole moment lies along the intermediate (b) momental axis and will not therefore lie along the unique momental axis at either extremity. This is well brought out in the Figure where the strong $37 \cdot 14 \text{ cm}^{-1}$, $(J = 0 \rightarrow 1)$, rotation line would apparently be forbidden for either symmetric top limit. In fact, because of the anomalous position of the dipole moment, the symmetric top selection rules given earlier become inappropriate. In fact the selection rules are more like $\Delta J = +1$, $\Delta K = +1$ but in interpreting these the parity conditions in Fig. 5.6 have to be borne in mind. The conclusion, for this "b-type" asymmetric rotor is that even in its two limiting symmetric rotor forms the spectrum would be quite complex. In fact it would, apart from a few frequency shifts, look remarkably like that of H_2O on its normal state. For a case "a" or case "c" asymmetric rotor, this would not be the situation and the spectrum in one limiting form would be the simple line spectrum of a symmetric rotor. The dipole moment of the water molecule is quite high (1·84 D) and its rotational transitions are therefore strong. In fact the transmission of the atmosphere in the millimetre/far-infrared regions is dominated by the pure rotation spectra of H_2O vapour [473]. Some of the lower frequency lines of H_2O are listed in Table 5.2. The first line, at 0·74 cm^{-1}, i.e. 22·235 GHz, was notorious during the evolution of wartime radar systems, for absorption in the atmosphere due to it made the expensively developed K-band radar systems virtually useless in high-humidity situations. Nevertheless this fiasco stimulated great interest in the rotational spectra of molecules and after the war when the freely available surplus microwave hardware was being used for the infant science of microwave spectroscopy this transition was extensively studied [474]. Some of the low-frequency pure rotation lines of $H_2{}^{16}O$ are shown in Fig. 5.7

TABLE 5.2

Pure rotational transitions in $H_2{}^{16}O$

Transition	Wavenumber cm^{-1}	Relative intensity	Transition	Wavenumber cm^{-1}	Intensity
$5_{23} \rightarrow 6_{16}$	0·741 683	VW	$4_{13} \rightarrow 4_{22}$	40·283	S
$2_{20} \rightarrow 3_{13}$	6·114	W	$2_{11} \rightarrow 2_{20}$	40·987	S
$4_{22} \rightarrow 5_{15}$	10·843	W	$6_{52} \rightarrow 7_{43}$	42·62	VW
$3_{21} \rightarrow 4_{14}$	12·682 0	W	$7_{34} \rightarrow 8_{27}$	43·24	VW
$3_{30} \rightarrow 4_{23}$	14·943 7	W	$5_{32} \rightarrow 6_{25}$	44.10	M
$1_{01} \rightarrow 1_{10}$	18·577 4	M	$5_{14} \rightarrow 5_{23}$	47·05	VS
$4_{41} \rightarrow 5_{32}$	20·704 3	W	$6_{33} \rightarrow 7_{26}$	48·060	W
$2_{02} \rightarrow 2_{11}$	25·085 1	M	$5_{42} \rightarrow 6_{33}$	51.434	W
$3_{31} \rightarrow 4_{22}$	30·560	VW	$4_{04} \rightarrow 4_{13}$	53·445	S
$4_{31} \rightarrow 5_{24}$	32·366	W	$2_{12} \rightarrow 2_{21}$	55·406	S
$1_{11} \rightarrow 2_{02}$	32·951	M	$1_{01} \rightarrow 2_{12}$	55.704	VS
$3_{03} \rightarrow 3_{12}$	36·605	VS	$2_{12} \rightarrow 3_{03}$	57·269	VS
$0_{00} \rightarrow 1_{11}$	37·137	M	$6_{24} \rightarrow 6_{33}$	58·772	S
$2_{21} \rightarrow 3_{12}$	38·465	M	$6_{42} \rightarrow 7_{35}$	58·895	W
$5_{41} \rightarrow 6_{34}$	38·643	W	$6_{15} \rightarrow 6_{24}$	59·871	S
$3_{12} \rightarrow 3_{21}$	38·791	VS	$7_{25} \rightarrow 7_{34}$	59·940	S

Apart from H_2O and its isotopic variants the other rigid asymmetric rotors of interest are ozone O_3, sulphur dioxide SO_2 and nitrogen dioxide NO_2, the first because it is a vital protective component of the upper atmosphere and its concentration there can be monitored via its pure rotation spectrum in emission and the latter two because they are important atmospheric poll-utants. In all three cases, it is found that the extremely complex line pattern shows considerable simplification at resolutions large compared to the line width. This is because the lines tend to bunch or blend and a low resolution spectrum shows isolated strong features which may even display regularity—like for example the so-called "Q-branches" of SO_2 [476]. This aggregation of lines make it feasible to measure concentrations etc. of these molecules with relatively low resolution. One interesting feature of the NO_2 spectrum is that the lines show splittings due to the unpaired electron spin. Some typical O_3, SO_2 and NO_2 spectra are shown in Fig. 5.8.

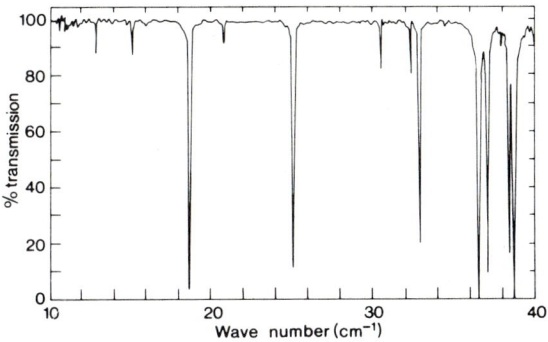

FIG. 5.7. Submillimetre absorption spectra of water vapour: pressure 18 torr, path length 203 mm, resolution 0.07 cm^{-1} (from Fleming and Gibson [475]).

The rigid rotor approximation is usually very good but there is a class of molecules for which even first order it fails, namely those for which there are two or more equivalent or nearly equivalent positions separated by a finite barrier. In classical mechanics this would be of no consequence, but in quantum mechanics the molecule can "tunnel" through the barrier to reach the alternative positions. The actual wave function then becomes a mixture of those for the equivalent positions and all the levels split into two or more components. The best known example is the pyramidal ammonia molecule NH_3 which can go over to an exactly equivalent (but inverted) position if the nitrogen atom passes through the plane of the three hydrogen atoms [478]. The barrier to inversion is that required to deform the molecule into the planar configuration: experimentally this is found to be 2077 cm^{-1}. The situation is illustrated in Fig. 5.9. The v_2 or bending vibration, which is the one which when sufficiently excited would take the molecule over the barrier, becomes split into two components, one symmetrical with respect to inversion and the other antisymmetrical. The ground state splitting is only 0.7934 cm^{-1} and transitions between the two components for a given J lead to a rich microwave spectrum at 23.786 GHz. This was in fact the first microwave absorption ever to be observed—by Cleeton and Williams in 1934. The spectrum is rich because the splitting increases with J. At much higher frequencies there is the pure rotation spectrum each "line" of which, apart from the first ($J = 0 \rightarrow J = 1$) at 19.096 cm^{-1}, is a doublet with a spacing of 1.59 cm^{-1}. The origin of this splitting is shown in the right inset of Fig. 5.9. The nuclear statistics of NH_3 are such that for $K = 0$ only one component of the doublet can exist since the overall wave function would be symmetric for the other. At very high resolution each component of the doublet breaks up into its K-substructure (due to centrifugal effects) and at ultrahigh, and in fact so far not achieved resolution there would be a further hyperfine splitting due to the ^{14}N quadrupole. In the first excited state ($\tilde{\nu} \sim 936 \text{ cm}^{-1}$) the splitting is much larger ($\sim 36 \text{ cm}^{-1}$) and since there will be a measurable population due to the non-negligible Boltzmann factor, lines may be observed in the spectrum due to transitions within this state. The result is a very complex spectrum—some exemplary sections are shown in Fig. 5.10.

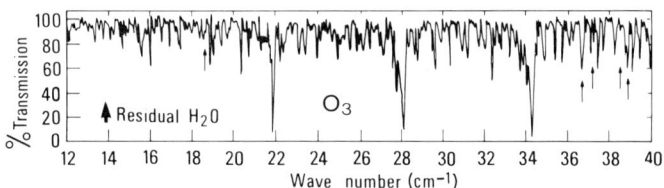

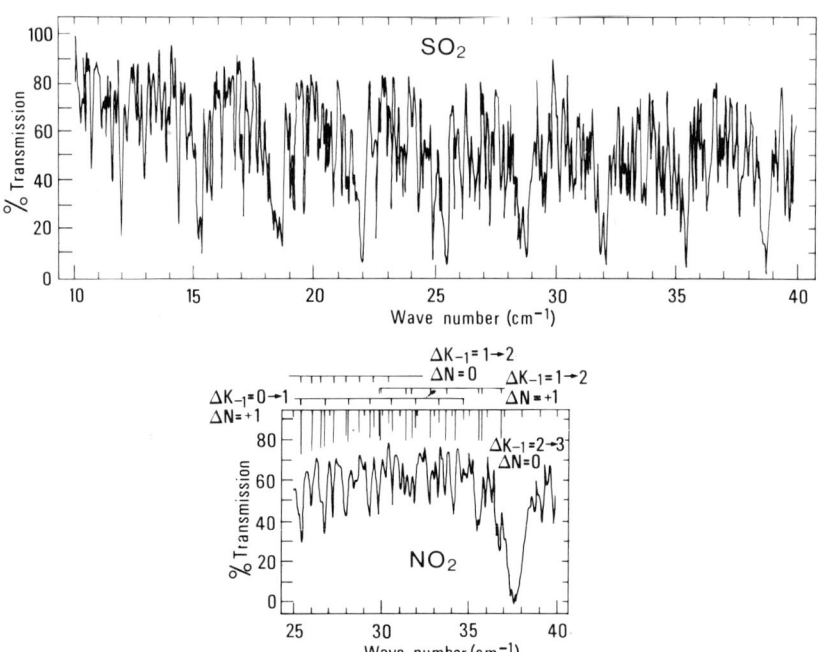

Fig. 5.8. Far-infrared spectra of ozone, sulphur dioxide and nitrogen dioxide: the path length was 933 mm for all the gases and the pressures were 18, 18.5 and 400 torr respectively. The resolution was 0.07 cm^{-1}. All the spectra are taken from the work of Fleming and his colleagues [476, 477].

The second type of molecule of this class is typified by methanol CH_3OH where the methyl group has three equivalent positions with respect to the OH group and the barrier separating them is low because all that is involved is rotation about a σ bond. The finite barrier then arises from non-bonded interactions and can range from virtually zero (as in CH_3NO_2) to quite high ($\sim 1\,200$ cm^{-1} as in CH_3CH_2F). For the reasonably high barrier cases one can think of a potential function similar to that of NH_3 but this time with three wells. The methyl group will therefore undergo torsional oscillations with a typical frequency of between 100 and 300 cm^{-1} in one of the wells but will occasionally tunnel through to continue the oscillation in an adjacent well. Just

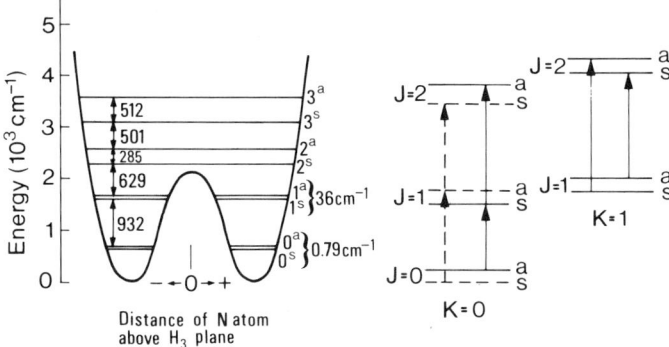

FIG. 5.9. Inversion doubling in the v_2 fundamental of NH_3. The right inset shows the rotational transitions $J = 0 \rightarrow J = 1$ and $J = 1 \rightarrow J = 2$. The levels and transitions shown dashed do not exist because of nuclear statistics for NH_3 but are present for ND_3.

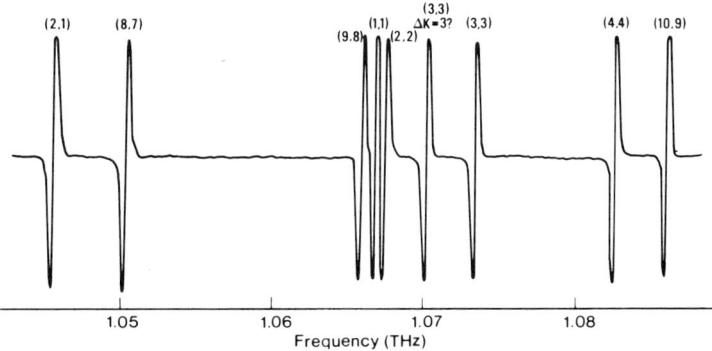

FIG. 5.10. Part of the inversion spectrum of ammonia in the $v_2 = 1$ state. The figures in brackets are the (J, K) values of the (a,s) pair involved in the transition. The spectra (taken from the work of Krupnov [101]) were observed by coherent microwave techniques and as usual are presented in first derivative form.

as for ammonia the tunnelling leads to a splitting of each of the rotational levels and the pure rotation spectrum becomes more complex. If, as in the case of methanol, the basic frame is an asymmetric rotor, the rotation spectrum will be complex to start with and when all the splittings due to the tunnelling are added it can present a major decipherment problem. For the commonest rotating top, viz. CH_3, the split states are to first order just doublets because the CH_3 group, having a threefold axis of symmetry, is itself a symmetric rotor and its rotational states can be divided into just A and E components (see section 5.2.3.1) where the E states are doubly degenerate. However, for many molecules the rest of the frame does not possess a threefold or higher axis; in other words the molecule is an asymmetric rotor, so the E-type double degeneracy is lifted. For molecules such as CH_2 DOH there is no symmetry at all and the lines would split into triplets having variable spacings and variable intensity ratios. The result would be a virtual forest of lines!

The complex structure of the rotational energy states of asymmetric rotors with free or merely hindered internal rotation is an interesting spectroscopic problem [479] but recently it has become of considerable technical importance. This is because of the introduction of optically pumped lasers (see sections 2.4.3 and 6.2.5) which for their operation require a near or exact coincidence of the frequency of a high-power pumping laser (e.g. CO_2 or N_2O) with a transition of the molecule to be pumped. Clearly the more lines there are per frequency interval, the better the chance of a coincidence. One of the best optically pumped lasers is that based on methanol and one reason for this is undoubtedly the very complex multi-line structure of the absorption bands of methanol in the 10 μm region. A simple calculation based on the ideas of Gebbie, Topping, Illsley and Dennison [480] would show that on average there will be a line every 100 MHz or so whilst if the laser line chanced to lie in a more densely populated region this might reduce to 20 MHz or so. Bearing in mind that the CO_2 laser produces a large number of lines (see section 6.2.2) and that it is available in isotopic and sequence band versions, the chance of a close enough coincidence becomes very high. Methanol is only slightly asymmetric ($A_0 = 5\cdot247\,18$ cm^{-1}, $B = 0\cdot823\,12$ cm^{-1}, $C_0 = 0\cdot792\,98$ cm^{-1}) so a meaningful labelling of its rotational states can be made in terms of the four quantum numbers, J, K, n and τ where n is the vibrational quantum number of the torsional oscillation and τ labels the three positions of minimum energy of the CH_3 top. In terms of the full rotational group [481] the rotational levels may be divided into the three classes A, E_1 and E_2 of C_3 and these are specified by the rules

$$A; \quad\quad \tau + K = 3N + 1$$
$$E_1; \quad\quad \tau + K = 3N$$
$$E_2; \quad\quad \tau + K = 3N + 2$$

where N is an arbitrary integer.

For an overall symmetric top the $\pm K$ states are, as usual, degenerate but for an overall asymmetric top, the A states split into A^+ and A^- components whose splitting increases slowly with J but decreases rapidly with K [482]. The asymmetry causes much wider splittings of the $\pm K$ components of the E_1 and E_2 states but time reversal symmetry demands that $E_1(K)$ and $E_2(-K)$ be degenerate and likewise $E_2(K)$ and $E_1(-K)$, so the E states remain doubly degenerate. This A, E_1, E_2 labelling is convenient`since, in terms of it, the symmetry selection rules take on the particularly simple form

$$A \longleftrightarrow A, \qquad E_1 \longleftrightarrow E_1, \qquad E_2 \longleftrightarrow E_2$$

The K selection rules are of two types; there is the usual symmetric top selection rule $\Delta K = 0$, which leads to the so-called "a-type" transitions and in addition there is the $\Delta K = \pm 1$ rule which gives the "b-type" transitions. It is the presence of these two types together with the internal rotation splittings which leads to the spectral complexity mentioned earlier. In the region of the CO_2 laser lines, the most prominent absorption band of CH_3OH is the v_5 fundamental (mostly C–O stretching) whose band centre is at $1033\cdot622$ cm^{-1}. The transition from $(J'', K'', n'', \tau'') = (16, 8, 0, 1)$ to $(J', K', n', \tau') = (16, 8, 0, 1)$, which is in the Q-branch of this band (see next section) is virtually coincident with the $9P(36)CO_2$ laser line at $1031\cdot478$ cm^{-1}. Transitions from the pumped v_5 (16, 8, 0, 1) state to lower states give the well-known CH_3OH laser lines

$$
\begin{aligned}
v_5(16, 8, 0, 1) &\rightarrow v_5(15, 7, 0, 2); &118\cdot834~\mu m \\
v_5(16, 8, 0, 1) &\rightarrow v_5(15, 8, 0, 1); &392\cdot069~\mu m \\
v_5(16, 8, 0, 1) &\rightarrow v_5(16, 7, 0, 2); &170\cdot576~\mu m
\end{aligned}
$$

Nearly all the strong lines of the CH_3OH laser have now been assigned [483] within this formalism. These transitions are, of course, not normally perturbed so one can use measures of their absolute frequencies to test the new and sophisticated internal rotation Hamiltonians which the theoreticians are starting to introduce [484].

5.2.3 Vibrational spectroscopy of molecules

It is usual to begin a discussion of molecular vibration by considering the simplest case—the diatomic molecule AB. The main reason for this is that it is easier initially to follow the development of the argument in one dimension but it turns out that all of the concepts which emerge can be transferred to help in the analysis of the more complicated polyatomic case in which it is necessary to work in an abstract space of many dimensions. The diatomic molecule has a single degree of vibrational freedom and this is completely specified by the value of the internuclear separation r. A schematic representation of how the potential energy varies with r is shown in Fig. 5.11. Although this curve is continuous, the actual molecule is constrained to have only discrete values of the energy and these energy levels are also indicated schematically by horizontal lines in the Figure. The form of the potential curve arises from a balance of two forces, the intense repulsion of the inner core electrons and of the nuclei and the attraction due to overlap of the bonding electron wave functions (see Fig. 5.3). The former leads to the very steeply rising edge at values only slightly less than r_e—the equilibrium separation—and the latter to the more gently rising edge for values greater than r_e. Strictly speaking the energy of interaction is zero at both infinite and at zero separation, the latter corresponding to the united atom, but this point is irrelevant for vibrational spectroscopy and potential functions, poorly behaved at $r = 0$, may neverthe-less give an excellent fit to the observed vibrational energy levels. The convention of a zero of potential energy at $r = \infty$ does, however, imply that the bonding energy $V(r_e)$ is negative and in fact equal to $-D_e$, where D_e is the dissociation energy. Potential energy curves such as 5.11 should in principle be calculable *ab initio* by the methods of quantum mechanics but this is, in general, impossibly difficult and reliable calculations have in fact been carried out only for H_2^+ and H_2. Numerical solutions to various degrees of approximation are possible for other cases, at the expense of considerable man

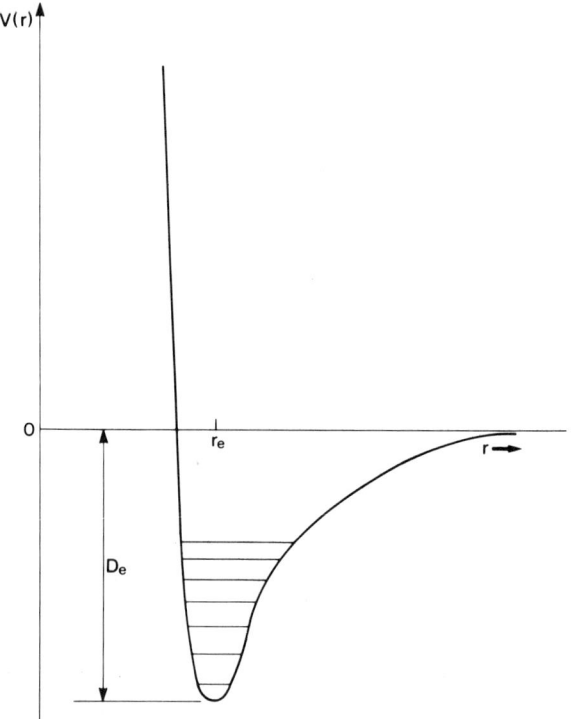

Fig. 5.11. Potential energy curve for a diatomic molecule.

and computer time, but these too have only been carried out for a few isolated cases. The more general approach is based on the Born–Oppenheimer approximation that is essentially on the separability of electronic and nuclear motion. One adopts an analytical form for $V(r)$ as an approximation to the true potential function and then, assuming that this is the correct solution for the electronic quantum mechanical problem one uses it as the starting point for a solution of the vibrational wave equation. In this way, provided one has been clever enough to choose an analytical form leading to a solvable wave equation, one can derive analytical expressions for the energy levels. The form chosen for $V(r)$ will usually be severely constrained, so if one has a reasonable number of adjustable parameters one will necessarily get a good fit to the true curve and the energy levels calculated will be in fair agreement with the experimental values. The simplest analytical expression is that of Lennard–Jones,

$$V(r) = D_{\mathrm{e}} \left[\left(\frac{r_{\mathrm{e}}}{r}\right)^{12} - 2 \left(\frac{r_{\mathrm{e}}}{r}\right)^{6} \right], \qquad (5.2.17)$$

which strictly speaking has no adjustable parameters since D_{e} and r_{e} can be determined experimentally. The Lennard–Jones curve for HCl ($D_{\mathrm{e}} = 7 \cdot 392 \times 10^{-19}$ J, $r_{\mathrm{e}} = 1 \cdot 2746 \times 10^{-10}$ m) is illustrated in Fig. 5.12. The Hooke's Law

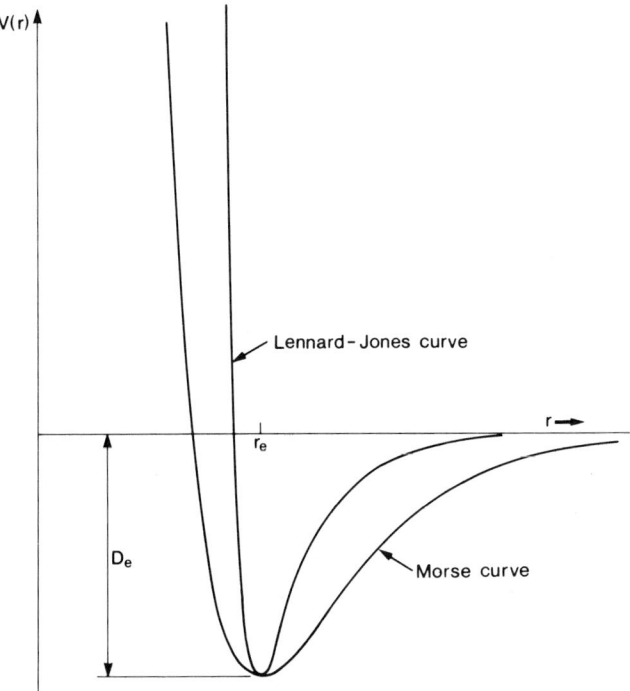

Fig. 5.12. Comparison of the Morse and Lennard–Jones Potential Curves for $H^{35}Cl$.

(or force) constant f of the bond is given by

$$f = \frac{d^2 V}{dr^2} = \frac{72 D_e}{r_e^2} \tag{5.2.18}$$

which for HCl is found to be $3 \cdot 276 \times 10^3 \, \text{N m}^{-1}$. The classical vibration frequency in the "zero-amplitude" (that is the harmonic oscillator) limit is given by

$$v_e = \frac{1}{2\pi} \sqrt{\left(\frac{f}{\mu}\right)}, \tag{5.2.19}$$

where μ is the reduced mass given by

$$\mu^{-1} = m_A^{-1} + m_B^{-1}. \tag{5.2.20}$$

The value for HCl is found to be 227 THz which in the units familiar to infrared spectroscopists is $7562 \, \text{cm}^{-1}$. This is considerably higher than the observed value ($\sim 3000 \, \text{cm}^{-1}$) so the Lennard–Jones formalism is not an accurate approach for HCl, but nevertheless it is interesting that the vibration frequency can be calculated to about a factor of $2 \cdot 5$ merely knowing the dissociation energy and the equilibrium bond length. A more frequently used empirical relation, however, is the *Morse curve*,

$$V(r) = D_e \left[\exp\left(-2a(r - r_e)\right) - 2 \exp\left(-a(r - r_e)\right) \right], \tag{5.2.21a}$$

which is sometimes written, with a conventional shift to $V(r_e) = 0$, as

$$V(r) = D_e \left[1 - \exp\left(-a(r - r_e)\right) \right]^2. \tag{5.2.21b}$$

This curve has a single adjustable parameter a. The value of a can be determined by requiring that the curve correctly gives any other observable spectroscopic quantity. Thus if $V(r)$ is required to have the correct second derivative at its minimum:

$$f = 2 D_e a^2. \tag{5.2.22}$$

For HCl again, the experimental value of f is $5 \cdot 16 \times 10^2 \, \text{Nm}^{-1}$ and this leads to a value of a of $1 \cdot 87 \times 10^{10} \, \text{m}^{-1}$. The Morse curve for HCl with this value of a is also shown in Fig. 5.12. The reason for the high force constant produced by the Lennard–Jones approach is seen from the Figure to lie in the higher curvature of this much "sharper" curve. The gentler Morse curve is in fact quite a good approximation to the observed curve for the case of HCl.

One can, as mentioned above, hope to calculate the vibrational energy levels corresponding to the chosen potential function by direct solution of the wave equation but since in practical spectroscopy one will always be working far down the potential curve it is usual, as a first approximation, to replace the potential curve by the parabola which fits it near r_e. This is the harmonic oscillator approximation. The solution of the harmonic oscillator problem has already been given in section 1.5.2 where it was shown that the energy levels are equispaced with their separation given by hv_e where v_e is the classical oscillation frequency (5.2.19). The lowest energy of the quantum oscillator is not however $-D_e$ but is instead one half quantum of the vibrational motion higher. This "zero-point energy" is important in chemical physics, especially for hydrogen containing molecules where the amplitude can be considerable. The second difference between the two systems of mechanics is that the quantum oscillator cannot have an arbitrary amplitude since its energy is fixed. Thus by identifying the quantum and classical energies one arrives at the result

$$\Delta r_n = \sqrt{\left[\frac{(n+\frac{1}{2})h}{2\pi^2 \mu v}\right]}, \qquad (5.2.23)$$

where Δr_n is the amplitude of motion in the nth vibrational state. For HCl in the ground state (i.e. $n = 0$) the value found is $\Delta r_0 = 1.096 \times 10^{-11}$ m, that is about 8 % of the equilibrium separation. This underlines what was said above about zero-point motion. For higher values of n the amplitude will become large and the harmonic oscillator approximation will become less and less appropriate. One can attempt to get a better polynomial fit to the potential curve by introducing cubic, quartic etc. terms and then one has an anharmonic oscillator. The energy levels of the anharmonic oscillator may be written

$$\mathscr{E}_n = hv_e(n+\tfrac{1}{2}) - x_e\, hv_e(n+\tfrac{1}{2})^2 + y_e\, hv_e(n+\tfrac{1}{2})^3. \qquad (5.2.24)$$

Because of anharmonicity, the energy levels are no longer equispaced and tend to crowd more and more together as the energy increases (see Fig. 5.11). It is always found that $v_e \gg x_e v_e \gg y_e v_e$ which explains the reasonable success of the harmonic oscillator model. In fact y_e is usually so small that it is only known for a few molecules. For HCl, experiment gives

$$
\begin{array}{ll}
n = 0 \rightarrow n = 1 & \tilde{v} = 2885.82 \text{ cm}^{-1} \\
n = 1 \rightarrow n = 2 & \tilde{v} = 2782.23 \text{ cm}^{-1} \\
n = 2 \rightarrow n = 3 & \tilde{v} = 2678.92 \text{ cm}^{-1}
\end{array}
$$

from which it follows that $\tilde{v}_e = 2989.74 \text{ cm}^{-1}$, $x_e = 0.0174$ and $y_e = 1.9 \times 10^{-5}$. Direct solution of the wave equation incorporating the Morse potential gives an energy level formula containing only the first two terms of (5.2.24). One has thus an analytical expression for x_e, viz.

$$x_e = hv_e/4D_e \qquad (5.2.25)$$

which for HCl is 0.0174. The perfect agreement with the experimental value is worth noting. Attempts have been made (see Herzberg [485] for details) to introduce modifications to the Morse potential to make it fit experiment even better but the success of the simple form makes these refinements only of academic interest.

For a transition to be allowed in infrared absorption or emission the motion in question must be associated with an oscillating electric or magnetic field but in the practical world of laboratory spectroscopy only an oscillating electric dipole (see section 2.5.1) is significant. In the case of the diatomic molecule one may expand the dipole moment function thus:

$$\mu = \mu_0 + \left(\frac{d\mu}{dr}\right)(r - r_e) + \tfrac{1}{2}\frac{d^2\mu}{dr^2}(r - r_e)^2 + . \qquad (5.2.26)$$

The simplest assumption one can make is that apart from μ_0 only the $(d\mu/dr)$ term is finite. This assumption of a linear dependence of μ upon r is rather oddly known as the approximation of "electrical harmonicity". For the diatomic molecule of course, $d\mu/dr$ can only be finite if the molecule is made up of two different atoms, HCl for example. If the oscillator is both mechanically and electrically harmonic, then the selection rule governing transitions between the energy levels is

$$\Delta n = \pm 1. \qquad (5.2.27)$$

If either of these requirements is not met, then overtones $\Delta n = \pm 2, \pm 3$ etc. may appear in the spectrum but they will always be weaker than the fundamental. The observed spectrum therefore consists of an intense band at the fundamental frequency accompanied by very much weaker bands at frequencies approximating the harmonics of the fundamental. Another effect of mechanical anharmonicity is that the transitions $1 \to 2$, $2 \to 3$, etc. become separated from the $0 \to 1$ transition and the effect in the absorption spectrum is that the strong fundamental will be accompanied by resolvable, much weaker, satellites. These satellites have strongly temperature sensitive intensities because their lower states are excited and for this reason are nearly always referred to as "hot-bands". Occasionally, however, and especially in the laser literature, they may be called "sequence-bands".

Having so far considered a non-rotating molecule we must now examine what happens when the molecule is allowed to both vibrate and rotate. The simplest assumption is separability of rotational and vibrational motion. Each vibrational state would have a set of rotational levels associated with it and the observed spectrum would be obtained by combining the vibrational selection rule (5.2.27) with the rotational selection rule (5.2.4). The result would be a series of lines with regular spacing of $2B$, those to higher frequencies than ν_0 being an *R-branch* with $\Delta J = +1$ and those at lower frequencies than ν_0 being a *P-branch* with $\Delta J = -1$. There would be a gap in the regular series at ν_0 itself. This absence of a *Q-branch* with $\Delta J = 0$ is characteristic of the parallel bands (i.e. those with the oscillating moment parallel to the axis) of linear molecules. It arises because photons have unit angular momentum and therefore cannot connect levels of the same J unless there is some other mechanism (see later) for conserving the angular momentum. Spectra like this are in fact observed (see for example, Fig. 6.1) but on closer examination two complications emerge. Firstly the lines are not quite regularly spaced, because of centrifugal distortion, as discussed earlier, but in addition the B values in the upper and lower states are not the same. This is of course not surprising since one cannot strictly separate rotational and vibrational motion. The best that can be done is to take B to be the average over the vibrational motion. If one does this one gets the relation

$$B_n = B_e - \alpha_e \ (n + \tfrac{1}{2}) + \text{higher terms.} \qquad (5.2.28)$$

In the light of this, all the B values discussed in section 5.2.2 should strictly be called B_0 values and the r_0 values derived from them will not exactly equal the r_e values. Thus for our standard example of $H^{35}Cl$ we have $B_e = 10 \cdot 5909$ and $\alpha_e = 0 \cdot 3019$, so $B_0 = 10 \cdot 4400$. The r_0 value is therefore $1 \cdot 284 \times 10^{-10}$ which is significantly different from r_e ($1 \cdot 275 \times 10^{-10}$ m). Bearing in mind the variation of B, the lines in the rotation/vibration spectrum of a diatomic molecule will now be given by

$$\text{R–branch } \tilde{v} = \tilde{v}_0 + (B' + B'') (J + 1) + (B' - B'') (J + 1)^2, \quad (5.2.29a)$$
$$\text{P–branch } \tilde{v} = \tilde{v}_0 - (B' + B'') J + (B' - B'') J^2, \qquad (5.2.29b)$$

which can again be regarded as a single series with one member missing. Depending on the value of $(B' - B'') = \Delta B$, one branch or the other will come to a "head", that is the lines will crowd together initially and then ultimately move in the opposite direction. The formation of "heads" is commonly observed in emission spectroscopy where high J values may be involved but for light molecules studied in absorption the head may not be obvious because the corresponding ground state J values may not be populated.

5.2.3.1 *Vibrational spectra of polyatomic molecules*

A polyatomic molecule containing N atoms, will be completely defined in space by a set of $3N$ cartesian coordinates. Of these, 3 will serve to locate the centre of mass and thus to define the translational motion whilst 3 more will serve to locate a set of inertial axes, fixed in the molecule, and thus to define the rotational motion. There remain $3N$-6 degree of freedom and these clearly serve to define the vibrational motion of the molecule. The normal modes of vibration can be calculated, to surprisingly good accuracy, by a simple model (see Appendix 6 for details) in which the molecule (now non-rotating and translating) is considered to be made up of a set of point masses (the atoms) connected by a set of Hooke's law springs. The resulting eigen frequencies will, in general, be distinct, but if the molecule possesses a three-fold or higher axis of symmetry pairs of frequencies may become identically equal, that is be degenerate. For the still higher symmetry of tetrahedral or octahedral molecules, groups of three frequencies may form degenerate triplets. The analysis is carried out using the concepts of Group Theory. The symmetry operations which send a molecule into itself from a group, and since there is always one point which is sent into itself by all the operations this group is a point group. The point group will have a set of irreducible representations, that is a set of primitive matrices with a multiplication table that exactly matches that of the group, and any property of the molecule can be assigned to one or more of these irreducible representations. Thus the set of atomic cartesian coordinates forms a reducible representation of the group and it may be factored down into its constituent irreducible components. The method of doing this makes use of the group characters, that is the traces or sums of the diagonal elements of the appropriate matrices. The characters for all the irreducible representations are listed for each possible point group in the standard works, for example Wilson, Decius and Cross [486]. The characters of the matrices making up the reducible representation have to be found by inspection, but this is not arduous since only atoms which are sent into themselves by the particular symmetry element in question contribute to the trace. As an example, we may consider the tetra-hedral molecule XY_4 whose point group is designated T_d. This group is made up of the 24 elements, E (the identity operation), $8C_3$ (threefold axes), $3C_2$ (two-fold axes), $6S_4$ (rotation-reflection axes) and $6\sigma_d$ (planes of symmetry). It will be seen from Table 5.3 that the three rotations and the three translations form degenerate triplets in F_1 and F_2 respectively. The dipole moment operator has the same symmetry as a translation and it follows therefore that only F_2 type fundamentals are allowed in infrared absorption. The six components of the polarisability tensor α are divided in various combinations amongst A_1, E and F_2 so it follows that fundamentals of any of these types will be allowed in the Raman effect. Taking now the representation formed by the 15 cartesian coordinates, Γ_x, one finds its characters to be

	E	$8C_3$	$3C_2$	$6S_4$	$6\sigma_d$
χ_{Γ_x}	15	0	-1	-1	3

TABLE 5.3
Character table for the tetrahedral group

T_d	E	$8C_3$	$3C_2$	$6S_4$	$6\sigma_d$	Primitive displacements	Spectroscopic parameters
A_1	1	1	1	1	1		$\alpha_{xx}+\alpha_{yy}+\alpha_{zz}$
A_2	1	1	1	-1	-1		
E	2	-1	2	0	0		$\alpha_{xx}+\alpha_{yy}-2\alpha_{zz}$
							$\alpha_{xx}-\alpha_{yy}$
F_1	3	0	-1	1	-1	$R_x R_y R_z$	
F_2	3	0	-1	-1	1	$T_x T_y T_z$	$\alpha_{xy},\alpha_{yz},\alpha_{zx}$
							μ_x,μ_y,μ_z

The decomposition of Γ_x is achieved now by means of the relation

$$n_j = \frac{1}{h} \sum_i \chi^i_{\Gamma_x} \chi^i_{\Gamma_j} \tag{5.2.30}$$

where n_j is the number of times irreducible representation Γ_j occurs in the decomposition, h is the order of the group (in this case 24), i signifies the elements of the group and $\chi^i_{\Gamma_j}$ is the entry in the group character table corresponding to Γ_j and element i. Applying the formula to the XY_4 case one has

$$\Gamma_x = A_1 + E + F_1 + 3F_2 \tag{5.2.31}$$

which after subtracting the translations and rotations gives

$$\Gamma_Q = A_1 + E + 2F_2 \tag{5.2.32}$$

for the representation formed by the normal coordinates. One has therefore the result that a molecule such as CCl_4 will have only four distinct normal modes all four of which will be active in the Raman effect but only two of which will be active in infrared absorption. The spectra of liquid carbon tetrachloride are shown in Fig. 5.13. It will be seen that broadly speaking these predictions are fulfilled but in the infrared several weaker bands corresponding to overtone, combination, difference or even forbidden fundamental transitions are observed as well. There are several mechanisms which could lead to this breakdown of the selection rules. Firstly, the assumption of electrical harmonicity is not as good as is that of a linear dependence of polarisability on the internuclear separation: consequently first-order forbidden bands are seen more commonly in the infrared spectrum than they are in the Raman spectrum. Secondly the selection rules are derived for the rather hypothetical case of a completely isolated molecule interacting solely with the electromagnetic field: they cannot be expected to hold so strictly for the case of molecules in condensed phases interacting strongly with one another. Thirdly, the vibrations of real molecules are anharmonic and this leads to relaxation of the harmonic selection rules: it also leads to the possibility of vibrational perturbation and to the "borrowing" of intensity by formally forbidden lines from allowed neighbours. It seems likely that all three mechanisms are operative for the case of liquid carbon tetrachloride.

It should be remarked, in passing, that for the linear molecule only two degrees of freedom are required to define its rotational motion. The third, rotation about the molecular axis, is to be regarded more properly as a form of electronic excitation. There are thus $3N$-5 modes for the N-atomic linear molecule but at least two of these will occur as a degenerate pair. Another point worth making is that if a molecule has a centre of symmetry, then any transition allowed in infrared absorption will be forbidden in Raman scattering and vice versa. This is the rule of mutual exclusion, which, because it is derived using parity arguments, holds rigorously even for the anharmonic oscillator.

In classical mechanics, one can have arbitrary excitations but it follows from the mathematical nature of the problem that these can always be resolved into linear combinations of the normal modes. In quantum mechanics, the only allowed excitations are those which can be resolved into integral combinations of the normal modes. Thus if an allowed excitation is resolved into

$$Q = n_1 Q_1 + n_2 Q_2 + \ldots + n_i Q_i + \ldots + n_{3N-6} Q_{3N-6} \qquad (5.2.33)$$

then the n_i will be integers—the vibrational quantum numbers. In classical mechanics, transitions between states like (5.2.33) could only take place by the

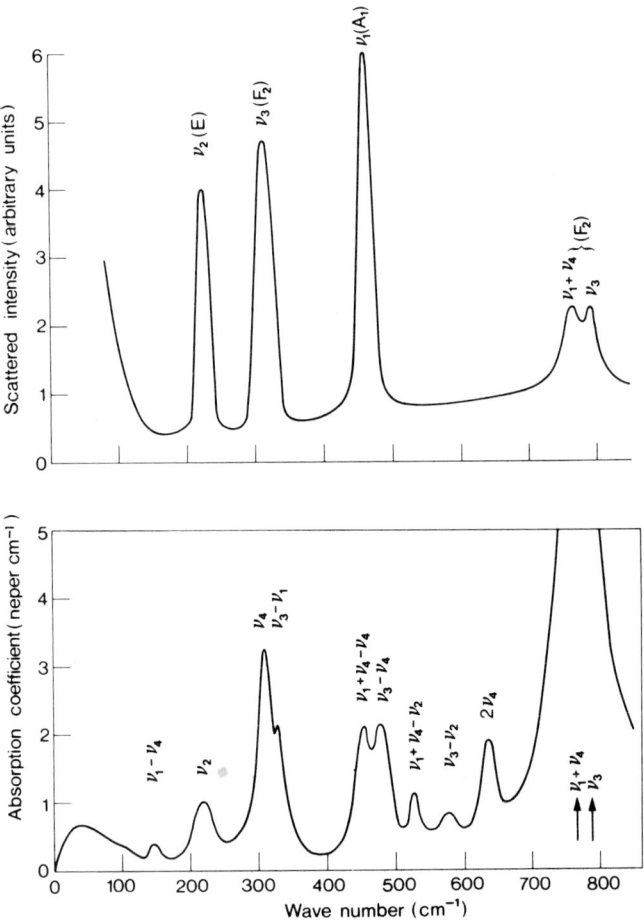

Fig. 5.13. Raman spectrum (upper) and infrared spectrum (lower) of liquid carbon tetrachloride.

molecule absorbing or emitting one of its eigen-frequencies and it follows by the Correspondence Principle that the quantum mechanical selection rule will be that transitions are only permitted between states if one and only one of the n_i changes and this by unity. What this means in practice is that ground state molecules can only absorb at their fundamental frequencies and only then if the symmetry rules mentioned above are also satisfied. Quite simple infrared spectra are therefore expected, especially for highly symmetrical molecules, and to a large extent this is what is observed, as was illustrated above for carbon tetrachloride. It is, in fact, this property of infrared spectra which makes infrared spectroscopy such a powerful analytical tool. The infrared spectrum of a quite complicated molecule will consist of a series of resolved bands spanning the whole region and this spectrum will be a unique "fingerprint" for that molecule. Even for the case of closely related molecules, the spectra will be sufficiently different for an unknown to be identified with certainty. A typical example is shown in Fig. 5.14. The method is not, however,

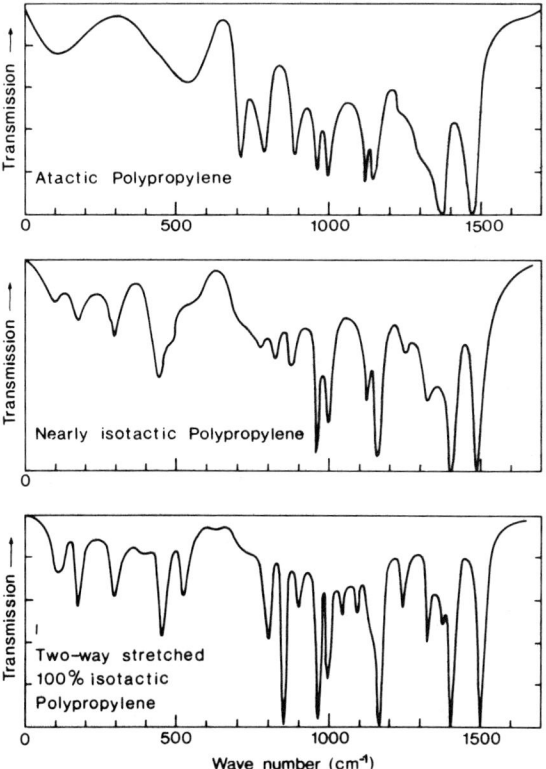

FIG. 5.14. Spectra of three different specimens of polypropylene. The samples merely differ in the way the macromolecules are disposed and in the conformation of the $(CH_2-C(CH_3)H)$ monomers with respect to one another.

restricted to the identification of pure substances since the spectrum of a mixture will consist of a linear superposition of those of its constituents. This area is a natural one for modern computer techniques and a medium-size computer with a good set of standard spectra in its file store can analyse quantitatively a mixture in a very short time. Another use of infrared spectroscopy is as a help in the deduction of the structure of a newly discovered or newly synthesised compound. This relies on the concept of "group frequencies" that is the one-to-one correspondence between certain structural groupings within the molecule and particular bands in the infrared spectrum. It is often possible to reduce the number of candidate structures down to a handful solely from a study of the infrared spectrum and then a few more tests will reveal the correct one. The method has been well described by Bellamy [5].

The mathematical problem of determining the eigen-frequencies and the normal modes of vibration of a polyatomic molecule has been solved, to the harmonic approximation, by the introduction of the Wilson FG matrix method, outlined in Appendix 6. This treatment can give an excellent account, to first order, not only of the vibrational frequencies, but also of the infrared and Raman intensities [487], the vibrational amplitudes [488], etc. To higher order, however, one must take account of anharmonicity. To do this one considers a 3N-6 dimensional space containing a set of potential wells of the form shown in Fig. 5.11. Each well is anharmonic in its own right, but in addition the wells perturb one another. The analogue of equation (5.2.24) is then

$$\mathscr{E}(n_1, \ldots \ldots n_{3N-6}) = \sum_i v_i(n_i + \tfrac{1}{2}) - \sum_{i=1}^{i=j} \sum_j x_{ij}(n_i + \tfrac{1}{2})(n_j + \tfrac{1}{2}) \quad (5.2.34)$$

$$+ \text{ higher terms}$$

where $\mathscr{E}$ is in frequency units and the anharmonicity parameters x_{ij} have been redefined so that they too are in frequency units. Anharmonicity has two principal effects in spectroscopy. Firstly the fundamentals, combinations and overtones are not found where they would be expected, but instead are slightly shifted from these positions. Secondly, because the quantum mechanical problem of the anharmonic oscillator is not separable when expressed in terms of the normal coordinates, the various states given by equation (5.2.34) may perturb one another and be found widely displaced from their calculated positions. The first effect is small and provided enough bands are observed the equilibrium or "bottom of the well" frequencies v_i and the anharmonicity constants can be observed. The second effect usually called "Fermi resonance" can occur whenever two states, not both fundamentals but having the same symmetry, lie, according to (5.2.34), very close to one another. Quantum mechanical mixing can take place and two new states be produced each of which takes on the characteristics of both of the unperturbed original states. The effect is most dramatic when the two original states are coincident for then the "resonance denominator" has its maximum effect and the displacement of the two resulting states—one up the other down—will be large. Thus for carbon tetrachloride (see Appendix 6) the two states v_3 (777 cm^{-1}) and $v_1 + v_4$ (774 cm^{-1}) are very close together and have the same symmetry F_2. A strong Fermi resonance results and the v_3 band is observed, not as a single feature, but as a virtually symmetrical doublet at 764 cm^{-1} and 790 cm^{-1} (see Fig. 5.13). The transition from the ground state to the combination level $v_1 + v_3$ is first-order forbidden and would therefore lead to a very weak absorption band in its own right, but because each of the resonating levels contains 50 % of the strongly connected v_3 state each band appears strongly in the spectrum. The positions of the resonating levels are given by the latent roots of a matrix which has the unperturbed energies as its diagonal entries and an interaction energy as its off-diagonal element. The magnitude of the interaction energy is determined by the cubic, quartic etc. terms in the potential energy and these also determine the x_{ij}, so it is possible, in principle, to use the positions of the unperturbed but nevertheless anharmonic levels together with the positions of the resonating levels to give a complete potential energy function. However this has been done so far only for a very small number of molecules.

Fermi resonance is a quite common phenomenon in molecular spectroscopy. Several examples are known of virtually perfect resonance with strong perturbation, whilst the cases of weak but nevertheless obvious perturbation caused by off-resonant interaction are legion. One of the best known examples of nearly exact coincidence is provided by the carbon dioxide molecule where the levels (10^00), i.e. v_1 and (02^00), i.e. $2v_2$, have almost the same energy and interact strongly. The unperturbed energy (in wave number units) is 1337 cm^{-1}, but the Raman spectrum shows no feature at this position; instead an intense doublet at 1285·6 cm^{-1} and 1388·3 cm^{-1} is observed. These states have A_{1g} symmetry and direct transitions to them from the ground state are forbidden by the dipole selection rules. Nevertheless they have a profound, if indirect, effect on the infrared spectrum and on the radiative transfer in carbon dioxide gas. Thus the allowed hot-band $v_2 - 2v_2$ is observed as a doublet and the radiative relaxation from either of the two resonating states is rapid since both can decay via the allowed cascade

$$(02^00) \rightarrow (01^10) \rightarrow (00^00) \tag{5.2.35}$$

In the absence of the resonance, one of the two states, viz. (10^00), would be metastable since there would be no dipole-allowed radiative path back to the ground state. The very important CO_2 laser (see section 6.2.2) operates on the difference band $(v_3 - v_1)$ and because of the resonance this band is a doublet and the laser therefore shows two branches, at 9·4 and 10·6 μm respectively. The rapid cascade relaxation (5.2.35) is vital to the maintenance of the population inversion in the continuous-wave CO_2 laser. All the "rungs" of the v_1 and $2v_2$ ladders will tend to coincide and the $(v_1, 2v_2)$ resonance will therefore repeated for successive stages of excitation. All pairs of levels of the type $(n0^00)$ and $(02n^00)$ are thus perturbed to a greater or lesser extent. This is not usually manifest in the absorption spectrum of CO_2 because these highly excited states have small populations at room temperature. However the effects of the resonances are clear in the emission spectrum of very hot CO_2 and also in the output of the so-called sequence band laser. In this laser, hot CO_2, contained in an intercavity cell, strongly absorbs the primary lines and thus prevents normal operation. The laser is therefore diverted to operation on the "hot" or sequence bands

$$(00^0n) \rightarrow \{ (n0^00) + (02n^00) \} \tag{5.2.36}$$

which, because of anharmonicity, are slightly shifted from the normal (i.e. $n = 1$) band positions. In this way, the CO_2 laser can be induced to give still more output wavelengths to add to those listed in section 6.2.2.

A good example of Fermi resonance perturbation without close approach is provided by the carbon disulphide molecule CS_2. Here v_1 (657 cm^{-1}) and $2v_2$ (792 cm^{-1}) are not in coincidence, but nevertheless there is a perturbation. $2v_2$ appears weakly in the Raman spectrum, due to intensity borrowing from the very strong v_1 band and, in the far infrared, the difference band $(v_1 - v_2)$ is observed [535]. That this is unusual and not due to anharmonic or non-linear electrical effects is clear from the non-appearance of the sum band $(v_1 + v_2)$ in the middle infrared spectrum. It is a theorem in quantum mechanics that the transition moments for corresponding difference and sum bands are the same, but in the observed spectrum the sum band will always be the more intense because of population effects. Thus the intensity observed for $(v_1 - v_2)$ must arise from borrowing from the strongly allowed hot band $(2v_2 - v_2)$. The CS_2 type of Fermi resonance does not give any nomenclatural problems because the mixing is so slight that one can meaningfully use the original labels. For the CO_2 and CCl_4 type, with virtually exact coincidence, this is not so. Each observed level is an almost exactly 50/50 mixture of the original states and to use either original label for either of the final states is misleading. Unfortunately, this does appear to be the present practice and the upper of the two CO_2 levels is usually labelled (10^00) and the lower (02^00). In the absence of a logical and universally agreed alternative, this practice will probably continue.

5.2.3.2 *Rotation–vibration spectra of polyatomic molecules*
The vibrational bands of polyatomic molecules in the gas phase also show rotational fine structure but in many cases this is much more complex than that observed for diatomic molecules. There are two principal reasons for this: firstly, new forms of rotation–vibration interaction are possible when the molecule contains more than two atoms and secondly the selection rules are often less restrictive. As a simple example of the latter, the selection rule $\Delta J = 0$ often applies and the observed band will then show a Q-branch in addition

to the P- and R-branches which are all that are observed for diatomic molecules.

We can begin by considering the linear molecule. Here the vibrational modes divide into two classes, those in which the atomic motions are confined to the molecular axis and which therefore feature dipole moment oscillations *parallel* to the axis and those in which the motions (and the dipole moment oscillations) are perpendicular to the axis. A state with parallel excitation has exactly the same simple rotational structure as the ground state and transitions to it from the ground state will lead to a simple band with line positions given by equation (5.2.29) as it stands. An example of such an excited state is provided by (04^00) of HCN as illustrated in Fig. 2.32. For molecules such as CO_2, with a centre of symmetry, the effects of nuclear statistics must be taken into account. All states with an even vibrational excitation have the odd J levels missing and vice versa. It follows that alternate lines will be missing in the spectrum and that the line spacing instead of being $\sim 2B$ will be $\sim 4B$. The perpendicular states are very different. This is because the bending modes are always doubly degenerate since the motion can take place equivalently in two planes at right angles. If both components of the degenerate motion were excited, one would have a Lissajous figure sort of resulting motion, that is there would be vibrationally induced angular momentum. Conversely, if one component were excited, then rotation of the molecule would tend to induce the other. It follows that all excited states of the perpendicular type will have a vibrational angular momentum which may be specified by the quantum number l. As usual the existence of two equivalent configurations leads to interaction and to a splitting of the rotational levels. This is called l-type doubling. It is illustrated schematically for the (11^10) state of HCN in Fig. 2.32. In the state notation, the value of l is usually given as a right superscript to the quantum numbers specifying the bending motion. The magnitude of the splitting is given by

$$\Delta = qJ(J+1) \qquad (5.2.37)$$

but since q is always very small (for OCS it is $6\cdot345$ MHz) the effects are not discernible in ordinary diffraction grating spectroscopy. In very high resolution spectroscopy and in double resonance spectroscopy, however, the existence of the l-doublets is obvious. The l-type doublet has one component of even parity and one of odd and it follows that rotational transitions within perpendicularly excited states will be doublets but the lines in a band arising from a transition from a parallel to a perpendicular state will be singlets. In such a band, the v_2 absorption of CO_2 for example, the effect of the vibrational angular momentum is made obvious by the presence of a Q-branch. Ignoring centrifugal effects, the lines in the Q-branch will be given by

$$\tilde{v} = \tilde{v}_0 + (B' - B'')J(J+1), \qquad (5.2.38)$$

so, since the difference in the B values will be small, the lines in the Q-branch pile up on top of one another leading to an intense line-like feature.

Doubling of the rotational levels due to rotation–vibration interaction will occur whenever one has a degenerate pair of vibrations whose species type contains a primitive rotation. A very common example is for the E-type vibrations of C_{3v} symmetric rotors. For the symmetric rotor, one can again define parallel and perpendicular motion with respect, this time, to the figure axis and one finds that the E modes are perpendicular. The parallel selection rules are simple except that when $K \neq 0$ a Q-branch is also allowed. One has therefore

$$K = 0, \quad \Delta K = 0, \quad \Delta J = \pm 1 \tag{5.2.39a}$$
$$K \neq 0, \quad \Delta K = 0, \quad \Delta J = 0, \pm 1 \tag{5.2.39b}$$

The observed spectrum will consist of the superposition of simple PQR sub-bands, each of which arises from a given value of K. The positions of the lines in a given sub-band will be given by the usual expressions (equations 5.2.29a and 5.2.29b) except that $\tilde{\nu}_0$ is replaced by $\tilde{\nu}_0^{sub}$ where

$$\tilde{\nu}_0^{sub} = \tilde{\nu}_0 + [\Delta A - \Delta B] K^2. \tag{5.2.40}$$

The displacement of $\tilde{\nu}_0^{sub}$ from $\tilde{\nu}_0$ will normally be small, the contributions from higher values of K will be attenuated by population effects and, of course, since K is necessarily less than J they will be still further attenuated. All these effects lead to an observed spectrum which even at moderate resolution looks remarkably like a simple PQR band. The E-type perpendicular bands on the other hand, look very different indeed. The main reason for this is the selection rule,

$$\Delta K = \pm 1, \quad \Delta J = 0, \pm 1, \tag{5.2.41}$$

which ensures that the sub-bands do not even approximately coincide. The subsidiary reason is the effect of the vibration–rotation interaction, mentioned above, which gives a very complicated rotational structure for states with the degenerate modes excited. The interaction between vibration and rotation in both linear molecules and symmetric rotors arises from a very general phenomenon in mechanics, Coriolis coupling. This is an apparent acceleration, at right angles to the velocity, when the motion of the system is referred to uniformly rotating axes. It is of some consequence in mechanics and ballistics because our normal observing frame, the surface of the Earth, is such a uniformly rotating reference system. In quantum mechanics, the Coriolis perturbation, like other forms of interaction treated by first-order perturbation theory, involves a resonance denominator and one can therefore usefully distinguish two cases, first-order Coriolis perturbation which involves two frequencies degenerate by symmetry and second-order perturbation which involves two distinct frequencies. The rule is that two modes may interact via a Coriolis perturbation when the product of their symmetry types contains one or more of the primitive rotations. The Coriolis perturbation in E-type states of C_{3v} symmetric rotors is first order and this leads to an additional term

$\pm 2A_i\zeta_iK$ to be included in the rotational energy level formula (5.2.3). Here the subscript i denotes the degenerate state in question and ζ_i is the Coriolis coupling constant. The two signs refer to the two ways of combining the rotational and the vibrational angular momentum and in fact correspond to the two values of the l quantum number. The ζ_i can have any value between $+1$ and -1 and are determined by the moments of inertia and by the force constants. It is possible therefore to use experimentally determined values of the ζ_i as additional information for determining the force-field. As usual, however, not all of the information is unique since the sum of all the ζ_i for a given symmetry class involves only simple combinations of the moments of inertia. The effect of this additional Coriolis term is to lift the double degeneracy of the $K \neq 0$ levels and one has an energy level scheme which looks like that of a slightly asymmetric rotor. The structure of perpendicular bands is thus very complicated indeed with a veritable forest of lines spanning the region of the band. Herzberg has given some elegant diagrams which analyse the make-up of these bands [466]. He also suggests a useful shorthand notation to identify each of the myriad lines making up the overall band. This is to denote the ΔJ value by the equivalent capital letter, to use a left upper superscript to denote ΔK and right lower index to denote K''. The value of J'' is then given in parentheses. Thus the line in the ν_6 band of methyl bromide which is coincident with the $P(20)$ line of the CO_2 laser is denoted $^P P_1(9)$ and arises from the transition

$$(J = 9, K = 1) \rightarrow (J = 8, K = 0)$$

The same convention can be used to describe the simpler line structure of a parallel band. Thus the two transitions in methyl fluoride pumped by the $P(20)$ line (see Fig. 2.33) would be denoted $^Q Q_1(12)$ and $^Q Q_2(12)$.

The set of rotational eigenfunctions of any molecule forms a group and, since the rotational operations form a sub-set of the total set of symmetry operations, this group is necessarily a sub-group of the molecular point group. In the case of a C_{3v} molecule, the rotational sub-group is C_3 which has only the two species A and E. All the rotational levels of a C_{3v} molecule can therefore be classified as being either of A or E symmetry. If the excitation in question is of the parallel type, that is it is non-degenerate, the classification depends solely on K, the rule being that if K is a multiple of 3 the levels are of A type and if not they are of E type. For perpendicular states affected by the first-order Coriolis interaction (that is *all* perpendicular states under C_{3v} symmetry) the split rotational levels are all of E symmetry if K is a multiple of 3 but if not then one l component will be A type and the other E-type. Despite this splitting, each rotational level is still a degenerate doublet provided K is not zero but in the case of the E component this degeneracy is absolute and may not be lifted by any internal perturbation. The A levels, however, have only a conditional degeneracy and internal perturbations, for example second-order Coriolis interaction, can cause a further splitting. This is known as K-doubling. It is always a very small effect and prior to the introduction of modern ultra-high resolution techniques, it lay purely in the realm of conjecture. Now several examples are known and the relevant spittings determined. The effects of K-doubling tend to be particularly obvious in laser/microwave double-resonance experiments [138]. The statistical weights of the A and E rotational levels are determined by the nuclear spins. For an XY_3 molecule with all nuclear spins zero, only the levels with K a multiple of 3 will exist. For other values of the nuclear spins, the $K \neq 3$ levels will exist but will have lower statistical weights. The general symmetry rules when applied to XY_3 molecules show that in a radiative or collisional transition only levels of the same kind may combine. One can therefore consider a gas of XY_3 molecules to be made up of two kinds an A type and an E type with a very low rate of interconversion.

Spherical tops have much simpler energy level schemes in their ground states because of the absence of the K structure and one has merely a set of levels specified by the quantum number J with each level $(2J + 1)^2$ degenerate. In vibrationally excited states, however, one can again get a first-order Coriolis interaction leading to a more complicated disposition of the rotational levels. The only types of spherical top to be extensively studied are the tetrahedral XY_4 of point group T_d and the octahedral XY_6 of point group O_h. In these only the F_1 and F_2 states are perturbed for T_d and only the F_{1g}, F_{1u}, F_{2g}, and F_{2u} perturbed for O_h. The perturbation, as before, is specified by the Coriolis coupling constant ζ_i and a sum rule exists which for XY_4 takes on the simple form

$$\zeta_3 + \zeta_4 = \tfrac{1}{2}. \tag{5.2.42}$$

For CH_4, as an example, the v_3 vibration (principally CH asymmetric stretching) has a ζ value of 0·05 and v_4 (principally HCH deformation) has a ζ value of 0·45. The perturbed levels being triply degenerate, however, one now has the situation that the splitting leaves one member of the triplet unaffected. Ignoring centrifugal effects, the rotational energy levels may be written

$$F^{(+)}(J) = B_i J(J+1) + 2B_i \zeta_i(J+1) \tag{5.2.43a}$$
$$F^{(0)}(J) = B_i J(J+1) \tag{5.2.43b}$$
$$F^{(-)}(J) = B_i J(J+1) - 2B_i \zeta_i J \tag{5.2.43c}$$

The rotational selection rule is $\Delta J = 0, \pm 1$, but the structure observed in an $A_1 \rightarrow F_2$ fundamental, for example, is much simpler than one might expect. This is because an additional selection rule

$$\Delta J = +1, \quad A_1 \rightarrow F_2{}^{(-)} \tag{5.2.44a}$$
$$\Delta J = 0, \quad A_1 \rightarrow F_2{}^{(0)} \tag{5.2.44b}$$
$$\Delta J = -1, \quad A_1 \rightarrow F_2{}^{(+)} \tag{5.2.44c}$$

applies. The R-branch therefore involves only the $F_2{}^{(-)}$ levels, the Q-branch only the $F_2{}^{(0)}$ levels and the P-branch only the $F_2{}^{(+)}$ levels.

The rotational sub-groups are T for T_d and O for O_h which have the irreducible representations A, E and F so the rotational levels can be classified into these three categories. As an example the seventeen components of the $J = 8$ level of the ground state divide into $A + 2E + 4F$. As before the statistical weights are determined by the nuclear spins and for a molecule in which all the atoms have zero spin, only the A levels will exist. Certain J values do not have any A-type components in their manifold and it follows that these J values would not occur in the spinless case. Just as for the symmetric top, only levels of the same kind can combine so it follows that a gas like methane will contain three kinds of molecule which can only slowly interconvert.

Second-order Coriolis perturbation has been mentioned several times so far but now needs to be discussed in some more detail. Basically all levels of the same J and the same parity can perturb one another even if they belong to quite different vibrational systems. However it is only when there is close approach that dramatic effects are observed. The results of distant perturbation hardly vary with J apart from the overriding rule that the effects are non-existent for $J = 0$, and all that happens is that there is a contribution to the rotation/vibration interaction constant α. With medium approach, the slight mixing of states can lead to the appearance of lines in the spectrum which would be forbidden in the absence of perturbation. This is similar in a way to Fermi resonance, but can readily be distinguished in practice because of the strong dependence on J. A very well-known example is provided by XY_4 molecules where the E modes are forbidden but are often observed due to Coriolis interaction with nearby F_2 modes. The origin of the forbidden intensity is obvious from the anomalous intensity distribution and in particular from the absence of transitions to the $J = 0$ level. The overall selection rule for Coriolis perturbation, sometimes called Jahn's rule, is that the product of the symmetry species of the two vibrational states in question should contain one or more of the primitive rotations. For an XY_4 molecule, interaction between v_2 (E) and v_4 (F_2) is allowed since

$$E \times F_2 = F_1 + F_2 \qquad (5.2.45)$$

and F_1 contains the three rotations R_x, R_y and R_z. In the case of very close approach the effects of the resonance denominators dominate and the perturbation can become very anomalous indeed with one or two lines in the spectrum strongly disturbed and the others not obviously affected. Coriolis perturbation of this sort is responsible for the population inversion in the HCN laser as shown in Fig. 2.32.

Much of what has gone before will now be put together in the analysis of a single but technically important example, the v_3 band of methane. Under low resolution, this band shows a simple PQR structure, as shown in Fig. 5.15. Under higher resolution, the "lines" break up into multiplets due to Coriolis interaction with other modes, e.g. v_2 and $2v_2$. At the best resolution available with dispersive spectrometers (slightly worse than 0.01 cm^{-1}) the multiplets are resolved into six components as shown in the lower inset of Fig. 5.15. For the case of P(7), these components occur at 2947·683, 2947·816, 2947·912,

2948·144, 2948·433 and 2948·489 cm⁻¹ respectively. Actually both the upper and the lower states are split, as shown in Fig. 5.16, the splitting of the ground state being due to centrifugal effects, as discussed earlier and the splitting of the upper state being due to Coriolis effects. The upper state splittings are, however, an order of magnitude larger. The significance of the P(7) line is that one component in the multiplet, that at 2947·912, lies within the tuning range of the important 3·392 μm helium/neon laser whose free running centre wavenumber is 2947·903 cm⁻¹. The laser can thus be used to pump this transition and double-resonance experiments [489] can be carried out. From these experiments, the magnitude of all the splittings in both the upper and the lower states can be deduced. Some of these are included in Fig. 5.16. These splittings are such that for all intents and purposes one can regard the ground state as a simple doublet with a spacing of 0·05 cm⁻¹ when carrying out conventional spectroscopy. The very large number of possible transitions which stem from the not very restrictive selection rule, namely, that the product of the species symmetry types should contain the F_2 irreducible representation, is thus

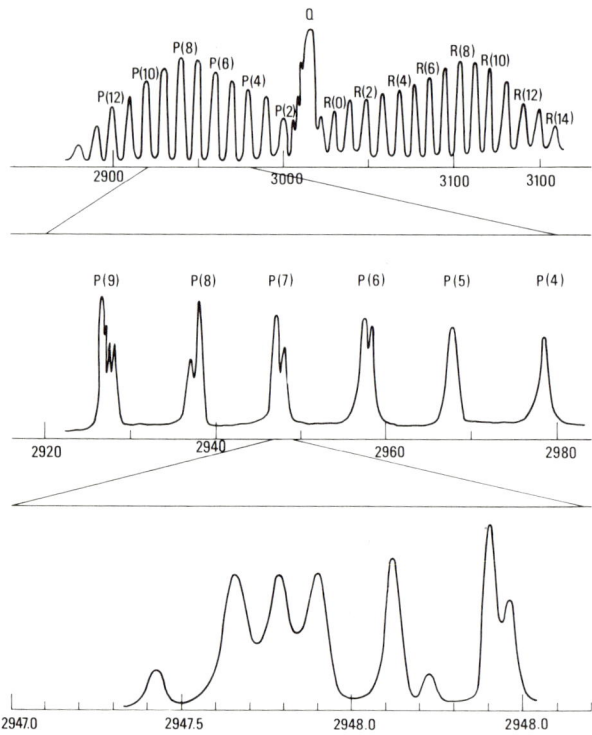

FIG. 5.15. The v_3 band of methane under low (upper), medium (middle) and high resolution (lower). The structure observed on the various lines is due mostly to second-order Coriolis perturbation.

reduced, in practice to a very much smaller number. In fact one can regard the spectrum as made up of five doublets each having a spacing of 0.05 cm^{-1} and, when these are convolved with a suitable line-shape function, one gets a result virtually identical with that observed. This near coincidence of the absorption line with the laser is close enough for the laser to be locked to the inverse Lamb-dip of the line [490]. Evenson and his colleagues [491] have shown that one can produce, in this way, an extremely stable frequency standard and they have used the harmonic multiplication technique to measure its absolute frequency. Their result $\nu_0 = 88.376\,181\,627$ THz, when combined with the measured corresponding wavelength, $\lambda_0 = 3.392\,231\,376$ μm, has provided us with the currently [2] best value for the speed of light, namely $c = 2.997\,924\,562\,10^8$ m s^{-1}.

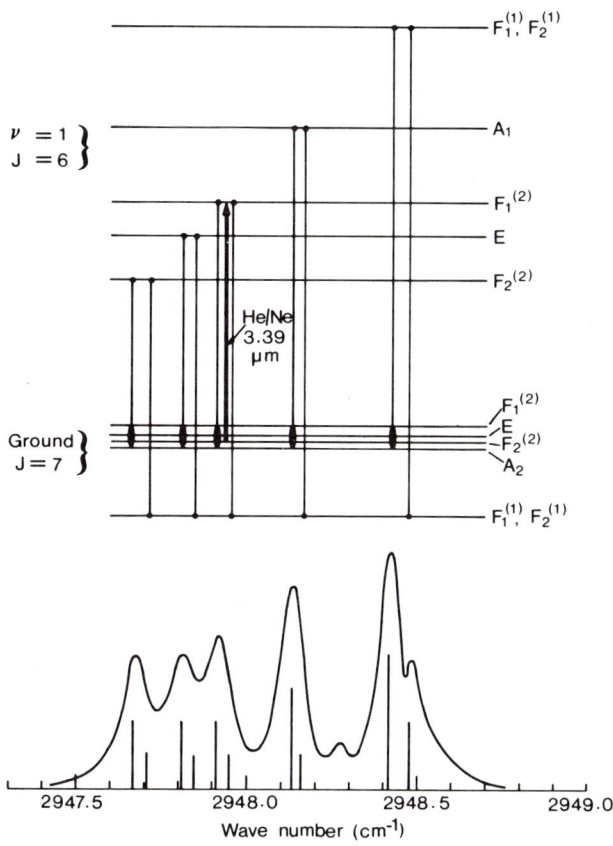

FIG. 5.16. Origin of the fine structure of the P(7) line of methane near 2948 cm^{-1}. The ground-state splittings are due to centrifugal effects (total span 1670 MHz) whilst the upper-state splittings are due to Coriolis effects (total span 22 497 MHz).

5.3 Infrared spectroscopy of solids

5.3.1 *Introduction*

The solid-state of matter covers a wide array of microstructures ranging from the virtually total disorder of glasses which are essentially ultra-viscous supercooled liquids, all the way to the virtually perfect order of single crystals which have long-range translational symmetry. The one property which all solids have in common is that the constituent atoms are not free to move away from their equilibrium positions and their motions are therefore restricted to rotations and vibrations about these positions. Large-angle rotational motion is rare, the only significant examples being provided by plastic crystals and especially simple solids such as crystalline hydrogen. Most solids consequently absorb electromagnetic radiation only by virtue of vibrational and electronic transitions. Infrared absorption by vibrational transitions in solids is virtually universal but absorption by electronic transitions is very rare, the only significant examples being furnished by the narrow band-gap semiconductors. It is therefore more convenient to postpone discussion of electronic absorption till section 5.4 where the other properties of semiconductors are drawn together. The treatment in this section will then be confined to the vibrational infrared spectroscopy of solids.

One can develop the theory of the vibrations of a solid in purely general terms in which one assigns coordinates to each atom and then decides the form of the potential linking each atom to every other atom in the specimen. Not surprisingly this general theory is not very useful. Much more progress can be made by dividing solids into various categories and then treating each category by appropriate approximate methods. It is conventional to begin with the treatment of perfect crystals and then to consider the case of a disordered crystal either by the use of perturbation theory or else by the use of statistical methods. In a perfect crystal, the coordinates of the equilibrium positions are defined so only the nature of the forces linking the atoms needs to be considered. There are several possible categories but some particularly important cases are:

a. Ionic crystals such as NaCl, which are made up of the two sublattices of the cations and the anions held together by long-range non-directional electrostatic forces.
b. Covalent crystals such as diamond, in which the atoms are held in place by short-range highly directional forces.
c. Molecular crystals in which the constituent molecules, themselves bound together by strong forces, are held in their places in the lattice by much weaker Van der Waals forces.

This division is rough because all kinds of intermediate cases and all kinds of mixed cases exist—one need only think of graphite, a macromolecular atomic crystal in two dimensions and a molecular crystal in the third to underline this, but nevertheless it is a useful basis for discussion. An outline of the vibrational theory with illustrations drawn from each of these three cases will now be developed in the following sections.

5.3.2 *Elementary theory of the vibrations of crystalline lattices*

In the crystalline state, the constituent entities, either molecules atoms or ions, are held in their equilibrium positions by a balance of attractive and repulsive forces. There is long range order which ideally extends throughout the macroscopic specimen and this order may be specified by the space group symmetry of the crystal lattice. If the crystal contains N atoms or ions then there will be $3N - 6$ (roughly $3N$ since N is large) vibrational degrees of freedom and the problem in determining the infrared absorption spectrum of a crystal is to decide the frequencies of these modes and their spectral activity. In general this problem cannot be solved exactly, even for the simplest lattices, but one can make a great deal of progress using a semi-phenomenological theory. Many workers have contributed to this topic but the simple exposition given here is inspired mostly by the work of Born and Huang [492] as summarised by Mitra and Gielisse [493].

It is usual to begin by discussing the idealised case of a simple linear diatomic lattice. We consider such a chain made up of two particles of masses M and m and connected by springs whose force constant is f. The model is shown schematically in Fig. 5.17 where the interatomic distance is a. To each lattice point is assigned a number: even $2l$ for the M sites and odd $2l + 1$ for the m sites.

If the displacements from the equilibrium positions are denoted by the coordinate u, the equations of motion are

$$M \cdot \ddot{u}_{2l} = f(u_{2l+1} + u_{2l-1} - 2u_{2l}) \tag{5.3.1a}$$

and

$$m \cdot \ddot{u}_{2l+1} = f(u_{2l+2} + u_{2l} - 2u_{2l+1}). \tag{5.3.1b}$$

Assuming simple harmonic solutions of the form

$$u_{2l} = A_1 \cos(2\pi vt + 2lka) \tag{5.3.2a}$$

and

$$u_{2l+1} = A_2 \cos(2\pi vt + (2l+1)ka), \tag{5.3.2b}$$

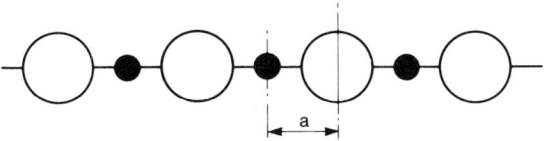

FIG. 5.17. Simple linear diatomic chain.

where k is used to denote the magnitude of the wave vector, and substituting in the equations of motion leads to two simultaneous linear equations for A_1 and A_2 and these equations have a solution other than $A_1 = A_2 = 0$ only if

$$\begin{vmatrix} 2f - 4\pi^2 v^2 M, & -2f \cos ka \\ -2f \cos ka, & 2f - 4\pi^2 v^2 m \end{vmatrix} = 0, \qquad (5.3.3)$$

which gives

$$v^2 = (1/4\pi^2)\{f/\mu \pm [f^2/\mu^2 - (4f^2 \sin^2 ka)/(Mm)]^{1/2}\} \qquad (5.3.4)$$

where μ is the reduced mass defined by

$$\mu^{-1} = M^{-1} + m^{-1}. \qquad (5.3.5)$$

Clearly, the minimum value of λ will occur when alternate lattice sites of a given kind are moving in phase, i.e. $\lambda_{min} = 4a$; $|\bar{k}|$ is therefore restricted to the range

$$-\pi/2a \leqslant |\bar{k}| \leqslant \pi/2a. \qquad (5.3.6)$$

The region between these limits of $|\bar{k}|$ is termed the first Brillouin zone. The lattice parameter a will be of the order 10^{-10} m, so the Brillouin zone has a width of the order 10^{10} m^{-1}.

The relation between the vibration frequencies and the wave vector is shown in Fig. 5.18 in the form of a dispersion diagram. This takes the form of two branches corresponding to the two signs of the radical in equation (5.3.4). Clearly both branches display considerable dispersion. The lower of the two branches is called the acoustic branch since it corresponds to the propagation of sound waves. For low values of $\bar{k}$ this branch is sensibly linear with a slope equal to the velocity of sound along the chain $v_s = [2a^2 f/(M+m)]^{1/2}$. At higher wave numbers the slope becomes less and the branch intercepts the zone edge at a frequency $v = (f/2\pi^2 M)^{1/2}$ for $M > m$. The upper branch is called

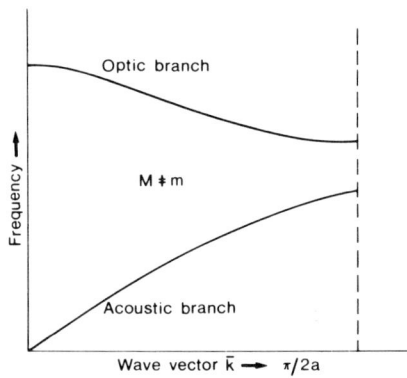

FIG. 5.18. Dispersion diagram for the vibrations of a simple linear diatomic chain.

the optical branch since for low $\bar{k}$ the phase-velocities are very high, corresponding roughly to light velocities. When $\bar{k}$ is zero, the phase velocity is formally infinite which is just another way of saying that the vibrations are everywhere in phase. Both the acoustic and the optical types of wave propagating along the chain are quantised in energy and one is led therefore to the concept of a "phonon". From what was said above, a typical phonon will have a wave vector with a magnitude of 10^8 cm^{-1}. An equivalent photon would lie in the X-ray region of the spectrum!

We now have to consider the absorption of radiation by this simple idealised system. In quantum mechanical language, this involves the absorption of a photon with the corresponding generation of one or more phonons and, as always, it is necessary that in the process both energy and momentum be conserved. The momentum of an infrared photon, $p = h\lambda^{-1} = h\bar{k}$, is very small compared with that of a typical phonon so absorption of radiation will be confined in the dispersion diagram to the region of very low $\bar{k}$. In fact solution of the relevant equations shows that there exists only one frequency for which both energy and momentum are conserved and this corresponds to the point of intersection of the optical branch with the dispersion curve for photons, i.e. a straight line through the origin of slope $c/2\pi n$. Since the optic branch is almost horizontal for low $\bar{k}$ values, it follows that the infrared spectrum of such an idealised system would consist of a single infinitely sharp line at a frequency given to a very good approximation by

$$v = (2\pi)^{-1} \sqrt{(2f/\mu)}, \tag{5.3.7}$$

which is the frequency of intersection of the optical branch with the ordinate axis. The intensity of this line will be determined by the rate of change of dipole moment and for this idealised system that would be simply proportional to the difference in charge of the two atoms making up the chain.

When we come to consider real three-dimensional crystals, the situation, not unexpectedly, is much more complicated but certain key concepts still remain: thus again we find optical and acoustic branches and the selection rule is still $\Delta\bar{k} = 0$. In three dimensions, however, it is possible to have transverse as well as longitudinal vibrations and so in general we expect to encounter *T*ransverse *O*ptical (TO), *L*ongitudinal *O*ptical (LO), *T*ransverse *A*coustic (TA) and *L*ongitudinal *A*coustic (LA) modes. The dispersion diagrams can get very complicated and for anisotropic crystals they will not necessarily be the same for different directions of propagation in the crystal.

The $\Delta\bar{k} = 0$ selection rule implies that one is considering modes of vibration in which all the symmetrically equivalent atoms throughout the crystal move with the same frequency, in the same direction and in phase. One need therefore consider merely the motion of the atoms making up a single unit cell since the motion in all the other unit cells will be identical. If one has N atoms in the unit cell, there will be $3N$ branches in the dispersion diagram, three of them being acoustic branches passing through the origin. For an alkali halide, for example, which has two atoms per unit cell there would be to first order (but see later) just two branches, one acoustic and the other optical, both triply degenerate. It is interesting, of course, to speculate what would happen if these two atoms were to be smoothly made identical since then there would be, in the limit, only one atom per unit cell and the optic branch would have to disappear. The answer to this apparent paradox is that in this limit the two old branches would meet at the zone edge and simultaneously the zone would double in size. The old diagram is therefore equivalent to the new one folded back on itself at the mid-point. These two branches of the old diagram become just one continuous branch of the new. There is likewise no paradox about the spectral activity since in the limit the transition which now goes to an intersection on the zone edge and therefore violates $\Delta \bar{k} = 0$ would have zero intensity because the dipole moment change during the vibration would be zero. This concept of the folding back of dispersion diagrams is very powerful, especially for polymer spectroscopy, where doubling or halving the repeat unit frequently occurs. The concept is strongly related mathematically to the "aliassing" which occurs in discrete Fourier transformation (see section 1.6.2).

The $\Delta \bar{k} = 0$ selection rule restricts infrared spectroscopy (or Raman for that matter since even an optical photon has negligible momentum when compared with a typical phonon) to observations of the frequencies of intersection of the optical branches with the ordinate axis. The amount of information forthcoming is thus rather small. One can, of course, estimate theoretically the form of the branches, knowing only the $\bar{k} = 0$ intersections, if one postulates some form for the interatomic force field but this is rather uncertain territory. Most lattice dynamicists prefer an experimental approach and for this one needs a probe not restricted to $\Delta \bar{k} = 0$ transitions. Thermal neutrons fill the bill admirably and to date neutron spectroscopy [494] has proved a very powerful tool for exploring the details of lattice dynamics dispersion diagrams. This topic has thus tended to be pursued most diligently in the large research centres such as the Institut Laue-Langevin at Grenoble where intense monochromatic neutron sources are available. However, the infrared contribution though minor is important in its own right and is, moreover, relevant to the use of solids as transmissive and reflective infrared materials so the vibrations of crystal lattices have nevertheless remained of considerable interest to the infrared technologist.

For real three-dimensional lattices, various extra refinements have to be introduced into the simple model in order to provide an adequate treatment. Most important of these is the acknowledgement that the atoms making up the lattice will be to a greater or lesser extent charged. For alkali halides this is very obvious; indeed the lattice is made up of ions, but even for solids which one might, to first order, think of as covalent, e.g. BN, ZnS etc, there is nevertheless some polarisation of the bonds and this leads to fractional charges on the atoms. The electrostatic forces between the charged atoms are very long range, falling off only as the inverse square of the separation, so the approximation of nearest-neighbour-only interaction which is adequate for covalently bonded entities would fail rather obviously for an ionically bonded lattice. The net result of this long-range Coulombic interaction is that modes of vibration which one would expect to be degenerate in the simplest treatment have their degeneracy lifted by the presence of the macroscopic electric field. The most well-known example is the splitting of the triply degenerate optical branch of alkali halides into a doubly degenerate TO mode and a non-degenerate LO mode. Good accounts of these electric field effects have been given by Mitra and Gielisse [493] by Moller and Rothschild [495] and by Hadni [496]. Following them one writes down the equation of motion for the forced oscillation of a cubic lattice made up of ions of charge $\pm q$ and reduced mass μ under the influence of an external field $E_0 \exp(i\omega t)$. One can take it without much loss of generality that the specimen is large in comparison with the wavelength and then, taking the natural resonant frequency of the lattice to be ω_0, the equation becomes

$$\frac{d^2 \hat{u}}{dt^2} + \gamma \frac{d\hat{u}}{dt} + \omega_0^2 \hat{u} = \left(\frac{q}{\mu}\right) E_{int} \exp(i\omega t) \qquad (5.3.8)$$

where u is a coordinate measuring the displacement of an ion from its equilibrium position, γ is a damping constant with dimensions of frequency and E_{int} is the actual or "internal" field which the ion experiences. The calculation of the internal field to full rigour is one of the more formidable problems in theoretical physics but, fortunately for lattice dynamics work, the simple classical approach stemming from Lorentz, Onsager and Clausius–Mosotti is perfectly adequate. One assumes that one can split the polarisation of the lattice, that is its dipole moment per unit volume, into two terms, one due to the displacement of the ions and the other due to the distortion of the electronic clouds. One can then write

$$P = Nqu/V + \varepsilon_0 (\varepsilon_\infty - 1) E_0, \qquad (5.3.9)$$

where N is the number of ion pairs in a volume V and ε_∞ is the relative permittivity at a frequency high enough for the lattice dispersion to be complete but not high enough for the electronic dispersion to have yet begun. It is a rather nebulous quantity but for all practical purposes can be taken to be the value of ε' at the near infrared permittivity minimum. Following Lorentz, one next assumes that the local field E_L will be given by

$$E_L = E_0 + P/3\varepsilon_0 \qquad (5.3.10)$$

where it has been taken that normal, i.e. transverse, electromagnetic propagation is being considered. Solving (5.3.10) using (5.3.9) gives

$$E_L = E_0 [(\varepsilon_\infty + 2)/3] + Nqu/3\varepsilon_0 V. \qquad (5.3.11)$$

Applying this entire procedure once more, allows us to calculate the true internal field in the form

$$E_{int} = E_L \left[(\varepsilon_\infty + 2)/3 \right]$$
$$= E_0 \left[(\varepsilon_\infty + 2)/3 \right]^2 + Nqu (\varepsilon_\infty + 2)/9\varepsilon_0 \, V. \tag{5.3.12}$$

Substituting this in (5.3.8) leads to a term

$$\omega_T^2 = \omega_0^2 - Nq^2 (\varepsilon_\infty + 2)/9 \, \varepsilon_0 \, \mu \, V \tag{5.3.13}$$

for the coefficient of u. The observed "resonant" frequency is thus reduced from its "true" value due to the effects of the macroscopic field. The final solution for u is

$$\hat{u} = \frac{q (\varepsilon_\infty + 2)^2 \, E_0 \exp i\omega t}{9\mu \left[\omega_T^2 - \omega^2 + i\omega\gamma \right]}, \tag{5.3.14}$$

and the polarisation due to this displacement is

$$\hat{P} = \frac{N}{V} q\hat{u} = \frac{Nq^2 (\varepsilon_\infty + 2)^2 \, E_0 \exp i\omega t}{q\mu \, V \left[\omega_T^2 - \omega^2 + i\omega\gamma \right]}. \tag{5.3.15}$$

Finally, writing

$$\hat{P} = \varepsilon_0 \left(\hat{\varepsilon}_r - \varepsilon_\infty \right) E_0 \tag{5.3.16}$$

gives

$$\hat{\varepsilon}_r = \varepsilon_\infty + \frac{Nq^2 (\varepsilon_\infty + 2)^2}{q\mu\varepsilon_0 \, V \left[\omega_T^2 - \omega^2 + i\omega\gamma \right]} \tag{5.3.17}$$

for the complex relative permittivity. This relation can also be written in the equivalent form

$$\hat{\varepsilon}_r = \varepsilon_\infty + \frac{\omega_T^2 (\varepsilon_s - \varepsilon_\infty)}{\left[\omega_T^2 - \omega^2 + i\omega\gamma \right]} \tag{5.3.17a}$$

where ε_s is the static (i.e. DC) value of the permittivity.

To analyse the physical content of equation (5.3.17) one can usefully begin by considering the limiting case when γ the damping parameter tends to zero. The dielectric response function is then pure real and has the form

$$\varepsilon_r = \varepsilon_\infty + \frac{\omega_T^2 (\varepsilon_s - \varepsilon_\infty)}{\omega_T^2 - \omega^2}. \tag{5.3.17b}$$

The vibration frequencies of the lattice always correspond to singular points that is infinities or zeroes of the dielectric response function. Clearly (5.3.17b) gives an infinity at $\omega = \omega_{TO}$, that is the transverse optic frequency. The zero occurs for the other optic branch namely the longitudinal optic, and its frequency will thus be given by

$$\frac{\omega_L}{\omega_T} = \sqrt{\frac{\varepsilon_s}{\varepsilon_\infty}}. \tag{5.3.18}$$

This is the celebrated Lyddane–Sachs–Teller relation [497]. The magnitude of the longitudinal optic frequency is therefore given by

$$\omega_L^2 = \omega_T^2 + Nq^2 (\varepsilon_\infty + 2)^2 / 9\mu \varepsilon_0 \varepsilon_\infty V. \qquad (5.3.19)$$

Equations (5.3.13) and (5.3.19) display the splitting of the originally triply degenerate vibration at ω_0 into a lower doubly degenerate mode at ω_T and an upper non-degenerate mode at ω_L. From (5.3.19) the total splitting is seen to depend on the square of the ionic charge thus reflecting its origin in the macroscopic electric field due to the ions.

From (5.3.17b) one has that ε_r rises from its static value ε_s as ω increases, that it becomes infinite at $\omega = \omega_{TO}$ and then becomes negative for ω infinitesimally greater than ω_{TO}: it remains negative until $\omega = (\varepsilon_s / \varepsilon_\infty)^{1/2} \omega_{TO}$, i.e. $= \omega_{LO}$.† The permittivity is therefore negative between $\omega_{TO} < \omega < \omega_{LO}$ and this implies that the refractive index is purely imaginary. The crystal is therefore perfectly reflecting (see equation 3.5.20) between these two limits.

The treatment just given is of course unrealistic in that no damping term is present in equation (5.3.17b) and we know from experience that the TO resonance is very broad, that is γ^2 is not negligible in comparison with ω_T^2. The effect of damping is to make n real everywhere, but it is nevertheless very small between ω_{TO} and ω_{LO}. Alkali halide crystals are therefore very highly reflecting in this region, a property which makes them rather valuable as reflection filters in the far infrared. Taking NaCl as our example, $\omega_{TO} = 164 \text{ cm}^{-1}$ and $\omega_{LO} = 259 \text{ cm}^{-1}$ and there is strong reflectivity between these bounds. The effect of the strong damping is, however, to make the reflectivity less than unity and to make it peak at a frequency near ω_{TO} on the high side. The peak in the reflectance spectrum was exploited by Rubens who succeeded in isolating far-infrared radiation by allowing the radiation from a hot body to undergo successive reflection from alkali halide plates. The resulting "residual rays" or "reststrahlen" were much studied in the early days of far-infrared spectroscopy and the name "reststrahlen" is still in common use to describe the absorptive and reflective properties of ionic crystals in the far infrared. By the converse argument, monatomic lattices such as that of pure diamond will not absorb at all in the infrared. It is this feature, together with its strength and durability, which makes diamond such a valuable window material. The longitudinal optic frequency can always be calculated from (5.3.18) but is not easily observed in normal experimental arrangements because the transverse electromagnetic field cannot excite a longitudinal mode. However, at grazing incidence it is sometimes possible to observe it [498]. The width of the TO resonance can be ascribed to weak multiphonon processes which violate the simple selection rules. These are the analogues of the hot bands observed in

---

† It should be noted that ε_s must exceed ε_∞ by Kramers–Kronig arguments since their difference is a measure of the integrated strength of the TO band. If this latter is zero, as it is for diamond, $\varepsilon_s = \varepsilon_\infty$ and $\omega_{LO} = \omega_{TO}$ as remarked earlier.

molecular spectroscopy so they get weaker as the temperature falls. At liquid helium temperature the TO resonance is very sharp indeed [499] but is shifted up in frequency because of the contraction of the lattice and the consequent increase in the force constant. The lattice frequency can be calculated from (5.3.7) as it stands but the force constant, f, is best regarded as an empirical parameter rather than as a quantity to be calculated from first principles. This is because the potential function is made up of two parts, an attractive Coulombic term, which can be calculated, and a quantum repulsive term which can only be estimated. The simple Lennard Jones estimation of the repulsive potential, for example, gives only an order of magnitude value for f whilst the much more sophisticated "Shell Model", which regards the overall interionic force as the sum of Hooke's law components which connect respectively electron shells, nuclei to shells and nuclei to nuclei [500] is still not able to give particularly good agreement. The reason for this is, of course, that f is the second derivative of V so a potential which gives the thermodynamic functions excellently may nevertheless give a poor result for the interaction frequencies. The real value of these calculations lies more in their being used the other way round, that is observed force constants being used to reveal fine detail in the interionic potential. What emerges is that for the alkali halides, the most studied group, the f values do vary from ion to ion but not much and that they also vary with interionic separation but again not much. The principal factor in determining the value of ω_{TO} for an alkali halide is the reduced mass. We may thus take a representative value of f, say 11 N m^{-1} and use this to calculate ω_{TO} values. The result is that for an average alkali halide the reststrahlen absorption will lie in the 100–200 cm^{-1} region. Some experimental [501] values are given in Table 5.4.

This simple classical approach gives the essential features of ionic crystal spectra, namely an intense absorption located at ω_{TO} but the further details

TABLE 5.4
Transverse-optic ($k = 0$) wave numbers for some ionic crystals

Substance	$\tilde{v}_{TO}$ (cm^{-1})	Substance	$\tilde{v}_{TO}$ (cm^{-1})	Substance	$\tilde{v}_{TO}$ (cm^{-1})
ZnO	413	InAs	210	KBr	117
MgO	403	KF	191	NaI	117
GaP	366	BaF$_2$	186	AgCl	103
ZnS	310	CaSe	185	KI	101
LiF	304	LiCl	165	CsCl	99
InP	304	RbF	158	AgBr	80
GaAs	266	NaCl	164	RbI	75
CdS	265	LiI	144	CsBr	74
CaF$_2$	261	KCl	141	PbS	65
NaF	246	NaBr	135	CsI	61
SrF$_2$	218	CsF	121	TlI	52
ZnSe	215	RbCl	119	TlBr	48

require a more elaborate theory [502]. The width of the feature is accounted for by anharmonic processes which conserve momentum by involving more than one phonon. The intensities of these multi-phonon bands, however, fall off rapidly as the number of phonons involved increases. Thus most alkali halides become quite transparent beyond a frequency corresponding to $\omega_{TO} + \omega_{LO}$ and are very transparent indeed beyond $3 \omega_{LO}$. Fine detail in the multiphonon spectrum tends to reveal the critical points at the zone boundary. With the very accurate radiometry now becoming available with the perfection of Fourier transform spectrometry it is becoming possible to analyse the broad multiphonon absorption and even to provide complete assignments [503].

Real crystals do not, of course, have the perfect translational symmetry implied in the above analysis. They will have defects where crystal planes slip with respect to one another or where an ion has moved into an interstitial position. In addition there will always be impurity ions present and maybe even vacancies. Two cases are usually distinguished: (a) where the defect or substitution is either so light or so strongly bound that its natural vibration frequency lies above the highest phonon frequency; and (b) where the defect has a mass or a restoring force which give it a vibration frequency lying within the phonon band. In the former case the vibration will be unable to propagate through the lattice and we have a local mode which will give a relatively narrow line in the absorption spectrum. In the second case the mode can propagate but will couple to all the modes of the host lattice and one can then get, in principle, all the lattice phonons activated. The spectrum will then reflect primarily the density of states function $g(v)$, i.e. the number of modes of vibration lying between v and $v + dv$, apart from some slowly varying functions which depend on the nature of the defect and on the shape of the phonon branches. Local modes and band modes provide powerful probes for investigating lattice dynamics and are useful supplements to neutron spectroscopy which sometimes cannot give the whole story, but so far as the use of crystalline solids as transmissive infrared media is concerned they are a considerable nuisance. Thus only the very pure type IIa diamonds are completely transparent, the type I diamonds which contain a high proportion of nitrogen are quite heavily absorbing and the type IIb which contain acceptor centres are also moderately absorbing in some regions. Impurity induced absorption in silicon severely restricts the range of this otherwise very useful material.

5.3.3 Soft-mode phenomena in perovskite crystals

Many compounds of the general formula ABX_3 where A and B are metals and X is oxygen or a halogen crystallise in the cubic system. The typical example is calcium titanate or perovskite $CaTiO_3$ and such materials are commonly referred to as "perovskites" and are said to adopt the perovskite structure. They are particularly interesting physically because below a certain transition temperature many of them become spontaneously polarized and are therefore ferroelectric [504]. This transition temperature is commonly around 50 K and additionally at a higher temperature there occurs another transition where the elastic constants of the crystal show anomalous behaviour. Strontium titanate $SrTiO_3$ undergoes this latter transition [505] at 110 K. The elastic constants of a crystal are intimately connected with the velocity of sound therein and anomalous behaviour of the elastic constants implies some extraordinary behaviour of the acoustic branches of the dispersion diagram. Cochran [506] has suggested, and subsequent calculations and experiments have confirmed, that the ferroelectric transition arises in some cases because the lattice becomes unstable with respect to the lowest frequency transverse optical mode. This mode has come to be known as a "soft" mode and intense interest has been shown in the lattice dynamics of perovskites because of this singular behaviour.

The perovskite structure has one formula unit per unit cell and the symmetry of the unit cell is octahedral with point group O_h. There are fifteen branches to the dispersion diagram but at the zone origin these form the representation

$$\Gamma_Q(\bar{k} = 0) = 4F_{1u} + F_{2u} \qquad (5.3.20)$$

to first order. However just as for the alkali halides the effect of the macroscopic electric field is to split the triple degeneracy of the F_{1u} modes into a doubly degenerate transverse mode and a non-degenerate longitudinal mode. One of the F_{1u} modes corresponds to zero frequency when $\bar{k} = 0$, i.e. is equivalent to an overall translation of the crystal, and therefore represents the acoustic branches. The net result is that we have three TO modes active in infrared absorption, three LO modes inactive and one triply degenerate F_{2u} mode which is also inactive. The infrared spectrum should show three lattice modes and there should be no first-order Raman effect. This is observed, for at room temperature the far-infrared spectrum of strontium titanate shows three bands at 100, 178 and 510 cm^{-1} and the Raman spectrum is entirely second order [507].

One would normally expect that the frequency of a lattice mode would rise as the temperature falls because the interatomic forces will increase as the lattice shrinks and the atoms come closer together. Such "blue-shifts" are indeed widely observed but, for the special case of ionic crystals, the opposite type of behaviour is possible since (see equation 5.3.13) the observed lattice frequency will be given by the difference of two terms and in general these will not have the same temperature dependence. In other words, the effect of the macroscopic electric field can be manifest as a "red-shift" and in extreme cases the observed lattice frequency can fall to zero. This type of behaviour seems to occur for the lowest frequency TO mode of perovskites which, say for $SrTiO_3$, has dropped to 40 cm^{-1} when the temperature has been reduced to 90 K. Cochran [506] has shown that if ε_s follows a Curie law, i.e.

$$\varepsilon_s = C(T - T_c)^{-1}, \qquad (5.3.21)$$

then the frequency of the lowest TO mode should be given by

$$v_{TO}^2 = K(T - T_C). \tag{5.3.22}$$

Both quantities follow these equations quite well with a Curie temperature T_C = 35 K. Cowley [505] has pointed out that at the ferroelectric transition in barium titanate the displacements of the ions away from their cubic positions are all in the same sense as the oscillations of the atoms in the lowest frequency TO mode of strontium titanate. It is most plausible to argue therefore that as the TO mode frequency falls, its amplitude to maintain quantisation must rise and eventually will be big enough to take the whole lattice over into the ferroelectric phase. The transition near 110 K has been explained as due to a near degeneracy of the lowest TO mode and the highest (i.e. longitudinal) acoustic mode over a wide range of wave vectors for this temperature.

This, in outline, is the currently accepted explanation for this remarkable phenomenon. The application [500] of the "shell model", which takes account of the polarizability of ions, has opened the way to reasonably good calculations of the dispersion relations, ω versus k, over the whole Brillouin zone and the recent rapid advances in inelastic neutron spectroscopy have permitted the accurate testing of these over a wide range of $\bar{k}$. Far- infrared spectroscopy, especially reflection spectroscopy, has enabled the $\bar{k} = 0$ intersections to be determined and their temperature and pressure variations explored. The current availability of high-power gas lasers for the visible region has increased the experimental reach of Raman spectroscopy enormously and nothing illustrates this more than the new technique of electric field induced Raman spectra. This phenomenon occurs when a perovskite crystal is exposed to a high electric field. There is no first-order Raman effect at zero field essentially because each ion is on a centre of symmetry of the lattice; when the field is present, the ions are slightly displaced and a first-order Raman effect occurs [508]. If the electric field is modulated at an audiofrequency, the first-order Raman effect can be picked out from the unmodulated second-order effect by the use of suitable narrow band amplifiers and phase detectors following the photomultiplier detector. The importance of this technique is that the induced Raman spectrum may show zone centre phonons which are not present in the far-infrared spectrum. The converse, i.e. electric field induced infrared absorption, has been observed with diamond [509]. A weak band appears at 1336 cm^{-1} coincident with the $\bar{k} = 0$ optical mode found in first-order Raman scattering.

5.3.4 *Lattice spectra of molecular crystals*

The lattice spectra of molecular crystals are much less intense than those of ionic crystals and are therefore much easier to study experimentally since they can be observed in transmission. There is no macroscopic electric field and, because of this, the unit cell symmetry or factor group method applies rigorously [510]. The simplest way of interpreting the spectrum of a molecular crystal is to assign the high-frequency bands, which occur more or less where those of the free molecule do, to purely intramolecular modes and to assign the low-frequency bands to the vibrations of the unit cell regarded as made up of a small number of rigid molecules. Even at moderate resolution, however, this approximation fails since the true normal modes of the unit cell are invariably mixtures of the two types and the failure is evident from the splitting of the "intramolecular" bands into several components and from the fact that some of the "intermolecular" bands have observable intensity even though the zero-order calculation would indicate vanishingly small matrix elements. The situation is quite akin to that in molecular vibration theory where any normal mode, of a given symmetry, is made up from a linear combination of the primitive internal coordinate modes of the same symmetry. The degree of mixing depends on the frequency separation, and we therefore expect that the lowest frequency "intramolecular" modes will show the largest factor-group or correlation splittings. The pure "rigid-molecule" lattice modes themselves can be regarded as mixtures of translational and librational motions of the constituent molecules. If these molecules, are situated on centres of symmetry, however, then the two types are rigorously separate and any given lattice mode belongs strictly to one type or the other, but even when this condition does not apply this division is useful in practice since modes are usually found to be either mostly translational or else mostly librational in character. With the assumption that the intermolecular and the intramolecular modes are separable, there will be $3N - 3$ translatory and $3N$ rotatory lattice modes for a unit cell containing N non-linear polyatomic molecules. Extensive investigations of the spectra of a range of molecular crystals by Hadni [511] and by Anderson and his colleagues [512] have confirmed this. A particularly interesting point which emerges from Anderson's work is the comparison of the lattice spectra of a hydrogen-containing compound with that of the corresponding deuterium compound. Crystalline hydrogen chloride for example gives [513] six absorption bands at 86, 109, 217, 296, 496 and 650 cm^{-1} whereas crystalline deuterium chloride absorbs at 89, 113, 169, 209, 258 and 328. The effective temperatures of the samples were close to 90 K. The two lowest frequency modes show little shift on deuteration and are therefore principally translational in character whereas the four higher frequencies arise from modes which are mostly librational. Analogous studies of the halobenzenes and their deuterated derivatives have been published by Fleming and his colleagues [514]. The symmetry of the unit cells are known for these compounds, unlike the crystalline hydrogen halides, and armed with this information Fleming has calculated the activity of the various modes and using this has assigned the observed spectra. A typical spectrum is shown in Fig. 5.19. Fifteen lattice modes made up of 6 translatory and 9 rotatory types are expected for the D_{2h} unit cell containing 8 molecules: five are observed for C_6H_5Cl, C_6H_5Br, C_6H_5I and their deutero derivatives and five are also observed for C_6H_5F which has a different crystal structure with four molecules per unit cell. The five lattice bands of C_6D_5Br shown in Fig. 5.19 occur at $33(T)\,cm^{-1}$, $44(L_y)\,cm^{-1}$, $66(L_x)\,cm^{-1}$, $88(T)\,cm^{-1}$ and $120(L_z)\,cm^{-1}$ where the letters in brackets indicate, translation assignment T or librational assignment L_x, L_y, L_z. Three intramolecular modes are also evident in Fig. 5.19. The lowest of these of B_2 symmetry occurs at $180\,cm^{-1}$ for the liquid phase and shows a large correlation splitting in the crystalline phase. The second lowest of B_1 symmetry occurs at 246 in the liquid and shows a very small crystal-state splitting. The third lowest of A_1 symmetry has not so far been resolved into split components. The broad band peaking at a wave number less than $40\,cm^{-1}$ for liquid bromobenzene is typical of the liquid state and bands of this type will be discussed in the next section.

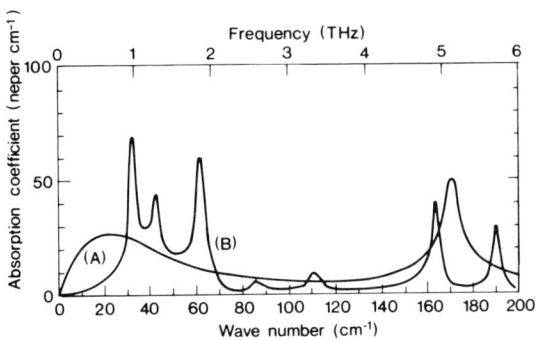

FIG. 5.19. Far-infrared absorption spectra of (A) liquid and (B) polycrystalline pentadeutero-bromobenzene at 300 K and 100 K respectively. In the crystal spectrum, the features at 33, 43, 62, 86 and $110\,cm^{-1}$ are lattice bands whilst the two bands near 160 and $190\,cm^{-1}$ are the crystal-field split components of the band observed at $170\,cm^{-1}$ in the liquid.

A very interesting application of crystal phase far-infrared spectroscopy is the detection of weak complex formation between molecules [515]. It is known, for example, that the crystals which separate from a mixture of chloroform and benzene on cooling are not pure crystals of either component but contain both. Are they simply solid solutions or are they crystals of stoichiometric complexes? The question is not readily answered by X-ray methods since single crystals would have to be isolated for study and it is possible that several types of complexes are being formed simultaneously. However, if stoichiometric complexes are being formed with defined crystal structures showing long-range regularity then the far-infrared spectra of polycrystalline samples may show new bands arising from the vibrations of unit cells which contain both chloroform and benzene. If this is observed, then this is conclusive evidence for the formation of stoichiometric complexes. Studies of the intensities of the new bands as functions of composition of the original mixture will indicate via the well known plots versus mole ratio, the exact stoichiometry of the various complexes present. Some results for chloroform and benzene are shown in Fig. 5.20. These spectra, in marked contrast with those of either pure crystalline benzene or chloroform, show several sharp lines whose intensities vary in a regular manner with composition of the original mixture. This evidence firmly establishes the existence of the two complexes $2CHCl_3 : C_6H_6$ and $CHCl_3 : C_6H_6$.

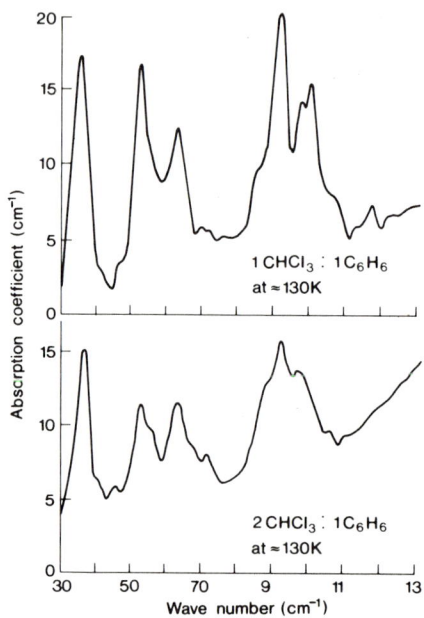

FIG. 5.20. Far-infrared spectra of the polycrystalline specimens produced by freezing two different liquid mixtures of chloroform and benzene.

Observations of crystal-phase spectra have some relevance to purely molecular spectroscopy either perforce because the spectroscopist is studying a compound which does not dissolve in any convenient solvent or else by design in that by choosing the crystalline phase the experimentalist might hope to observe fundamentals which would be forbidden for the free molecule. In the first case great care must be taken to ensure that the various components of a split fundamental are not misidentified as separate fundamentals and the spectroscopist has also the considerable problem of identifying which observed bands are molecular fundamentals and which are lattice bands. If the crystal can be studied at various temperatures or at various hydrostatic pressures this identification can be made more easily since in general the lattice modes show much bigger shifts than do the intermolecular modes. The observation of forbidden fundamentals, in the crystal phase, arises because the selection rules in this phase are dictated by the symmetry of a molecule's environment (i.e. the site symmetry) rather than by the molecular symmetry itself and frequently a molecule will occupy a site whose symmetry is lower than that of the molecule itself. This has been demonstrated [516] for napthalene $C_{10}H_8$ where the lowest frequency A_u mode, which is of course forbidden for the free molecule, does appear at medium intensity in the far-infrared spectrum of the crystal at 210 cm^{-1}.

It was remarked earlier that the infrared spectrum can give only a limited amount of information so far as the density of states function $g(v)$ is concerned. Specific heats are, however, often discussed in terms of approximate forms of $g(v)$ and the classic example is that chosen by Debye [517] who postulated that the density of states function for the acoustic branches was a quadratic function of $\tilde{v}$ up to a limiting wave number $\tilde{v}_m$ and was zero for all higher $\tilde{v}$. At low temperature, when by the Boltzmann factor only the acoustic phonons are significant, the molar specific heat is then given by

$$C_v = (12\pi^4 R/5)(k T/hc \, \tilde{v}_m)^3. \qquad (5.3.23)$$

The Debye temperature θ_D is that temperature for which $k\theta_D = hc \, \tilde{v}_m$. This equation satisfactorily accounts for the behaviour of many solids at cryogenic temperatures which is surprising in view of the crude model used in its derivation, but if $\tilde{v}_m$ is reinterpreted as the wave number of the centroid of the density of vibrational states, as suggested by Plendl [518], then the treatment becomes physically much more acceptable. It is therefore most interesting to note that for crystalline HCl, $\tilde{v}_m$ is $89{\cdot}6 \text{ cm}^{-1}$ and that this is close to the mean of the two translational mode frequencies. For NaCl as another example $\tilde{v}_m$ is 196 cm^{-1}—quite close to the reststrahlen frequency (164 cm^{-1} at 300 K, 174 cm^{-1} at 90 K).

5.4 Infrared properties of semiconductors

5.4.1 Introduction: electronic solid-state physics

It would be difficult to overstress the importance of semiconductors to infrared science and technology since these materials provide the infrared worker with some of his most flexible sources, some of his most useful optical materials and some of his most sensitive detectors. It would be equally difficult to overstate the good fortune of the infrared practitioner in having all these invaluable materials together with the appropriate very high technology fabrication techniques made available to him as a spin-off from the multi-billion dollar microelectronics industry. To make sense, though, of a discussion of semiconductor physics does require some familiarity with the basic concepts of solid-state physics, a much more extensive topic, which includes semiconductor physics as a special case. Solid-state physics is an intellectually rewarding discipline but some of its ramifications can be rather daunting. Fortunately it can be approached in terms of two theories, each of them relatively accessible which between them give a good account of most of the elementary parts of the topic. These theories stress respectively the physical and the chemical aspects of the materials and it is convenient to use these terms to label them. Both theories assume that the solid in question is a near perfect crystal since this makes for considerable mathematical simplifications. It is important to note, however, that mathematical expedience and physical reality are not necessarily concurrent and that several features which emerge from these theories are nevertheless commonly manifest in amorphous solids and even liquids. Thus glass is an excellent insulator, amorphous silicon is a good semiconductor and mercury is a perfectly good metal. These facts do not give the lie to the theories, merely that the mathematical framework could be generalised without physical loss. In the next two sections each theory will be developed and then the two will be brought together as a basis for a discussion of the optical properties of semiconductors.

5.4.2 *The physical approach—electron waves in periodic lattices*

The physical approach stems from the classical "free-electron" theory originally devised to account for the properties of metals. This theory had some success especially in accounting for the optical and electrical properties but, as is well known, it failed completely to account for the thermal properties. The quantum version of this theory developed originally by Sommerfeld [519] but much elaborated in later years by Bloch [520], Peierls [521], Mott [522] and others [523] has had quite remarkable successes when one considers that it is basically a very simple theory. It starts from the classical result that for a free electron there is only kinetic energy so

$$\mathscr{E} = (1/2)\, mv^2 = p^2/2m = \hbar^2 k^2/2m. \tag{5.4.1}$$

The dispersion diagram is therefore very simple—just a parabola. The next point is to introduce the quantum idea of electron waves propagating through the lattice. Such waves will be diffracted by the lattice points and if in a given direction the spacing is d, the propagation of the waves will be entirely impeded when the Bragg condition

$$n\lambda = 2d, \tag{5.4.2a}$$

that is

$$k = n\pi/d, \tag{5.4.2b}$$

is satisfied. The periodicity of the lattice therefore introduces a set of forbidden points on the dispersion diagram corresponding to k values given by (5.4.2b). These points correspond to standing rather than propagating waves. The next step is to acknowledge that the diffracting centres are not points but extended interacting entities—atoms in fact. One now must solve the Schrödinger equation with these atomic potentials included. This is quite impossible because (a) one doesn't know the potential and (b) one couldn't solve the equation even if one did. An approximate approach is therefore indicated based on some more tractable model. Kronig and Penney [524] introduced the concept of deep "wells" to represent the atoms and in particular did some elegant calculations for the particular case of square wells. If the depth of the wells is V_0 the width is b and the electron energy is $\mathscr{E}$, two parameters,

$$\alpha = (2m\mathscr{E}/\hbar^2)^{1/2} \quad \text{and} \quad P = mV_0bd/\hbar^2, \qquad (5.4.3)$$

may be introduced and in terms of these Kronig and Penney show that the stability condition may be written

$$P[(\sin \alpha d)/\alpha d] + \cos \alpha d = \cos kd. \qquad (5.4.4)$$

This equation has real solutions for k only if α lies within certain restricted ranges and the point discontinuities of the simple model broaden out into forbidden bands for k as shown in Fig. 5.21. This has the effect that corresponding ranges of $\mathscr{E}$ are not allowed either and one finishes up with the concept of forbidden bands separating the allowed bands. This is one of the most important concepts in modern solid state physics. The theory in this form is sometimes called the "nearly free" electron model.

An electron in an allowed band is in principle (but see later) permitted to flow under the effect of an applied electric field and to interact with an impressed electromagnetic field but it will not behave exactly like an electron in free space. In particular it will exhibit an effective mass and an effective velocity

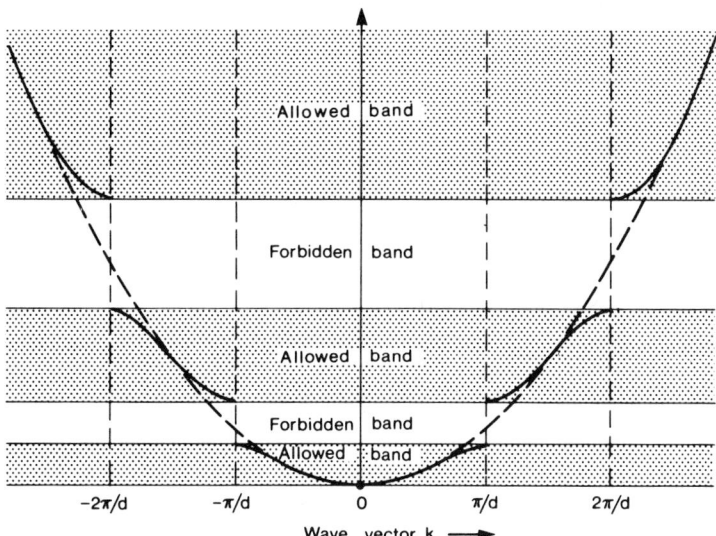

FIG. 5.21. Origin of bandstructure according to the "nearly-free" electron theory.

both different from what one might expect. Both these effects can be introduced rather nicely by differentiating (5.4.1). Once gives

$$v = (2\pi/h)\,(d\mathscr{E}/dk) \tag{5.4.5}$$

and twice gives

$$1/m = (4\pi^2/h^2)\,(d^2\mathscr{E}/dk^2). \tag{5.4.6}$$

For a free electron these would be identities and of course $\mathscr{E}$ would be a parabolic function of k but for the electrons in the band structure they can be taken as definitions of the effective velocity and of the effective mass:

$$m^* = h^2/4\pi^2\,(d^2\mathscr{E}/dk^2). \tag{5.4.7}$$

Thus since both $(d\mathscr{E}/dK)$ and $(d^2\mathscr{E}/dk^2)$ vary with $\mathscr{E}$ (Fig. 5.21) the mass and the velocity of the electron will both vary with its degree of excitation. The velocity is a maximum in the middle of the band and zero at the top and bottom. The effective mass is positive in the lower half of the band but negative in the upper half. At the point of inflection, it becomes formally infinite. The effective velocity defined by (5.4.5) explained away nicely the observed phenomenon that electron velocities in metals are much higher than might be expected from equipartition of energy. If one has a single electron in an otherwise empty band and then applies an electric field the electron will be accelerated and gaining energy move up the band but when it gets into the upper part of the band its effective mass will change sign and it will be accelerated in the *opposite* direction; the electron will therefore oscillate with mean energy equal to half the mid-band energy. In quantum mechanical terms this would be interpreted as another manifestation of Bragg reflection. In a real medium the electron will suffer inelastic collisions with the atoms and the momentum will be redistributed. The net effect is that the electron will spend more time in the lower part of the band and a net current will flow. The electron will therefore have a *drift velocity* which is an externally measurable quantity and this drift velocity ($\sim 1\ (\mathrm{m\,s^{-1}})/(\mathrm{V\,cm^{-1}})$) will be very much less than the average velocity of the electron between collisions. If one considers the opposite limit—that is a completely full band—one sees at once that no current can flow since if an electron is to be accelerated its energy must increase and there is no unoccupied higher energy level to accommodate it. The startling conclusion therefore is that the electrons in a full band are electrically invisible! If one now imagines a single electron removed from the top of a full band it will leave behind a "hole" and, since one has removed a negatively charged particle of negative mass, the hole being the complete opposite will be a positively charged particle of positive mass. There are obviously strong connections here with the Dirac theory of the positron [454]. This is particularly well brought out if one considers photons incident upon the non-conducting medium but having energy greater than the width of the forbidden gap. An electron can now be promoted into an empty band where it is, of course, perfectly "visible" and simultaneously a hole will appear—both as if from nowhere. The electron and the hole can both conduct electricity and we have the phenomenon of *photoconductivity*. This phenomenon is the basis of some of the most sensitive infrared detectors currently available. The electron and the hole will, however, in general, have different effective masses, will consequently have different drift velocities and hence different mobilities. The mobility μ is defined by

$$\mu = v_{dr}/E, \tag{5.4.8}$$

where v_{dr} is the drift velocity. In the general case where we have concentration n_e of electrons, n_h of holes and where the mobilities are μ_e and μ_h, the conductivity will be given by

$$\sigma = e[n_e \mu_e + n_h \mu_h]. \tag{5.4.9}$$

So far we have tacitly assumed a one-dimensional crystal but real crystals are three dimensional and even if they have cubic symmetry their properties in different directions need not be the same. It is possible therefore for the forbidden gap along one crystal direction to coincide with an allowed band in a different direction. In other words it is possible, because of directional effects, for the allowed bands to overlap and for the forbidden band separating them to disappear. It is also possible that the minimum energy position in the first allowed excited state does not occur for $\bar{k} = 0$ as it must for the ground state. Vertical electron promotion, which, by the Born–Oppenheimer approximation, is strongly favoured can then only take place with either the annihilation or the creation of a phonon—otherwise momentum would not be conserved. Transitions of this type are called *indirect*. The converse when the minima do lie vertically above one another involves *direct* transitions.

This concept is of particular importance when one comes to consider the converse of photoconductivity, namely the emission of photons produced by the recombination of the holes and electrons produced by the passage of a current through a suitable material (for details see later). This mechanism is the basis for LED action and for the corresponding semiconductor laser action, both of inestimable value in infrared technology. However the mechanism is orders of magnitude more efficient for direct materials such as gallium arsenide then it is for indirect materials such as silicon.

One needs also to consider temperature effects when trying to relate this model to real materials. At temperatures above 0 K, electrons can be *thermally* promoted from a filled band, across the gap and into an empty band. One therefore expects that for insulators there will be always a small conductivity and that this will increase rapidly (in fact exponentially) as the temperature rises. This behaviour is in stark contrast with that of metals where the conductivity *falls* as the temperature rises. The explanation of this latter phenomenon is, of course, that conduction due to electrons in a partially full band will be impeded by thermally induced atomic disorder as the temperature goes up. Equivalently one could say that the electron waves are scattered by both the phonons and the lattice itself at finite temperature whereas only the lattice can scatter at absolute zero. The actual distribution of the electrons amongst the available levels is determined by the statistical properties of electrons which, being identical particles of half integral spin, obey the Fermi–Dirac statistics whose distribution function is

$$N(\mathscr{E}) = N(0)[\exp[(\mathscr{E} - \mathscr{E}_F)/kT] + 1]^{-1}, \qquad (5.4.10)$$

where $\mathscr{E}_F$ is the so-called Fermi energy or more colloquially the Fermi level. For an insulator, $\mathscr{E}_F$ has of necessity to lie exactly in the middle of the forbidden gap in order to achieve the results that promotion can only take place *across* the gap and that charge must be conserved.

5.4.3 *The chemical approach—the tight-binding model*

The chemical approach starts from almost the opposite extreme by considering in detail what exactly are the entities making up the lattice and how exactly do they interact with one another. One begins, as in section 5.3.1, by dividing the different kinds of crystalline solid into the three categories:

a. Molecular crystals
b. Ionic crystals
c. "Macromolecular" atomic crystals

From an infrared viewpoint, the first two categories are mostly of interest on account of their lattice vibrations—as discussed in section 5.3. They are nearly always insulators with band gaps in excess of 5 eV so their electronic properties do not become manifest until well into the ultraviolet. The third category, however, includes many materials which have band gaps of 2 eV or less and which are therefore of considerable infrared interest. These materials can be elemental, e.g. germanium or compound, e.g. indium antimonide, but their electronic properties are broadly similar so the discussion will begin by outlining the simpler case where all the atoms are identical.

These atoms are located in a cubic lattice where each atom is surrounded tetrahedrally by four nearest neighbours. One can imagine such a structure smoothly extended until the N atoms making up the lattice are widely separated but still maintaining the basic cubic symmetry of their disposition with respect to one another. The atoms will then be essentially non interacting so the electronic energy levels will be identical to those of the isolated atoms but, of course, they will now be N-fold degenerate. The atoms are then allowed to approach one another smoothly until they finally achieve their correct positions on the actually observed lattice. As they approach one another, the atoms begin to interact and the N-fold degeneracy is removed. Each level now breaks up into N separate levels and since these will have finite widths (because of the Uncertainty Principle), overlap will occur and for macroscopic specimens where N may well be in excess of 10^{20} one will finish up not with discrete energy levels but with continuous bands. This is shown schematically in Fig 5.22. The bands are perfectly continuous and an electron, in the solid, is permitted to have *any* energy which lies within the band. The bands are, however, separated by forbidden gaps of width $\mathcal{E}_g$—energies lying within such a gap do not correspond to stationary solutions of the overall wave equation. Each discrete level of the isolated atoms yields a corresponding band for the crystal but only the outer electrons give bands of significant width. The lower lying electrons are too well shielded for effective interaction to occur. In the particular case of infrared physics it is only the outer valence electrons and the bands formed by the overlap of their atomic orbitals which are of significance. One has then two important bands: the *valence* band which is constructed from bonding orbitals and the *conduction* band which is constructed from antibonding orbitals. This is the basic idea of the "tight-binding" model. It should be noted carefully that the concept of allowed bands separated by a forbidden gap is common to both models but the interpretation is very different. All this means physically is that we are not telling the whole truth. Sometimes one approximate way of looking at things will be appropriate, sometimes the other. The real difficulties in solid-state physics arise from those cases where neither model is completely adequate for the discussion.

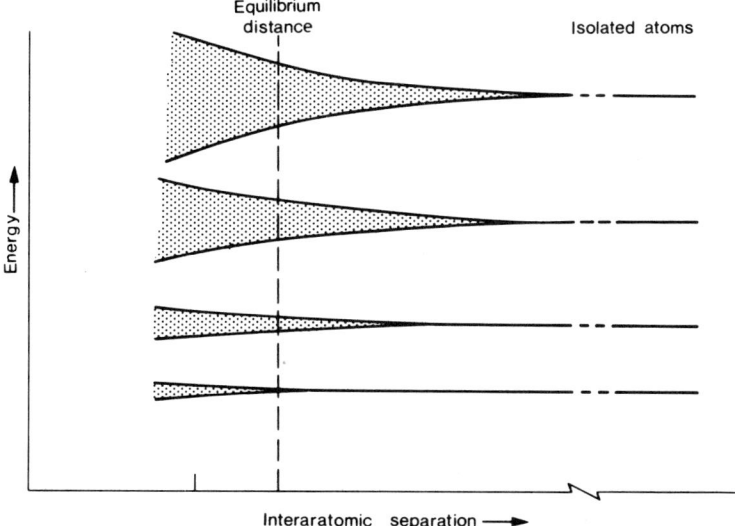

FIG. 5.22. The formation of continuous energy bands when a crystalline solid is formed from its components. The bands are separated by forbidden gaps of width $\mathcal{E}_g$.

To continue the development of this model it is best to discuss a concrete example and an excellent one is provided by the second row of the periodic Table, that is the elements, sodium, magnesium, aluminium, silicon, phosphorus, sulphur, chlorine and argon. For sodium which has a single outer electron in the 3s orbital, the valence band will be exactly half full and it follows that the element will be metallic. For magnesium, on the other hand, for which the isolated atom structure is $(1s)^2 (2s)^2 (2p)^6 (3s)^2$, one might expect to have an insulator since two electrons per atom will fill the valence band. However what happens is that the bonding orbitals are formed from hybrids; in fact sp^3 hybrids and therefore the s and p bands overlap. Hybrid orbitals are necessarily highly directional so this band overlap can be thought of as equivalent to the directionally induced overlap present in the physical model. One should therefore think of four orbitals pointing out tetrahedrally from each atom and overlapping with the equivalent orbitals from the next nearest neighbours to form four bonds. This bond structure extends throughout the macroscopic specimen and, if there are less than two electrons per bond, an electron can "hop" from a bond into a vacancy and thus propagate macroscopic distances. The network of interlinking bonds is thus the more detailed chemical interpretation of the valence band. In the case of magnesium one has on average one electron per bond so the element is metallic. The same is true of aluminium where one has 1·5 electrons per bond but when one comes to silicon with two electrons per bond one expects to find an insulator. This is true at absolute zero, but since the band gap (~ 1 eV) is relatively small it shows semiconducting properties at ambient temperatures. Moving on now to the interesting case of phosphorus one might imagine that, with the fifth electron having to go into the non-bonding conduction band, one would be back to metallic properties, but instead the whole lattice collapses and one gets an insulating molecular crystal made up of P_4 molecules. The same is true of sulphur with S_8 molecules and of chlorine (Cl_2) and argon (A) at sufficiently low temperatures. It thus seems that the strongly destabilising effect of having non-bonding electrons is just too much and the solid flips over into the more stable molecular crystal form. For heavier elements, new phenomena are possible, d-orbital hybridisation for example, and the forbidden gaps get progressively smaller. Overlap, leading to metallic properties, gets therefore easier and easier and metals are found, amongst the typical elements, further and further to the right of the Table. Thus tin has two forms, white which is metallic and grey which is semiconducting, whereas lead is entirely metallic.

The condition for forming an insulating or intrinsically semiconducting giant lattice is that there be on average four electrons per atom. This can obviously be satisfied for the Group IV elements C, Si, Ge, Sn, themselves and for the compounds formed amongst them, but it can also be satisfied for the compounds formed from the adjoining Groups III and V. Thus GaAs is analogous to Ge and InSb is analogous to Sn. The process can be continued to specify some II-VI compounds such as ZnSe, CdTe etc, but the differences between the atoms soon become manifest in terms of bond polarisation effects and the true structures tend to be a cross between the macromolecular type and the ionic lattice type. This is not necessarily detrimental for device purposes since it opens up new electromagnetic possibilities (deliberate use of the dispersion near the reststrahlen frequency for example). II-VI compounds involving light atoms have large band gaps and are of more use for their insulating properties than they for their electrical properties—they make excellent infrared optical materials for example but those made from heavier atoms can be very useful semiconductors, $Hg_x Cd_{1-x} Te$ for example. In this connection it is interesting to note that the heavy atom effects mentioned above lead to the possibility of variable band gap materials. Thus $Pb_x Sn_{1-x} Se$ is, according to the value of x, either semi-metallic or semiconducting.

5.4.4 *Semiconductors*

The phenomenon of semiconductivity has been mentioned several times in what has gone before but now it needs to be discussed in rather more detail. One must say at once that any division of non-metals into insulators and semiconductors is arbitrary: one just has materials of progressively lower and lower conductivity. As new measurement techniques are developed (and one can now buy femtoammeters!) so materials previously thought of as insulators are shown to be merely very high resistors. However, a division, arbitrary though it is, is valuable and one bases this division naturally enough on the technological semiconducting devices which work at ambient temperature and with reasonably high currents. One therefore thinks of any material with a band gap less than say 50 kT as a semiconductor and any material with a larger band gap as an insulator. 50 kT for $T = 300$ K is 1·29 eV which is equivalent to $1·04 \times 10^4$ cm^{-1}, that is approximately 1 µm, so this definition means that any material with a band-gap wavelength

$$\lambda_g = hc/\mathscr{E}_g \qquad (6.5.11)$$

lying in the infrared will be a semiconductor. The relevant parameters for some technically important semiconductors are given in Table 5.5.

The infrared properties of intrinsic semiconductors are basically simple, especially at low temperatures. At wavelengths longer than λ_g the material is transparent but in a short range straddling λ_g the absorption increases enormously and at wavelengths shorter than λ_g the absorption is so high ($\sim 10^4$ neper cm^{-1}) that the material behaves as though it were a metal. This is the explanation of the well-known metallic sheen displayed by silicon and germanium crystals. The onset of absorption is so sudden that it is conventional to refer to it as the "absorption edge". This absorption edge phenomenon can be used in the construction of long-wave passing filters or long-wave rejecting reflection filters. The edge becomes less abrupt and shifts as the temperature rises. In their transparent regions, semiconductors make very useful optical materials. Germanium is very widely used for high-grade optics for the 10·6 µm region and zinc selenide is rapidly establishing itself as a most useful general purpose infrared optical material. All these substances have quite high refractive indices, this being a Kramers–Kronig consequence of the existence of the intense absorption processes at higher frequency. In general the larger is λ_g the larger is the value of ε_r (or n) in the transparent region lying beyond. Intrinsic semiconductors are therefore very valuable optical materials for use in the infrared since they offer a combination of properties not available from other candidates—alkali halides for example. They are also very valuable as photoconductive detectors, especially for the shorter wavelengths.

TABLE 5.5
Properties of some important semiconductors and insulators

Substance	Formula	$\mathscr{E}_g$(eV)	λ_g(μm)	ε_s	m^*(m$_e$)
Diamond	C	5·47	0·277	5·8	0·20
Zinc sulphide	ZnS	3·68	0·337	8·1	0·40
Aluminium arsenide	AlAs	3·02	0·411	—	0·15
Zinc selenide	ZnSe	2·58	0·481	9·1	0·16
Cadmium sulphide	CdS	2·42	0·512	—	0·21
Gallium phosphide	GaP	2·26	0·549	10·75	0·82
Cadmium selenide	CdSe	1·70	0·729	—	0·13
Cadmium telluride	CdTe	1·56	0·796	9·7	0·11
Gallium aluminium arsenide	Ga$_{1-x}$Al$_x$As	1·42 + 1·25x	—	—	—
Gallium arsenide	GaAs	1·42	0·875	12·56	0·067
Indium phosphide	InP	1·35	0·919	12·4	0·077
Silicon	Si	1·12	1·108	11·4	0·98$_L$, 0·19$_T$
Gallium antimonide	GaSb	0·72	1·724	15·7	0·042
Germanium	Ge	0·66	1·880	15·36	1·64$_L$, 0·08$_T$
Lead sulphide	PbS	0·41	3·027	—	0·25
Indium arsenide	InAs	0·36	3·447	14·6	0·023
Lead telluride	PbTe	0·31	4·003	—	0·17
Indium antimonide	InSb	0·165	7·521	17·9	0·014
Grey ( ) tin	Sn	0·082	15·133	—	—
Lead tin telluride	Pb$_{1-x}$Sn$_x$Te	(0·1 for x = 0·2	Variable	428	—
Mercury cadmium telluride	Hg$_{1-x}$Cd$_x$Te	(0·25 for x = 0·3	Variable	18	—

Useful though intrinsic semiconductors are, it has to be admitted that the extrinsic types are even more valuable. In fact most of the technologically important properties of semiconductors are due to their having been carefully "doped" with appropriate impurities. Up till very recently, semiconductors, being uncommon and rather exotic materials, were only known in very impure condition. These impure specimens were sometimes shown to have remarkable properties—the ability of a metallic point (or cats whisker) impinging on a galena (PbS) crystal to demodulate radio-waves being a well-known example—but the real advances came when the chemists devised ways of making very pure semiconductors which could then be doped in a controlled manner. The purification techniques commonly used include vapour-phase preparation from carefully purified liquid compounds followed by zone refining of the resulting solid-state material. Techniques have also been devised to manufacture large single crystals of most of these substances. The purity situation is now completely reversed. Semiconducting elements such as silicon are being routinely prepared in a purer condition than that of any other class of substance. Impurity levels of 1 part in 10^{12} can be achieved but routine manufacture probably achieves a figure nearer 1 part in 10^9. Levels less than 1 part in 10^9 are probably beyond chemical detection but the presence of the impurities can still be detected by the much more sensitive electrical methods. It should be noted that an impurity level of 1 part in 10^9 nevertheless implies the presence of 10^{20} impurity atoms per cubic metre. Although, then, one cannot have truly intrinsic material it nevertheless remains a useful practice to use this term to describe the very pure forms and to reserve the terms "extrinsic" or "doped" to describe the forms made by deliberately adding impurities to the very pure starting material.

The impurities can be of two types. One can have adjacent atoms from the periodic table which, being of about the right size and having the appropriate chemical properties, can fit into the giant valence-bonded lattice. These would be called *substitutional* impurities. One can also have smaller atoms which can squeeze themselves in by distorting the lattice—transition metal atoms are good examples. These are called *interstitial* impurities. The modifications of the electromagnetic properties of the host medium caused by the presence of the impurities presents a difficult theoretical problem but fortunately one can give a very reasonable account in terms of a rather simple approximate theory in which the additional electron (or hole) is regarded as entirely hydrogenic. Thus if one had antimony in germanium, one would assign four of the five electrons to the job of bonding the lattice but the fifth would be regarded as localised around the antimony nucleus and in fact executing Bohr orbital motion around it. The energy levels of a hydrogenic atom have been given previously (5.1.5) but that equation applied to free space. In a material with relative permitivity ε_r the Rydberg constant is reduced by the factor ε_r^2 and the electron mass is reduced to its effective value m^* so that for an impurity electron in germanium, one would have an ionisation energy $\mathscr{E}_\infty$ equal to 0·0053 eV. The electron, if ionised, would be in the conduction band so it follows that the ground state must lie in the forbidden band at a distance of 10^{-2} eV or so below the conduction band. Electrons in such levels can readily be ionised thermally at ordinary temperatures so the levels are commonly called shallow *donor states*. A semiconductor containing donor states is said to be *n-type*, the "n" denoting negatively charged carriers. In an entirely similar manner if one had indium impurities in germanium one would have shallow *acceptor levels* lying about 10^{-2} eV above the top of the valence band. The material would then be called "p-type". Shallow donor and acceptor levels of these types are indeed found but usually not at quite the energies predicted by the simple "hydrogen-atom" theory. This is because of the operation of two factors. Firstly, electrons in the s orbitals of hydrogenic atoms have wave functions which have finite amplitudes at the origin, i.e. at the nucleus. The electron energy is therefore sensitive to the exact atomic number of the impurity. Secondly, the crystal field around the impurity does not have the full spherical symmetry of free space, rather it has the symmetry imposed by the tetrahedral disposition of the surrounding host atoms and characterised by the tetrahedral point group T_d. As a result of these twin effects, the s states split into a closely spaced E, T_2 (also denoted F_2) pair and a much lower singlet A_1 level. One thus has the so-called chemical shifts, valley–orbit interactions or "central-cell-corrections". The $1s(A_1)$ level shows by far the largest depression and chemical sensitivity, both being due to the relatively high amplitude at the nucleus for the wave function of this particular state.

Due to the presence of the impurities, the otherwise clear region between the plasma edge and the band gap will show discrete absorption features. These are of considerable interest to the theoretician since, by studying the lines due to known impurities, it is possible to test the validity of the effective mass formalism [525]. A well-known early example of such a hydrogen-like spectrum, due to Fisher and Fan [526], is shown in Fig. 5.23(a). This shows the Rydberg sequence, i.e. $1S \to nP$, leading up to the ionisation continuum. Nowadays, with the perfection of Fourier transform spectroscopy and the availability of supersensitive cryogenically cooled detectors, it is possible to produce superb Rydberg spectra [527]. Figure 5.23(b) is taken from the elegant work of Jagannath and Ramdas [528] at Purdue. It clearly shows the splitting of the degeneracy of the p orbitals due to the cylindrical, rather than spherical, symmetry of the wave functions near the minimum of the conduction band energy for semiconductors such as silicon and germanium. In equivalent language one could say that the effective mass is tensorial rather than scalar. To get properly resolved spectra, it is essential that the specimen be very cold, but then only the 1S Rydberg (or "Lyman") lines can be seen. A very useful way round this difficulty is optical excitation with black-body radiation. This populates all the various levels of the possible "ladders" and many more lines appear in the spectrum. Stradling [525] quotes the example of germanium in which, with optical excitation, nine lines are observed spanning the range from 5.64 to $10.4 \, \text{cm}^{-1}$. These can be assigned to hydrogen-like transitions ranging from $2p_{+1} \to 4d_0$ to $3p_0 \to 3d_{+1}$. The inverse of this type of investigation, that is observe the lines and infer the nature of the impurities, is very important technically. Afsar and Button [529], for example, have been able, using a combination of dispersive Fourier transform spectroscopy and superb magnet facilities, to deduce the nature of most of the residual impurities in semi-insulating gallium arsenide,

Magnetic, and particularly cyclotron, resonance spectroscopy of semiconductors is an important topic in its own right; some further details are given in section 5.6. Excitons, that is electron-hole bound states, occur commonly in optically excited semiconductors and effects due to them are readily ob-

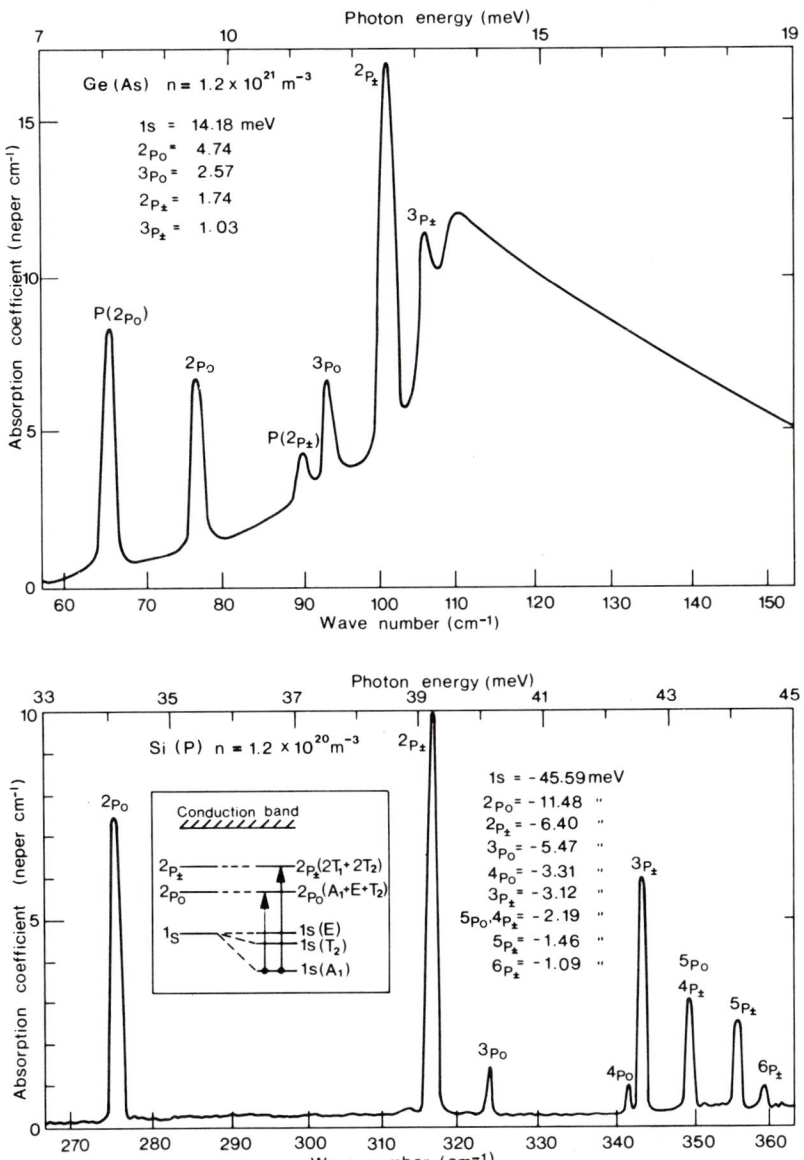

Fig. 5.23. Two examples of hydrogen-like spectra due to impurity donors in elemental semiconductors. In (a) is shown the spectrum of arsenic in germanium [526] whilst in (b) is shown the spectrum of phosphorus donors produced by neutron irradiation of a pure silicon specimen [527].

servable since the binding energies usually lie in the far infrared. Materials such as silicon, which have indirect band gaps, are of special interest because at high levels of excitation the free excitons undergo a phase transition into a metallic electron-hole plasma. This plasma, almost at once, decays to a lower energy state made up of spherical "drops" surrounded by virtually insulating material [530]. Spectroscopic studies of these systems can be very informative [531]. Thus transitions between the discrete levels of the excitons, free-carrier absorption and magneto-plasma oscillations of the drops, can all be distinguished [532] from one another.

An important consequence of the high permittivity of semiconductors is that the radii of the Bohr orbits are enormously increased and impurity atoms, which are widely separated, can nevertheless interact with one another. Thus at a certain level of doping the discrete lines due to the impurities will broaden out into a half-filled band and the material will develop metallic properties. As an example, germanium, as normally made, is a quite good conductor even at low temperatures. It is interesting, in this connection, to note the profound difference between metals and semiconductors *vis-à-vis* impurities. For metals these impurities can depress the conductivity whereas for semiconductors they can increase it by orders of magnitude. However the number density of electrons is much less than that which prevails for normal metals ($\sim 10^{29}$) and the characteristic plasma frequency (2.2.3) is shifted into the far infrared. One thus has the interesting situation of a material which is metallic up to far-infrared frequencies, then becomes a rather lossy dielectric and then resumes metallic behaviour at frequencies greater than that of the absorption edge. Plasma edges in the far infrared have been studied by Birch and his colleagues [533]. Apart from their theoretical interest, overlap phenomena are important in determining the spectral response and hence the performance of infrared semiconductor detectors. Thus, by careful choice of doping, detectors can be made which have a high responsivity over a chosen waveband but which do not respond significantly anywhere else.

These are just a few examples of the power of infrared spectroscopy to tell us things about semiconductors that would probably not be forthcoming by any other technique. Already it is being applied to actual semiconducting devices, for example MOS structures, where one has essentially a two-dimensional plasma [534]. The results are not only interesting in their own right, they will undoubtedly prove invaluable to the engineers who are designing the next generation of practical devices.

5.5 Infrared spectroscopy of liquids

5.5.1 *Introduction*

At low resolution and at a cursory glance, the infrared spectrum of a pure liquid looks very similar to that of the corresponding gas or vapour. However, at higher resolution or on closer inspection, some significant differences appear. Thus at a sufficiently high resolution the bands in the spectrum of a low-pressure vapour or gas break up into a large number of discrete very sharp lines, the rotational fine structure, whereas the bands in the liquid-phase spectrum are always diffuse. The continuous absorption bands displayed by a liquid have typically a bell-shaped profile whose centre frequency is close to, but significantly different from, v_0 for the corresponding gas-phase band. The relative integrated intensity of the band is also close to that of the gas-phase band but again there are differences which are sometimes dramatic. Perhaps the best illustration of this is the frequent appearance in a liquid-phase spectrum of bands which are forbidden by some symmetry selection rule for the isolated molecule and which therefore would not be expected in the gas-phase spectrum. These differences reflect the differing environment of a gas phase molecule and a liquid-phase molecule, the former spending most of its time in isolation with only an occasional collision, whilst the latter is in a state of more or less continual collision. Careful study of how the spectrum changes as the specimen passes from the gas into the liquid can therefore tell us a great deal about the microdynamics of the liquid state but it should be emphasised, as remarked above, that the differences are small and that to quite acceptable accuracy one can do most analytical and diagnostic infrared spectroscopy on pure liquids and solutions without any need to go to the considerable difficulties of working in the vapour phase.

Each particular liquid has its own idiosyncrasies but there are sufficient common features to make it worthwhile to quote a single illustrative example and we choose carbon tetrachloride, CCl_4, a simple XY_4 type tetrahedral molecule. From what was said in section 5.2.3.1 it will be seen that of the four modes of vibration $v_1(A)$, $v_2(E)$, $v_3(F_2)$ and $v_4(F_2)$ only the latter two are permitted in infrared absorption for an isolated molecule. Indeed the gas phase spectrum shows normal absorption only near $314\ cm^{-1}$ (v_4) and in the $776\ cm^{-1}$ region (v_3). However, in infrared spectroscopy, anomalous effects are almost the norm and carbon tetrachloride is no exception. Thus the v_3 fundamental is split by a strong Fermi resonance involving the virtually coincident combination level $v_1 + v_4$ and two bands of almost equal intensity are observed instead of one. From what was said earlier it is meaningless to attach the labels v_3, ($v_1 + v_4$) to either of these and we will adopt the neutral notation v_{3+} (for the upper level) and v_{3-} (for the lower). In addition to this very obvious anomaly, weak absorption showing a very anomalous dependence on J is seen in the v_2 region. This arises from Coriolis perturbation chiefly by the neighbouring v_4 level. The energy level diagram and the corresponding spectrum are shown in Fig. 5.24. The spectrum of a thin liquid layer looks much like that of the gas but that of a thicker specimen (see also Fig. 5.13) shows a rich array of features. These are assigned in the Figure with the exception of the broad feature at $45\ cm^{-1}$ which is thought to arise from "liquid-lattice" motion, a topic discussed further in the next section. It will be seen from Fig. 5.24 and from the insert spectrum that the Fermi resonance at 776 cm^{-1} manifests itself by a doubling of observed bands in other spectral regions and thus further complicates the spectrum. In addition it seems very likely that the overtone $2\nu_4$ is weakly involved in the resonance since its intensity seems higher than might otherwise be expected.

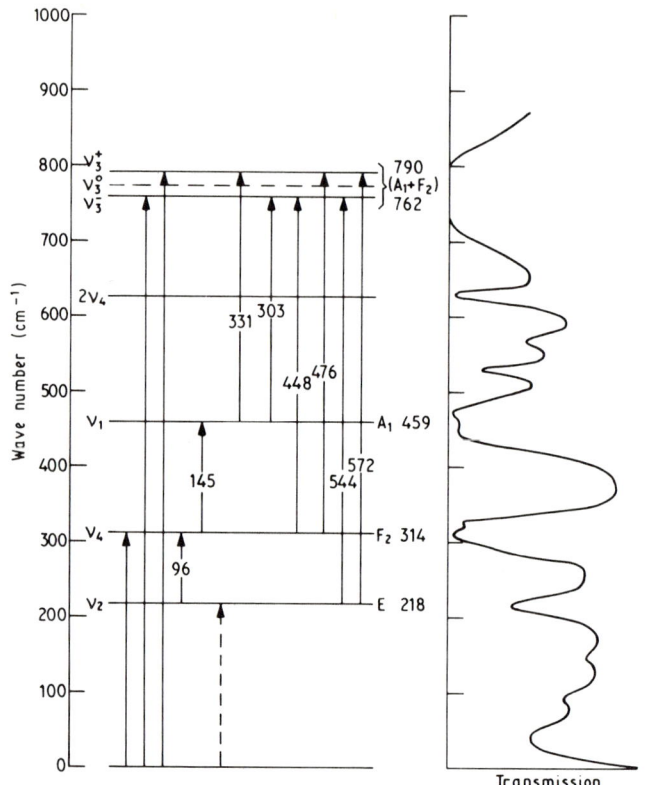

FIG. 5.24. Energy levels, some transitions between them and the resulting far-infrared spectrum of carbon tetrachloride.

Many of the features observed are difference bands of the form $\nu_i - \nu_j$. Such bands would be expected to be weak for two reasons: firstly because the "one quantum number only to change" selection rule is violated and secondly because of population effects. The population of a non-degenerate state of energy $\mathscr{E}_i$ is given by the Boltzmann expression (see section 2.2.3)

$$N_i = N_0 \exp(-\mathscr{E}_i/kT) \tag{5.5.1}$$

where N_0 is the population of the ground state. So, either as $\mathscr{E}$ increases or T falls the population of the state will diminish. In the case of CCl_4, the lower vibrational levels have energies comparable to kT ($\sim 200\,\text{cm}^{-1}$ for room temperature) and they have significant populations at ambient temperature. When, however, CCl_4 is cooled considerably below room temperature the difference bands weaken considerably and the spectrum simplifies. This apart, the absolute intensities of the difference bands do seem to be high. The gas phase spectrum has not so far been observed with sufficient precision to show whether the difference bands appear there too, but it seems probable that specifically liquid-state effects are involved in making these difference bands so obvious. This is underlined by the apparent absence of the corresponding sum bands. It is a quantum mechanical theorem that the transition moment for a difference band equals that of the corresponding sum band if the mechanism for giving them a finite value is the presence of non-linear terms in the expansion of the dipole moment as a function of molecular distortion. Population effects always favour the sum band so, if difference bands are observed due to this mechanism, the corresponding band should be still more obvious. Difference bands (unlike sum bands) may not be involved in Fermi resonance since they do not correspond to real stationary states of the molecule but in the case of CS_2, where there is a weak (v_1, $2v_2$) Fermi resonance, the difference band $v_1 - v_2$ becomes enhanced [535], due to intensity borrowing from the "hot" band $2v_2 - v_2$ and this effect is certainly intramolecular since the difference band also appears in the gas-phase spectrum.

The main liquid-state effects, as mentioned above, however, are the removal of all rotational fine structure and the appearance of the "liquid-lattice" band. These have their origin in the spatial and temporal randomness characteristic of the liquid state and in the strong interactions between molecules in a condensed phase. When, as sometimes happens, these two phenomena are relatively less important, for example solutions of HCl in SF_6 or argon, residual pure rotational absorption is observed [536], but for all other cases it is not. One then has two effects. First, because of the spatially random instantaneous positions of the molecules, each of them will be vibrating in a Van der Waal's force field which is different from that of any other and which, moreover, is spatially non-isotropic. There will therefore be a large spread of lattice-type vibration frequencies (see section 5.3 for a discussion of the ordered crystalline case). Secondly, because of the temporally random nature of the molecular collisions and because of the rapid variations of the interaction with distance, each of the component lattice vibrations will be considerably broadened. The result is a featureless "liquid-lattice" band in the same spectral region (i.e. the submillimetre band) where occurs the true lattice spectrum of the corresponding perfect crystal. It also follows that a given molecule will not be able to rotate freely and one can either say that one is now dealing with classical rotational diffusion or else, and equivalently, say that one is dealing with a rotational spectrum so broadened that all discrete structure is lost. Essentially the molecules are being bombarded on a time scale (of the order of tens of picoseconds) which is short in comparison with the rotational time and the quantised structure is blurred out. A treatment in terms of the classical Langevin equations modified to include inertial terms therefore becomes appropriate and this will be outlined in broad detail in the next section.

5.5.2 *The absorption of far-infrared and microwave radiation by polar and non-polar molecules in the liquid state*

The study of submillimetre and microwave absorption in non-polar liquids is a relatively new field but the study of the corresponding absorptions in polar liquids has been going on—at least theoretically—for most of the century. The main impetus came from the work of Debye [537] in the 1920s who extended his own treatment of the static permittivity to the case where the applied field was varying in time. The full problem is formidable since one is dealing with the solution of 10^{23} coupled Newtonian equations under conditions where one has only sketchy ideas of the forces and of how they vary with distance. This would not, of itself, be a serious drawback in discussing the *equilibrium* properties of the liquid since powerful statistical methods are then available but these are of much more limited application in discussing the non-equilibrium, irreversible, processes associated with the dissipation of absorbed radiant energy. Clearly, a full "head-on" attempt to solve the dynamical problem is not going to get anywhere, so three very different approaches have evolved, each of which, in its own way, has made major contributions to our understanding of the liquid state. These are:

a. Analytical approaches which, via the use of fairly drastic approximations, lead to solvable equations of the Langevin or Liouville type [538].
b. The use of new formalisms, developed within the framework of classical statistical mechanics, which constrain the possible solutions into narrow regions and thus give a feel for the possibly fruitful areas of exploration.
c. The use of high-speed computers to simulate the motion of the molecules making up the liquid [539].

Each of these will now be briefly outlined and then an attempt will be made to combine them in a sort of global picture of our present understanding of long-wave absorption in liquids.

5.5.2.1 *Approximate analytical treatments*
The analytical approaches all stem from Debye [537] who considered the motion of a large solute sphere of radius a, and mass m, having a dipole μ embedded in it and suffering random Brownian bombardment from the smaller solvent molecules. The Brownian forces can be averaged into an effective viscous force and as the polar molecule attempts to rotate in synchronism with the applied sinusoidally varying field, the viscous force will

cause a delay and energy will be absorbed from the field and be transferred to the liquid. Because of the energy loss, the permittivity becomes complex and Debye, by making some deft approximations, derived an explicit expression for the frequency dependence of the complex permittivity $\hat{\varepsilon}$. The assumptions he made were:

a. That the effects of the Brownian motion could be replaced by a torque force equivalent to "white noise"—this means the observing frequency must be less than kT/h.

b. That "dipole–dipole" interactions could be ignored—in other words that the system is in the limit of virtually zero concentration.

c. That there are no inertial effects—in other words that the polar molecule can be accelerated instantaneously. For most cases this is tantamount to insisting that ω be less than τ^{-1}.

d. That there are no attractive short-range forces between the molecules— in other words that only collisional interactions need be considered. This restricts the observing frequency to be much less than the natural lattice frequencies.

With these assumptions, the Langevin equation [540], which describes the motion of the test particle, can be solved and all the macroscopic properties including the complex permittivity can then be derived in simple terms. This equation can be written [541] in the form

$$m\dot{v}(t) = -(m/\tau_R)v(t) + F_N(t),\qquad(5.5.2)$$

which separates the total force into two terms, the first being a frictional retardation proportional to the velocity and the second a randomly varying force due to the Brownian impacts of the solvent molecules on the moving particle. The time constant, or relaxation time τ_R, specifies the frictional properties of a particular solute/solvent system. Debye's solution for $v(t)$ led to an analytical expression for $\hat{\varepsilon}(\omega)$ which, in modern notation, would be written

$$\frac{\hat{\varepsilon}(\omega) - \varepsilon_\infty}{\varepsilon_s - \varepsilon_\infty} = \frac{1}{1 - i\omega\tau_D},\qquad(5.5.3)$$

where $\tau_D = \tau_R$ is the Debye relaxation time, ε_s is the static permittivity and ε_∞ is the limiting high-frequency permittivity, assumed by Debye to be equal to the square of the visible refractive index n_D. From the model of viscous drag, it follows that the relaxation time τ_R is related to the macroscopic viscosity η by the relation

$$l(l+1)\tau_R = [8\pi\eta a^3/kT] + C,\qquad(5.5.4)$$

where l is the order of the appropriate Legendre polynomial (1 for infrared absorption, 2 for light-scattering etc.) and C is a constant usually taken to be zero. The appropriate expression for dielectric relaxation is then

$$\tau_D = 4\pi\eta a^3/kT.\qquad(5.5.4a)$$

At first sight it might be thought that so formidable a list of approximations might make the Debye approach of limited utility but in fact quite the opposite is the case. By direct substitution into (5.5.4a) it follows that Debye relaxation times are typically of the order 10^{-11} s and therefore that the characteristic frequencies are of the order of tens of GHz, i.e. that they lie in the microwave region. There are thus no difficulties in meeting points (a) and (d) and point (b) can usually be met approximately so radio frequency and microwave data for dilute solutions should fit equation (5.5.3) well. In fact they do but only if τ_D and ε_∞ are allowed to be adjustable parameters. The values for τ_D thus found, do agree with the predictions of equation (5.5.4a) to within a factor of two or so whilst the values of ε_∞ are very much closer to the square of n_D. More detailed theories which take account of the microstructure of the liquid can deal with these discrepancies and Hufnagel [542], for instance, has proposed an empirical relation

$$\tau_D = \tau_0 \exp\left(\sigma r_{eff}\right), \tag{5.5.5}$$

where τ_0 is the mean time between collisions, σ is a structure parameter characteristic of the solvent and r_{eff} is a measure of the effective radius of the solute molecule. This relation has proved an interesting new tool [543] in the physical chemistry of dilute solutions of organic molecules but since the Debye equation is used much more widely, it is more convenient in the present context to take τ_D to be an adjustable parameter. With this modification, the basic theory has proved very valuable in the analysis of radio frequency and microwave dielectric dispersion in polar liquids [544]. The dilute solution restriction has turned out to be not so important in practice and many pure liquids obey the Debye equation. Robert and Kenneth Cole [545] showed that equation (5.5.3) could be rewritten alternately in the form:

$$\left[\varepsilon' - (\varepsilon_s - \varepsilon_\infty)/2\right]^2 + (\varepsilon'')^2 = (1/4)(\varepsilon_s - \varepsilon_\infty)^2, \tag{5.5.6}$$

which is the equation of a circle in the complex plane. The centre of the circle lies on the real axis at $(\varepsilon_s + \varepsilon_\infty)/2$ and the radius is $(\varepsilon_s - \varepsilon_\infty)/2$. Plots of ε'' versus ε' (usually known as Cole–Cole plots) are widely used in dielectric work but since absorption alone is measured, only the semicircle in the right upper quarter plane is experimentally observable.

Experimental results do follow the general sense of (5.5.6), but nearly always there are departures from the ideal arc even for the low-frequency regions where inertial effects make no contribution. The plots frequently produce skew arcs, depressed arcs or even both. Attempts to deal with this situation have led to some purely empirical modifications of the Debye equation. Thus Cole and Cole [545] suggested

$$\frac{\hat{\varepsilon}(\omega) - \varepsilon_\infty}{\varepsilon_s - \varepsilon_\infty} = \frac{1}{1 - (i\omega\tau_D)^\beta}, \tag{5.5.7}$$

Cole and Davidson [546] proposed

$$\frac{\hat{\varepsilon}(\omega) - \varepsilon_\infty}{\varepsilon_s - \varepsilon_\infty} = \frac{1}{(1 - i\omega\tau_D)^{1-\alpha}} \tag{5.5.8}$$

and Navriliak and Negami [547] took the process to its logical conclusion with

$$\frac{\hat{\varepsilon} - \varepsilon_\infty}{\varepsilon_s - \varepsilon_\infty} = \frac{1}{[1 - (i\omega\tau_D)^\beta]^{1-\alpha}}. \tag{5.5.9}$$

In the time domain, these empirical expressions can be considered [548] to be tantamount to the existence of a spread of relaxation times. A completely different interpretation has been given by Jonscher [549] who postulates the existence of a "universal" dielectric response. A physical model for this has been given by Hill and Dissado [550].

The Debye equation and its empirical descendents serve well for the analysis of data restricted to the frequency region $\omega < \tau_D^{-1}$, that is time frequencies less than ten GHz or so. Infrared spectroscopists obviously cannot meet this restriction and to work out what will be the dielectric properties of liquids in this region where $\omega\tau_D > 1$ the difficulties so neatly sidestepped by Debye have to be tackled. The resulting equation (the Langevin equation plus inertial terms) cannot be solved analytically, but Sack [551] and others have shown that recurrence relations may be derived and from these the solution can be expressed as continued fractions. One simple (and slightly approximate) solution is

$$\frac{\hat{\varepsilon} - \varepsilon_\infty}{\varepsilon_s - \varepsilon_\infty} = 1 - \cfrac{i\omega\tau_D\gamma}{i\omega\tau_D\gamma - \cfrac{\gamma}{1 - i\omega\tau_D\gamma + \cfrac{2\gamma}{2 - i\omega\tau_D\gamma + \cfrac{3\gamma}{3 - \cdots}}}} \tag{5.5.10}$$

Where $\gamma = i/\tau_D^2 k T$ and I is the relevant moment of inertia. For all molecules of interest γ is of the order of a few per cent. The theory of continued fractions [552] is not so well known as is the corresponding theory of infinite series and the question of convergence is certainly more difficult but just as one often terminates an infinite series after a finite number of terms, so one can do the same with the continued fraction. These approximations are usually known as convergents. The successive convergents of equation (5.5.10) are

$$1, \quad \frac{1}{1 - i\omega\tau_D}, \quad \frac{1}{(1 - i\omega\tau_D)(1 - i\omega\tau_D\gamma)}, \quad \text{etc.}$$

and because γ is small, these represent good approximations for the regions $\omega\tau_D \ll 1$, $\omega\tau_D < 1$, $\omega\tau_D > 1$ etc. These expressions represent, respectively, no dispersion, Debye dispersion and Rocard dispersion, named after Rocard [553], who derived this expression, using a dynamical model in 1933. Just as for the Debye relation, there is a tendency to generalise the Rocard result and to write it in the form

$$\frac{\hat{\varepsilon} - \varepsilon_\infty}{\varepsilon_s - \varepsilon_\infty} = \frac{1}{(1 - i\omega\tau_D)(1 - i\omega\tau_F)} \tag{5.5.11}$$

where τ_F is regarded as a parameter to be determined experimentally. It is usually called the friction time. Later, in 1948, Powles [554] proposed an alternative solution,

$$\frac{\hat{\varepsilon} - \varepsilon_\infty}{\varepsilon_s - \varepsilon_\infty} = \frac{1}{\left[1 - i\omega\tau_D - (1 - \lambda)i\omega\tau_F - \tau_F\tau_D\omega^2\right]}, \tag{5.5.12}$$

where λ, lying between 0 and 1, is an adjustable parameter. Setting $\lambda = 0$ gives the Rocard result, whereas setting it equal to 1 gives the result

$$\frac{\hat{\varepsilon} - \varepsilon_\infty}{\varepsilon_s - \varepsilon_\infty} = \frac{1}{(1 - i\omega\tau_D - \tau_F\tau_D\omega^2)}. \tag{5.5.13}$$

This complex of equations is often referred to as the "Rocard–Powles" solution. McConnell [555] has discussed Sack's exact solutions and shows that their second convergent gives (5.5.13) and not the Rocard result (5.5.11). His own work [556] on the other hand which is based upon power series expansions in the small quantity γ, does lead to expressions of the Rocard type. For example the spherical rotor is found to be described by the equation

$$\frac{\hat{\varepsilon} - \varepsilon_\infty}{\varepsilon_s - \varepsilon_\infty} = \frac{1}{\left[1 - (1 - \frac{1}{2}\gamma)i\omega\tau_D\right]\left[1 - (1 - 2\gamma)i\omega\tau_F\right]}. \tag{5.5.14}$$

This difference is more apparent than real, however, because γ is so small and, as will be seen later, there is no chance of experimentally deciding between the two versions because of the intervention of the intense Poley process.

One thing all of these equations do is to remove one very obvious defect of the Debye theory. If one computes the imaginary component of the Debye function one gets

$$\varepsilon'' = \frac{(\varepsilon_s - \varepsilon_\infty)\omega\tau_D}{1 + \omega^2\tau_D^2}, \tag{5.5.15}$$

and hence ε'' goes as ω^{-1} for sufficiently high values of ω. Now since $\alpha = \omega\varepsilon''/cn$ (equation 3.4.10) and n will be constant at high frequency, it follows that α will reach a limiting high-frequency value,

$$\alpha_\infty = (\varepsilon_s - \varepsilon_\infty)/\varepsilon_\infty^{1/2}c\tau_D, \tag{5.5.16}$$

the so-called Debye plateau. This is not in agreement with observation since all polar liquids eventually recover transparency. The Rocard–Powles expressions, and all the higher convergents, go more rapidly than ω^{-1} at high frequencies and the calculated values of α fall eventually to arbitrarily low values and one has acceptable behaviour. McConnell [557] has shown that this is true for all classes of inertial rotor.

5.5.2.2 Approaches based on abstract formalisms

Debye derived his equation in terms of a specific model—namely viscous drag—but nowadays the topic is usually approached via the generalised concept of a correlation function $f(\tau)$. This is defined [558] for the molecular dipole case by the relation

$$f(\tau) = \frac{\langle \vec{\mu}(t)\,\vec{\mu}(t+\tau)\rangle}{\langle \vec{\mu}(t)\,\vec{\mu}(t)\rangle}, \tag{5.5.17}$$

where $\vec{\mu}(t)$ is the dipole vector at time t and $\vec{\mu}(t+\tau)$ is its value at a later time $t+\tau$. The function $\mu(t)$ is assumed to be a stationary ergodic stochastic process (i.e. it is produced by the equivalent of Debye's "white-noise" driving force) and $f(\tau)$ is by definition its autocorrelation function. Invoking the stationarity, one can write the correlation function as a Taylor series, thus

$$\mu^2 f(\tau) = \langle \vec{\mu}(t)\,\vec{\mu}(t)\rangle + \tau\langle \vec{\mu}(t)\,\vec{\mu}^1(t)\rangle + (\tau^2/2)\langle \vec{\mu}(t)\,\vec{\mu}^2(t)\rangle + \cdots \tag{5.5.18}$$

where the superscripts on the μ's indicate the order of differentiation. Now, since $f(\tau)$ is *not* a function of t (because of the assumed stationarity), it follows that all the differential coefficients with respect to t will be zero. If one therefore progressively differentiates (5.5.17) and sets all the results equal to zero, one can derive general relations amongst the coefficients appearing in (5.5.18). Thus it follows [559] that

$$\langle \vec{\mu}(t)\,\vec{\mu}^{2n+1}(t)\rangle = 0 \tag{5.5.19a}$$

and that

$$\langle \vec{\mu}(t)\,\vec{\mu}^{2n}(t)\rangle = -\langle \vec{\mu}^n(t)\,\vec{\mu}^n(t)\rangle \tag{5.5.19b}$$

for all n. Substitution into (5.5.18) gives the final result

$$\mu^2 f(\tau) = \langle \vec{\mu}(t)\,\vec{\mu}(t)\rangle - (\tau^2/2)\langle \vec{\mu}^1(t)\,\vec{\mu}^1(t)\rangle + (\tau^4/4!)\langle \vec{\mu}^2(t)\,\vec{\mu}^2(t)\rangle - \cdots \tag{5.5.20}$$

The correlation function thus has perfect time-reversal symmetry, a fact which may be underlined by writing its expansion in the simpler notation.

$$f(t) = 1 - at^2 + bt^4 - \cdots \tag{5.5.21}$$

The main value of the correlation function, to the spectroscopist, lies in the so-called "fluctuation dissipation" theorem [560] which states that the results of externally applied stimuli decay in exactly the same way as do those resulting from purely internal spontaneous excitations, provided one is within the linear regime. Kubo has shown that this theorem leads to the relation [561]

$$\frac{\hat{\varepsilon}(\omega) - \varepsilon_\infty}{\varepsilon_s - \varepsilon_\infty} = -\int_0^\infty \dot{f}(t)\exp(i\omega t)\,dt, \tag{5.5.22}$$

that is that the complex permittivity is determined by the one-sided Fourier transform, or equivalently the pure imaginary Laplace transform of the time derivative of the correlation function. By applying the rule for integration by parts, this equation can also be written

$$\frac{\hat{\varepsilon}(\omega) - \varepsilon_\infty}{\varepsilon_s - \varepsilon_\infty} = 1 + i\omega \int_0^\infty f(t) \exp(i\omega t)\mathrm{d}t. \qquad (5.5.23)$$

Thus by knowing only the form of the time decay of an initial orientation in the liquid, one can deduce all its dielectric properties. It follows that if two or more models lead to an identical form for the correlation function, then spectroscopic measurements will be unable to distinguish between them. The fluctuation–dissipation theorem also shows that the correlation function must be a continuous monotonic function which, for sufficiently long times, must approach zero exponentially. The function must be such that all its derivatives exist and it must be expressible as an even Taylor's series for all values of the time. These conditions are such as to ensure that $f(t)$ is a deterministic function and that $\mu(t)$ is also a deterministic function. All this not surprising result really says is that if one wants to know everything about the irreversible processes going on in the liquid one will have to solve the full dynamical problem and that we know is impossible. Nevertheless these general arguments are useful because, although we cannot evaluate all the coefficients in (5.5.21), we can nevertheless compare suggested approximate functions and see how close they come to the desired ideal. Thus in Fig. 5.25(a), a hypothetical ideal function is shown and this compared with the simple exponential decay correlation function, which corresponds to the Debye equation. In simple terms, this equation derives from the explicit form $f(t) = \exp(-t/\tau_D)$, but to ensure the time-reversal symmetry one must write it in the form

$$f(t) = \exp -[|t|/\tau_D]. \qquad (5.5.24)$$

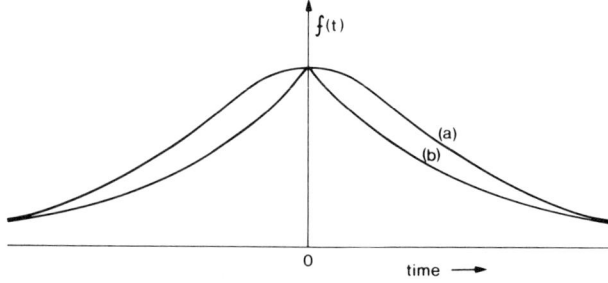

FIG. 5.25. Allowed form for a physically acceptable correlation function. The physically unacceptable simple exponential decay is shown for comparison.

This function (Fig. 5.25(b)) has a cusp at $t = 0$ and its Taylor series therefore does not exist for this value of the time. This is a general result: the autocorrelation functions of non-deterministic functions do not have Taylor expansions at the origin [559]. The cusp is not, of course, acceptable on simple physical grounds since, apart from any general arguments, it implies instantaneous rotational acceleration.

The problem which faces any attempt to use the correlation function formalism for real problems is that solvable models lead to correlation functions which do not meet all the requirements and the purely empirical functions which do (e.g. $f(t) = \operatorname{sech}(t/\tau_D)$) do not correspond to any obvious physical picture. Thus all the functions which have been suggested to deal with high-frequency dispersion are unsatisfactory in that eventually odd terms appear in the expansion [562]. As an example, the function

$$f(t) = \exp(-t) \exp\left[-\alpha^{-1} \exp(-\alpha t - 1)\right] \qquad (5.5.25)$$

suggested by Lewis [563], is well-behaved near the origin but is not even. If it is forced to be even by replacing t by $|t|$, then it is no longer well-behaved at the origin! However, from a purely pragmatic spectroscopic viewpoint, it seems that provided $f(t)$ has a horizontal tangent at $t = 0$ and behaves like $\exp(-t/\tau_D)$ for large t, it will lead to a spectrum which always looks, albeit to first order, like that observed. The Debye function is therefore only wrong near $t = 0$ and the spectral consequence of this is that it will start to fail only at correspondingly high frequencies. This is, of course, the point made above where the theory was restricted to $\omega\tau_D < 1$, which in practice means frequencies less than a few tens of GHz. All functions which have a zero slope at the origin lead to spectral functions which show an eventual recovery of transparency at high enough frequency, in other words a "roll-off" of the Debye plateau. The simplest way of showing this is to apply successive integration by parts to equation (5.5.22). Scaife shows [559] that this gives

$$\frac{\hat{\varepsilon}(\omega) - \varepsilon_\infty}{\varepsilon_s - \varepsilon_\infty} = -\frac{f^1(0)}{i\omega} + \frac{f^2(0)}{(i\omega)^2} - \frac{f^3(0)}{(i\omega)^3} + \ldots \qquad (5.5.26)$$

where the superscripts on the fs indicate the order of differentiation. Because of the time-reversal symmetry and because of the smoothly monotonic behaviour of $f(t)$ and all its derivatives, the odd terms in (5.5.26) must necessarily be zero and therefore at a sufficiently high frequency where the infinite series (5.5.26) certainly converges, it follows that the LHS will be pure real. The imaginary part, and hence the absorption coefficient, must then be zero.

To deal with this high-frequency region where the roll-off and other phenomena are becoming manifest, it is necessary to extend the abstract formalism to cover time-dependent properties. This is equivalent to generalising the Langevin equation, which can then be written [541]

$$m\dot{v}(t) = -m \int_0^t V(t')M_0(t-t')dt' + F_N(t), \qquad (5.5.27)$$

where $M_0(t)$ is a memory function which describes the past history of the friction. This memory function is itself a correlation function and therefore has its own memory function, $M_1(t)$, which, in turn, has its own memory function and so on. One thus generates an infinite series of memory functions, any two adjacent members of which are related by

$$\dot{M}_n(t) = M_n(t)^*M_{n+1}(t). \qquad (5.5.28)$$

The hierarchy of memory functions is, of course, only of value if an early member of the series degenerates into a delta function and thus terminates the progression. Thus if $M_0(t)$ is a delta function, one immediately recovers the Debye equation. In general experimentalists only use two, or three at the most, in their attempts to analyse experimental data [564].

An elegant way of handling the progression of memory functions was introduced by Mori [565], who applied Laplace transforms to the basic equation (5.5.28). Thus if one writes

$$M_0(t)^*f(t) = f'(t) \qquad (5.5.29)$$

then, applying the convolution theorem, equation (1.6.42), gives at once

$$\mathscr{L}_p(M_0{}^*f) = \mathscr{L}_p(M)\mathscr{L}_p(f) = \mathscr{L}_p(f') \qquad (5.5.30)$$

and hence, using the analogue of (5.5.21), it immediately follows that

$$\mathscr{L}_p(M_0)\mathscr{L}_p(f) = f(0) - p\mathscr{L}_p(f), \qquad (5.5.31)$$

where $f(0)$ is, of course, unity but has been left in this form to preserve generality. Equation (5.5.31) can be rewritten in the form

$$\mathscr{L}_p(f) = f(0)[p + \mathscr{L}_p(M_0)]^{-1}, \qquad (5.5.32)$$

and it follows, by applying exactly the same arguments, that an analogous fraction, viz.

$$\mathscr{L}_p(M_0) = M_0[p + \mathscr{L}_p(M_1)]^{-1} \qquad (5.5.33)$$

can be written for $\mathscr{L}_p(M_0(t))$. The process can obviously be continued indefinitely. Now, by substituting (5.5.33) in (5.5.31), and continuing the process, one will generate a continued fraction

$$\mathscr{L}_p(f) = \cfrac{f(0)}{p + \cfrac{M_0(0)}{p + \cfrac{M_1(0)}{p + \dots}}} \qquad (5.5.34)$$

This is the celebrated Mori expression which plays a central role in much modern theoretical analysis of liquid-state phenomena. The practical use of the Mori continued fraction lies in truncating it so as to produce, as in the treatment mentioned above, analytical expressions containing only a small number of adjustable parameters. The justification for approximating the Mori continued fraction by its convergents has been hotly debated [566] and some purists even point out that the convergence of the infinite fraction itself has not been established, but nevertheless early truncations of the Mori expression are likely to remain popular as analytical tools for dealing with the twin problems of interial roll-off and damped oscillatory motion in the high-frequency region where the simple Debye approach is invalid.

5.5.2.3 The Poley absorption

This advance of the theory to include inertial effects gives, as described above, a "roll-off" of the Debye plateau as shown schematically in Fig. 5.26 but the actually observed spectra look completely different. Thus in Fig. 5.26 it will be seen that the observed absorption coefficient for chlorobenzene is twice as high as the Debye plateau instead of being less! This is usually interpreted by saying that there is an additional or Poley absorption band present as well as that due to relaxation of the orientational polarisation. This sort of absorption gains its epithet from J. Ph. Poley who suggested [567] its existence, very bravely for the time, from slight discrepancies in the microwave data. The Poley band is often attributed to vibrations of the "liquid-lattice" and corresponds to a smeared out lattice vibrational spectrum [568, 569, 570]. In the case of non-

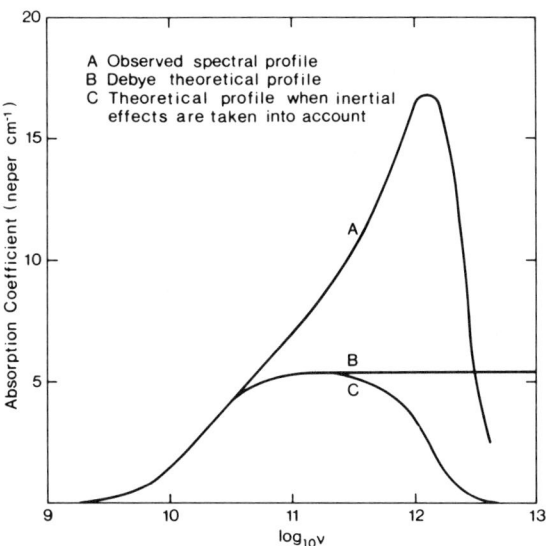

FIG. 5.26. Microwave and far-infrared absorption spectrum of chlorobenzene.

polar liquids there is no Debye process so only the Poley band is observed. This two-component theory of the total microwave/far-infrared absorption has received some support from studies of inertially similar—but electrically different—pairs of molecules such as methyl chloroform (CH_3CCl_3) and carbon tetrachloride [571]. An interpretation of the spectrum of methyl chloroform in terms of a Debye process with roll-off and a liquid-lattice band is shown in Fig. 5.27. The resonant component is derived by scaling up the Poley absorption of CCl_4 so that it matches that of CH_3CCl_3 in the 100 cm^{-1} region where presumably non-resonant absorption is negligible. When the presumed resonant component is subtracted from the overall profile, one is left with a residuum which looks very much like a rolled-off Debye process. In particular its peak absorption agrees very well with the value of α_∞ derived from the application of equation (5.5.16). An interesting consequence of this interpretation is that because of the lack of a Debye process a non-polar molecule will have a higher frequency for maximum absorption than will a polar molecule of similar mass and moments of inertia.

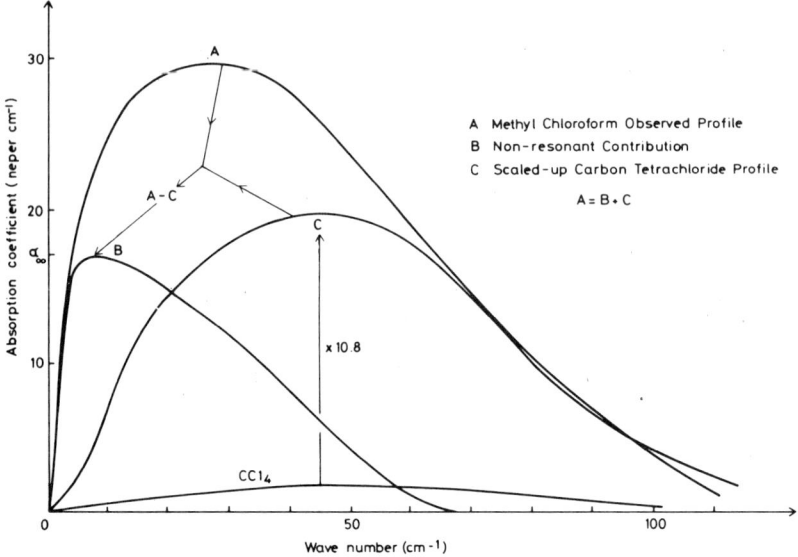

FIG. 5.27. Decomposition of the observed absorption profile for methyl chloroform into a "rolled-off" Debye process plus a resonant "liquid-lattice" contribution.

The intensity of the overall absorption for polar molecules is proportional to the square of the dipole moment and the band can to first order be regarded as a single molecule effect. Thus the absorption divided by concentration for chlorobenzene, dissolved in cyclohexane, has been shown to depend only slightly on the concentration [572]. The absorption for non-polar molecules on the other hand appears to depend almost entirely on intermolecular interactions and is greatly enhanced if the sample is subjected to high hydrostatic pressure [573].

Bands similar to those observed for liquids are present in the far-infrared spectra of many amorphous solids—they are particularly obvious in the spectra of amorphous polymers. Thus it is possible to make a continuous series of copolymers of ethylene and vinyl acetate in which the crystallinity steadily falls as the proportion of the vinyl acetate increases. This series also demonstrates an analogous transition from sharp lattice bands to a broad continuum which looks exactly like the Poley band of a liquid [574]. Pure homo-polymers which are capable of existing with a varying degree of crystallinity, e.g. polyethylene terephthalate, can show [575] most interesting effects as an initially amorphous sample is slowly annealed. Altogether, the existence of these "Poley" bands for amorphous polymers is taken as further strong evidence for the validity of the liquid lattice theory in general [576].

Attempts to provide analytical expressions for the Poley band-shape have, as mentioned earlier, mostly been carried out within the framework of the Mori formalism. The step from the Mori correlation function, equation (5.5.32), to the complex permittivity, and hence the spectrum, is conceptually easy; all one needs to do is to equate a Laplace transform with the corresponding Fourier transform via the simple substitution $p = i$. Pure mathematicians worry about this high-handed procedure since there is no guarantee that an integral which exists with a pure real argument will also exist with a complex, or pure imaginary, one. Nevertheless the physicists tend to press on regardless [577]. One then has

$$\mathscr{F}\left[f(t)\right] = \mathscr{L}_{-i\omega}\left[f(t)\right], \tag{5.5.35}$$

and from the Kubo relation, equation (5.5.22), it follows that

$$\frac{\hat{\varepsilon}(\omega) - \varepsilon_\infty}{\varepsilon_s - \varepsilon_\infty} = 1 + i\omega \, \mathscr{L}_{-i\omega}\left[f(t)\right]. \tag{5.5.36}$$

To illustrate this, one takes the simplest possible case of truncation, namely setting $M_0(t) = \tau_D^{-1}\delta(t)$. This expression has the Laplace transform τ_D^{-1} and equations (5.5.32) and (5.5.36) give at once the Debye equation. The physical content of this result is that the Debye equation corresponds to a model in which everything is instantaneous and there is no memory of previous events.

Physically more acceptable models have to introduce finite time-dependences, that is that the present state of the system depends on its previous history, but to make the treatment tractable, they have, as stressed several times before, to make some non-negligible approximations. Two widely used models, both due to Gordon [578], are the M-diffusion model in which instantaneous collisions lead to infinite torques at the moment of impact and all angular velocity directional correlation is lost and the J-diffusion model in which, in addition, the magnitude of the angular velocity is also randomised. These correspond to the truncations

M-diffusion $\qquad\qquad M_0(t) = M_0(0)\exp(-t/\tau_D)$ $\qquad\qquad$ (5.5.37a)

J-diffusion $\qquad\qquad M_0(t) = M_0(0)\exp(|t|/\tau_D)$ $\qquad\qquad$ (5.5.37b)

These expressions have found some use in the interpretation of infrared and Raman bandshapes (see later) but are not so useful in the analysis of Poley absorption. Much better here is the higher-order truncation,

$$M_1(t) = M_1(0)\exp(-t/\tau_D), \qquad\qquad (5.5.38)$$

introduced by Evans and his colleagues [579]. This has the Laplace transform

$$\mathscr{L}_p[M_1(t)] = M_1(0)[p + \tau_D^{-1}]^{-1}. \qquad\qquad (5.5.39)$$

Applying this truncation to Mori's continued fraction, gives

$$\mathscr{L}_p[f(t)] = \frac{p^2 + p\tau^{-1} + M_1(0)}{p^3 + p^2\tau^{-1} + p[M_1(0) + M_0(0)] + \tau^{-1}M_0(0)}, \qquad (5.5.40)$$

and from the general relation

$$\alpha(\omega) = \frac{\varepsilon_s - \varepsilon_\infty}{n(\omega)c}\omega^2 \, \text{Re}\,\{\mathscr{L}_{-i\omega}[f(t)]\} \qquad\qquad (5.5.41)$$

one finally derives

$$\alpha(\omega) = \frac{(\varepsilon_s - \varepsilon_\infty)\omega^2\tau_D^{-1}M_0(0)M_1(0)}{\tau_D^{-2}[M_0(0) - \omega^2]^2 + \omega^2[M_0(0) + M_1(0) - \omega^2]^2} \qquad (5.5.42)$$

for the expected band profile. This expression, with a suitable choice of parameters, fits the observed Poley band in CCl_4 very well indeed as is shown in Fig. 5.28. A similar application to vibrational bandshapes might be expected to be equally successful.

It must be stressed, however, that, despite these successes, the Mori formalism is merely a convenient way of doing curve-fitting. The parameters are equilibrium averages and in the averaging all the dynamic information is lost. The physical problem is no different. If one wants to have a completely valid theory then one needs to have the whole hierarchy of memory functions and one is back with the original problem where, to understand completely the dynamics of the liquid, one needed to have all the coefficients in the infinite series expansion of the correlation function. To show the close relationship, it need only be noted that if the correlation function expansion is written in the form

$$f(t) = 1 - (a_1/2)t^2 + (a_2/4!)t^4 - (a_3/6!)t^6 + \ldots \qquad (5.5.43)$$

then [579], $M_0(0) = -a_1$, $M_1(0) = a_1 - (a_2/a_1)$, $M_2(0) = (a_2^2 - a_1a_3)/a_1$ $(a_2 - a_1^2)$, etc. Of course, one does not need to go about things in this direction only. The absorption coefficient is, in essence, related to the correlation function by a Fourier transform so one could take the inverse Fourier transform of the spectrum and hence derive the correlation function. Experimental error limits the ultimate utility of this approach but, nevertheless, it has proven useful to compare these experimental functions for different systems and thus derive some information about how one molecule in one

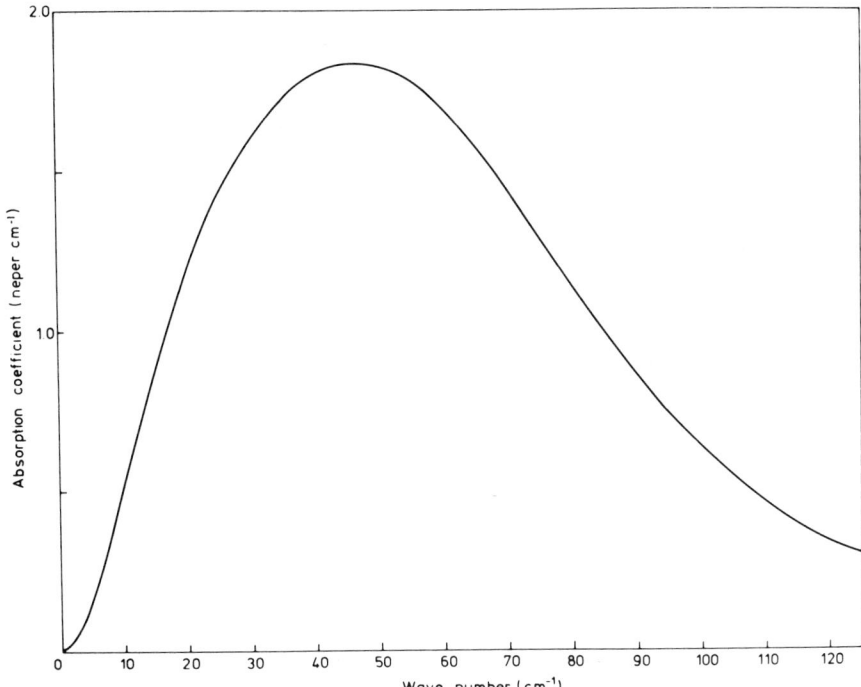

FIG. 5.28. Theoretical absorption curve for liquid CCl_4 derived using the theory of Evans and Evans [579]. This curve agrees almost exactly with the experimental one when the latter is corrected for intramolecular absorption.

environment differs from another in a different environment. This is a familiar pattern: chemists have never been inhibited, by their lack of fundamental knowledge, in making meaningful empirical comparisons. The most useful way of going about this task is in terms of the moments of the spectrum. The coefficients a_n are related to the various even moments of the absorption by

$$a_n = \frac{4^n \pi^{2(n-1)} c^{2n}}{(\varepsilon_s - \varepsilon_\infty)} \int_0^\infty \tilde{v}^{2(n-1)} \alpha(\tilde{v}) d\tilde{v}. \tag{5.5.44}$$

So, from an experimental record determined with sufficient accuracy, one could, in principle, determine as many of the coefficients as one wished.

5.5.2.4 *Corrections for the internal field*

One ought to remark, in this connection, that, just as for the case of ionic crystals, it is necessary to apply corrections for the internal field. The reason is the same, the relaxing molecular dipole is surrounded by a polarisable medium and its interaction with the medium leads to its experiencing an electric field different from that applied externally. The problem has been tackled successively by a number of authors but it is now generally agreed that the most successful approach has been that of Fatuzzo and Mason [580]. Their result has been extended by Klug, Kranbuehl and Vaughan [577], by Rivail [581] and generalised by Nee and Zwanzig [582]. Following these authors, one writes

$$\frac{(\hat{\varepsilon}(\omega) - \varepsilon_\infty)(2\hat{\varepsilon}(\omega) + \varepsilon_\infty)\varepsilon_s}{(\varepsilon_s - \varepsilon_\infty)(2\varepsilon_s + \varepsilon_\infty)\hat{\varepsilon}(\omega)} = \int\limits_0^\infty \dot{f}(t)\exp(i\omega t)\,dt \qquad (5.5.45)$$

as the currently accepted best form for the Kubo relation. It should be stressed, however, that for dilute solutions and for pure liquids which are not too polar, equation (5.5.45) hardly differs from the simpler equation (5.5.22). This is basically because ε_s is much closer in value to ε_∞ for a dipolar liquid than it is for an ionic crystal. This explains how it is that the simple Debye equation has been used so successfully for so many years to analyse experimental results. Even so, one should, strictly, draw a distinction between the macroscopic relaxation time which one derives from (5.5.45) and the true or microscopic one. These two need not necessarily be the same. Equation (5.5.45) is derived on the assumption that the molecules are spherical with a point dipole embedded at the geometrical centre. This is obviously not a very realistic model and recently some new work [583], inspired by attempts to understand the quantitative aspects of far infrared absorption, has led to expressions which take account of the anisotropy of the molecular polarisability. This will undoubtedly prove an expanding field of interest to physical chemists in the future, especially when it is backed up by computer simulations.

5.5.2.5 *Computer simulations of molecular motions*

With the development of the first high-speed digital computers in the 1950s, physicists quickly realised that sheer size of a "number-crunching" problem no longer meant that that problem was necessarily impossible. Problems like the time evolution or the thermodynamics of systems made up of large numbers of interacting particles thus moved into the realm of the possible. The dynamics of the liquid state was an obvious example and was, in fact, one of the first to be tackled. The rigorous proof of the existence of a fluid/solid transition, at a specific temperature, in a system made up of hard spheres was an early triumph. After that, however, the subject tended to languish because the available computers were not big enough or fast enough to deal with a realistic number of particles. Now with far better machines available and most workers having reasonable access to such a machine via a remote terminal the subject is thriving again.

The basic idea is that if one cannot follow the motions of the molecules analytically, one can follow them numerically using the computer as a simulator. The simulation has two main strands, the physics and the computation. The physics of the problem is inserted in the form of the interatomic or intermolecular potential and also in the thermodynamic boundary conditions, e.g. constant energy, constant volume, etc., and the computer then works out how a given starting configuration will evolve in time. Of course, even the biggest and fastest modern machines cannot deal with physically realistic numbers of "molecules", the numbers actually used being only of the order of a few hundred up to, at most, a few thousand. Also only a limited number of "time" steps can be computed. A typical computation might involve 512 molecules over a total time interval equivalent in the real world to some tens of picoseconds. To make the treatment more realistic and also to deal with confinement problems, it is extremely convenient to use cyclic (or periodic) boundary conditions. Here one takes a unit cube containing the small number of molecules and then tells the computer to assume that this is one of an infinite number of identical cubes filling all of space. In this way, not only does the number of molecules in the cube stay constant, the number of molecules whose motion is being considered is essentially infinite. The assumption of periodic boundary conditions works well when the inter-molecular potential is short-range, Van der Waal's forces for example, but it is not so good for long-range forces. Thus computations of the static permit-tivity, where Coulomb forces are involved, have not proven satisfactory and this has stimulated attempts to avoid the use of periodic boundaries [584]. The intermolecular potentials which are used in practice fall into several categories: thus one has the hard-sphere (or HS) model, the "rough" hard-sphere variant, the Lennard–Jones potential (equation 5.2.17), the Stockmayer potential (hard-spheres plus embedded point-dipole), etc.

Almost from the beginning the topic of computer simulation divided into two main techniques [585]. Firstly, there was that where the phase-space of the molecules was sampled via a random-number generator. The analogies with roulette led to its being called the "Monte-Carlo" method. Secondly, there was the technique where the trajectory of a particle, in phase-space was followed. This is the method of "molecular dynamics". The Monte-Carlo technique concentrates on the energy of the system and works by rejecting any randomly generated configuration which does not satisfy the Boltzmann criterion. The thermodynamic properties are then computed as phase-space averages. The molecular-dynamics method concentrates on the forces involved and deduces the physical properties from the appropriate time averages. The molecular dynamics method is the more general in that it can deal with non-equilibrium properties as well as the equilibrium ones.

In the early days of simulation, the practitioners were adamant in their refusal to "traffick in real molecules" but now simulations of quite complicated molecules are routinely being attempted. As an example one might quote the work of Powles and his group [586] on HCl. With this advance, the simulations can prove of inestimable value to the liquid-state theorist. Thus Bellemans [585] remarks that simulations apart from doing their original job of supplying the solutions to analytically intractable problems, can now tell us just how reasonable a "reasonable physical hypothesis" actually is. They can help greatly in constructing effective intermolecular potentials and they can provide "experimental" data which more orthodox experiments are incapable of supplying at the moment. The place of computer simulation in the experimentalists armoury is now well established and comparisons of computer generated far-infrared spectra of liquids [587] with those actually produced in the laboratory are telling us much about the subtle microdynamics of the liquid state.

5.5.3 Vibrational band-shapes in liquid-state spectra

As remarked earlier, the overall absorption curve for a liquid-state specimen consists of a set of more or less overlapping bell-shaped individual profiles, each of which corresponds fairly closely in position and in intensity with the corresponding vapour-phase band. The bands have lost their resolvable rotational fine structure because the molecules are suffering strong collisions on a time scale short in comparison with the rotation time and the phase-coherence is therefore lost. As a result and because the bandwidths are small in comparison with (kT/h), it is possible to discuss the form of the observed profiles in purely classical terms. The simplest approach, and the one most widely used, is to choose the best fit to a Lorentzian function (see Appendix 4) for each band. In this way one can derive good estimates for the integrated intensity, the frequency of maximum absorption and the line width parameter. These quantities can then be interpreted in terms of the various theories of the molecular vibration and of how it is perturbed by the strong interactions typical of the liquid state. The Lorentz function has many attractions, not least being its smooth decline towards zero for frequencies much larger than the centre frequency but it is nevertheless fundamentally unsound [135] in that it corresponds to collisions which do not preserve a Boltzmann distribution. The Van-Vleck and Weisskopf line-shape function is more acceptable on this count and has been widely used in microwave spectroscopy. It correctly reduces to the Debye function when the centre frequency tends to zero but similarly is afflicted with a "plateau" problem because of the incorrect treatment of rotational inertia.

The close relationship between vibrational line-shapes and long-wave dielectric phenomena has led to the development of a new branch of physical chemistry in which these line-shapes are carefully determined and their forms are then interpreted in much finer detail than would be possible with the traditional Lorentz-fitting approach. The development owes much to Gordon [578] but in more recent times major contributions have come from many other workers, for example Zwanzig [588], McClung [589], Evans [590] and Bratos [591]. The basic idea is to Fourier transform the normalised line-shape function in order to obtain the correlation function and then to compare this with correlation functions derived from various theoretical models. This comparison in the time domain is much more sensitive and, of course, much more convenient to the theorist whose primary data are correlation functions rather than spectra. The technique can be applied to both infrared and Raman spectra, but although the infrared spectra of simple diatomic molecules dissolved in liquid rare-gas solvents have proved informative, the modern tendency is definitely inclining towards Raman spectroscopy which has many advantages, not least that it is an emission technique. The combined spectroscopic techniques, when combined with other techniques, e.g. NMR and light-scattering, are showing us that the randomisation of molecular dipoles in solution seldom follows a simple monotonic exponential path: nearly always there is some measure of oscillatory motion. This is just one example of the exciting flow of information which is emerging from this rapidly developing area and hopefully we may well be on the verge of a breakthrough in liquid-state physics which will parallel that which occurred in the much easier solid-state physics in the middle years of this century.

5.6 Magnetic resonance spectroscopy

5.6.1 *The Zeeman effect*

When a magnetic field is applied to an isolated system, such as an atom, ion or molecule, which has narrow absorption lines, these lines are often observed to shift, to break up into multiplets or to do both. This phenomenon is known, after its discoverer, as the Zeeman effect. It was first noticed for the case of atoms in the gas phase, where it is very common, but examples are now known for atoms in condensed phases and for molecules in the gas phase. The essential criterion for the observation of a Zeeman effect is that the systems under study have unpaired electrons in either the upper, the lower or both states of the transition. Unpaired electrons can be thought of as executing orbital motion and therefore represent a circular current. This current produces a permanent magnetic dipole moment. When an external magnetic field is applied, the isotropy of space is removed and the energy of the system will depend on the relative orientation of the magnetic dipole moment and the applied field, that is on the quantum number m (or M). For weak fields, the energy level shifts are given in the simple semi-classical approach by the linear relation

$$\Delta \mathscr{E} = m\beta B \qquad (5.6.1)$$

where β is the Bohr magneton (sometimes denoted by μ_B) given by

$$\beta = eh/4\pi m_e$$
$$= 9\cdot274 \times 10^{-24} \text{ J T}^{-1}$$
$$= 0\cdot4669 \text{ cm}^{-1} \text{ T}^{-1} \qquad (5.6.2)$$

For the case of a single p optical electron, (5.6.1) would give a splitting into three levels corresponding to the allowed values of m, namely $+1$, 0 and -1. There are two complications to this simple picture. Firstly, at high fields the linear relation between $\Delta\mathscr{E}$ and B fails and is replaced by a quadratic dependence. This is the Paschen–Back effect. Secondly, for very many systems the Zeeman splittings are neither quantitatively nor qualitatively in agreement with the expectations of (5.6.1)—i.e. they are anomalous. To deal with this situation it is usual to introduce a multiplicative factor g called the Landé splitting factor and to rewrite (5.6.1) in the form

$$\Delta\mathscr{E} = gm\beta B. \qquad (5.6.3)$$

The Landé splitting factor can always, at least in principle, be calculated by quantum mechanical methods so it is not merely a "fudge-factor" but there are only a few cases where this calculation can be done exactly. Hydrogenic atoms provide an example and one can write

$$g_j = 1 + \frac{j(j+1) - l(l+1) + s(s+1)}{2j(j+1)}. \qquad (5.6.4)$$

Examination of this equation shows that the anomalous Zeeman effect arises from electron spin, since were s to be zero, g would be unity. For free electrons or electrons in S states, it follows that g will be 2, since l will be zero and j will equal s. The observed value of g, for free electrons, is indeed close to 2 but not exactly so. This is due to small quantum electrodynamics (QED) corrections which lead to a value

$$g = 2\cdot002\,319\,304 \ldots \qquad (5.6.5)$$

The QED corrections, as usual, can be expressed as a power series in α. The result is

$$a_e = (1/2)(\alpha/\pi) + C_4(\alpha/\pi)^2 + C_6(\alpha/\pi)^3 + \ldots \qquad (5.6.5a)$$

where a_e is equal to $(g-2)/2$ and the coefficients C_i are increasingly complicated sums of integrals which have to be evaluated using the Feynman graph technique. The coefficients C_6 and C_8 require for example the summation of 72 and 891 graphs respectively. Fortunately, however, the series (5.6.5a) very rapidly converges, because α for electrons is so small, and present levels of measurement precision require merely that the summation be taken as far as C_6. The number so produced for a_e, namely $0\cdot001\,159\,652\,222 \ldots$, agrees, as far as we know, exactly with the measured value: another illustration of the remarkable power of the QED approach.

For atoms in which Russell–Saunders coupling is a good approximation, one can calculate the g values by taking (5.6.4) and replacing the lower case letters by their upper case equivalents, i.e.

$$g_J = 1 + \frac{J(J+1) - L(L+1) + S(S+1)}{2J(J+1)}. \qquad (5.6.6)$$

Clearly we will have a "normal" Zeeman effect when $S = 0$. Otherwise one will not and this is the value of the Zeeman effect since from (5.6.6) it follows that the splittings will differ in the upper and lower levels (since some of the J values will differ) and the observed line will show splittings. In the past this was of considerable theoretical value since it gave a way of investigating the electronic structure of atoms but more recently it has also become of considerable practical value since it gives a way of tuning the otherwise fixed frequency atomic gas lasers. The He/Xe laser for example at 3.5 μm can be tuned over as much as 6 GHz by the application of an axial magnetic field. This laser line is in close coincidence with some absorption bands of formaldehyde and the magnetic tuning has been used [592] to explore the fine structure of these bands.

5.6.2 *Laser magnetic and cyclotron resonance spectroscopy*

More usually, however, the magnetic field is applied to the *specimen* in order to bring *its* absorption lines into coincidence with a fixed frequency laser. One then has the relatively new but very powerful technique of laser magnetic resonance spectroscopy (or LMR). This is particularly valuable in the study of free radicals which, unlike closed-shell molecules, have large magnetic moments. Since radicals are usually produced under very low pressure conditions, their absorption lines will be Doppler limited and at exact resonance there will be high sensitivity. The sensitivity can be still further enhanced by having the gaseous specimen enclosed in a Brewster-windowed cell which is mounted *inside* the laser cavity. LMR spectrometers commonly feature a CO or CO_2 laser since many radicals have absorption bands in the 5 μm and 9–11 μm bands but some valuable and interesting studies have been made of rotational spectra at longer wavelengths using optically pumped lasers [593]. The powerful CO_2 laser has the advantages of having a large number of lines available, so the chance of the necessary near coincidence is improved and also of readily saturating the lines so very high resolution, Lamb-dip spectroscopy is available. This is particularly desirable in the study of radicals since the hyperfine structure, when resolved, can give a very great deal of information about the electronic structure of the radical. The method has been used amongst others to study NO [594], NO_2 [595], NF_2 [596], NH_2 [597] and the short-lived excited-state atoms Cl(^{2}P) and Hg(^{3}P$_0$) [598]. An excellent example of the use of the technique to explore the ro-vibrational states of a free radical is provided by the work of Johns, McKellar and Riggin

[599] on the ν_2 fundamental band of HCO at 9·25 μm. Many coincidences were observed amongst them that of the $N_{KaKc} = 1_{10} \leftarrow 1_{11}$ and $1_{11} \leftarrow 1_{10}$ lines with the R (28) line of the laser at 1083·479 cm^{-1}. The observed spectra arising from the first of these and the underlying magnetic variation of the energy levels are shown schematically in Fig. 5.29.

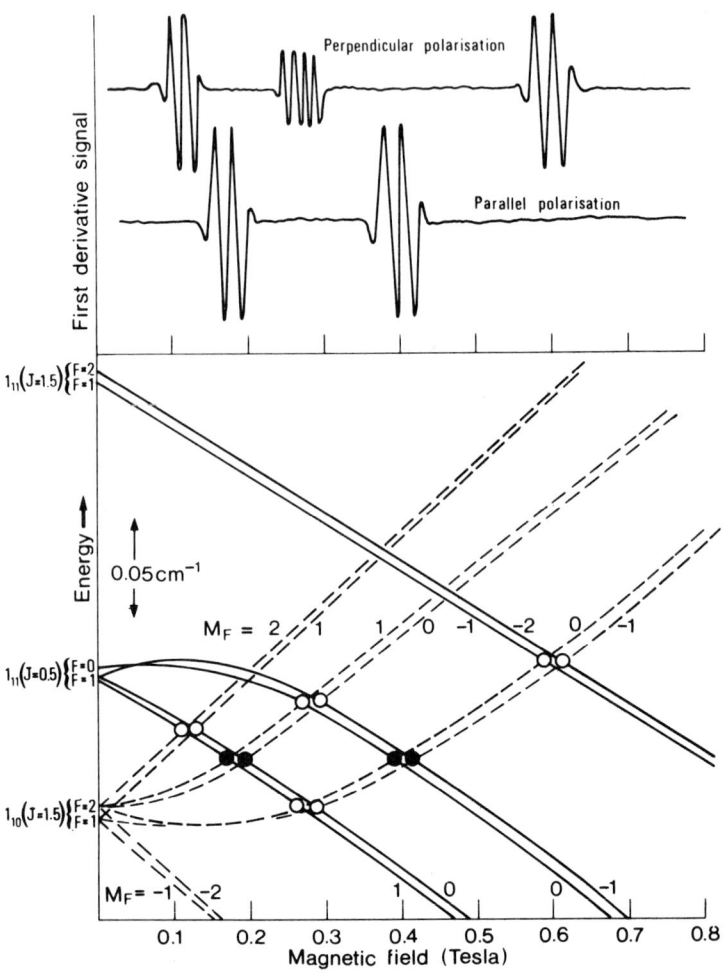

FIG. 5.29. Zeeman splitting of some of the states of HCO and the resulting resonances with the R(28) CO$_2$ laser line. The excited state levels have been shifted down by the laser wave number (1083·479 cm^{-1}) so that crossings of solid (i.e. ground-state) lines with dashed (i.e. excited-state) lines indicate possible laser magnetic resonances. Those indicated with open circles are expected in perpendicular polarisation ($\Delta M = \pm 1$) whilst those indicated by filled circles are expected in parallel polarisation ($\Delta M = 0$). After Johns, McKellar and Riggin [599].

The magnetic fields readily available with normal metal solenoids are restricted to the range below 10 T. Superconducting solenoids can extend this range upwards to 40 T but to go further demands the use of pulse techniques in which the solenoids sustain enormous currents for necessarily very brief ($\sim$ 1 μs) times [600]. This latter type of technology, which can produce fields of up to 100 T, is only available in a few very specialised research centres, so everywhere else magnetic resonance spectroscopy is restricted to fields less than about 40 T. It follows from (5.6.2) that the expected Zeeman splittings will be of the order of 10 cm^{-1} or less: in practice very much less! The power of laser magnetic resonance is that we can exploit close coincidences so that small Zeeman splittings, of this order, are sufficient to bring a transition from a lower split component to an upper split component into resonance with the laser. Transitions between the split levels themselves are not normally of interest to the infrared/ near-millimetre-wave spectroscopist because, with conventional magnet technology, the observing frequencies will be too low: in the microwave region for electron paramagnetic resonance EPR [601] and in the radio frequency region for nuclear magnetic resonance or NMR [602]. The main exceptions are found firstly in cyclotron resonance spectroscopy of semiconductors where the very small effective masses (see section 5.5) lead to much enhanced effective values for the Bohr magneton and hence to much higher observation frequencies and secondly in antiferromagnetic and ferrimagnetic resonance spectroscopy of crystalline transition metal and rare earth compounds where the natural fields produced by the magnetic ions can be an order of magnitude larger than the maximum fields which can be applied artificially.

Cyclotron resonance was briefly mentioned in Chapter 2. Basically a "gas" made up of free carriers has a continuum of possible energy states but, in the presence of a magnetic field B, the continuum "freezes out" into a set of discrete Landau levels which are separated by the frequency interval

$$\nu_{CR} = qB/2\pi m^* \qquad (5.6.7)$$

where q and m^* are the charge and effective mass of the carriers. This equation reduces to (5.6.3) for free electrons where $g = 2$ and $q = e$. The system can undergo allowed electric dipole transitions between any two (i.e. $n \to n + 1$) adjacent "rungs" of the Landau ladder and will therefore absorb electromagnetic energy at a frequency given by (5.6.7). This is cyclotron resonance absorption. The free carriers in semiconductors can be studied very conveniently by cyclotron resonance and it remains the best possible technique for determining effective masses near the band gap [603]. Taking InSb as an example it is found that a resonance with the 357 cm^{-1} (28 μm) line of the water vapour laser occurs at an applied field of 6·8 T. The effective mass of the carrier involved is thus $0·0178 m_e$.

Studies at these infrared wavelengths are important for many reasons but three major ones are:

a. The width of the resonance is determined by the ratio of the time to carry out a cyclotron resonance orbit compared to the carrier–carrier collision time. In fact the resonance can only be observed at all distinctly provided

$$2\pi v_{CR}\tau_e \gg 1. \tag{5.6.8}$$

The collision time τ_e is determined mostly by the concentration of carriers so for semiconductors such as silicon and germanium it is possible to prepare samples pure enough ($n_e < 10^{20}$ m^{-3}) that (5.6.8) can be satisfied at microwave frequencies but for all other semiconductors τ_e is very much less than 10^{-11} s so submillimetre observing wavelengths are essential. It is also worth remarking that even for an electric-dipole allowed transition one soon runs out of signal as the number of interacting carriers falls, so even for Si and Ge it is better to use a higher level of doping and a higher observing frequency.

b. Experimentally it is far preferable to work in transmission and this is only possible for wavelengths shorter than the plasma edge.

c. Full resolution of the detail in the cyclotron resonance spectra demands that $hv_{CR} > kT$ but one cannot always satisfy this by lowering the temperature since one may wish to observe high n states.

These are the reasons why cyclotron resonance, which has been known since the 1950s and which was pursued heroically with such devices as carcinotron sources [604], had to await the introduction of the infrared laser before it could fully flower.

The high resolution is necessary because the cyclotron resonance spectra observed with real semiconductors are very complex and, moreover, vary with direction of observation. The principal reason for this is that the simple picture of equispaced Landau levels does not even apply to the conduction band since the spacing is found to be n dependent—a kind of electronic "anharmonicity" arising from band non-parabolicity. It applies even less to the valence band where the states of even the simplest elemental semiconductors such as germanium are very complex indeed. The valence band of germanium is, to zero order, six-fold degenerate at $\bar{k} = 0$, but the effect of the spin–orbit coupling is to split the degeneracy leading to a four-fold degenerate state at the top of the band and a doubly degenerate state which is sufficiently far below the top to be ignored. The levels of the four-fold state interact with each other to lift the degeneracy for low quantum numbers. The result is that at high temperatures, where states with high n are involved, the cyclotron resonance spectrum is made up of lines due to conduction-band electrons together with two broad resonances which are ascribed to the "light" and "heavy" hole respectively. Furthermore quantum effects [605] lead to splittings, of the observed lines, which can be interpreted in terms of the presence of several distinct Landau ladders.

Cyclotron resonance is thus a complicated phenomenon but the spectra, when fully analysed, can give us enormously detailed information about the band-structures in semiconductors [606]. So far, however, only some twenty semiconductors have been investigated [603] but improvements in experimental technique are already opening up new possibilities. Thus the very high field pulsed technique mentioned above permits the investigation of materials such as GaP [607] which cannot so far be obtained even approximately pure (n_e $\sim 10^{23}$ m^{-3}). Alloy systems such as In As$_{1-x}$P$_x$ [608] can now be investigated with the conventional submillimetre laser systems and the results have shown interesting non-linear dependences of the electronic properties on the stoichiometry. Finally the development of reflection spectroscopic techniques has permitted the investigation of cyclotron resonance in highly conducting semiconductors like PbTe and HgSe. With these high-mobility specimens spectra are observed [609] which contain several very puzzling features but undoubtedly their elucidation will shed new light on the complicated phenomena characteristic of compound semiconductors.

5.6.3 *Antiferromagnetic and ferrimagnetic resonance*

The phenomenon of ferromagnetism, that is a bulk specimen displaying permanent magnetisation, has been known since prehistory but an explanation for it was only forthcoming this century when Heisenberg applied the newly developed quantum theory of electron spin to a lattice made up of atoms or ions, each of which had unpaired electrons and hence possessed a magnetic moment. Heisenberg evaluated the interaction energy between the ith and jth atoms of the lattice and found it to have the form [610]

$$\mathscr{E}_{ex} = -eJ\vec{s}_i \cdot \vec{s}_j, \qquad (5.6.9)$$

where J is the value of the exchange integral and $\vec{s}_i$ and $\vec{s}_j$ are the spin vectors. For ferromagnets J is positive and the resulting negative exchange energy compels all the spin vectors to align parallel to one another and there is thus an observable macroscopic magnetisation provided the temperature of the specimen is less than the order/disorder transition temperature, otherwise known as the Curie temperature. The phenomenon was first observed with iron, hence the epithet, and since the analogy with the electric polarisation is complete, it is usual to refer to the corresponding electric effect as "ferroelectricity" even though this phenomenon would not be possible in a conductor such as iron. In principle it is possible to flip over one of the spins with respect to all the others and this would lead to energy absorption but selection rules [610] prohibit this from happening and ferromagnetic resonance is therefore not observable.

If J is negative, the spins have to align themselves antiparallel to lower the energy and then one gets the magnetic atoms or ions arranged on two interpenetrating lattices such that the net moment of one sub-lattice exactly cancels that of the other. This is antiferromagnetism. Again there is a characteristic Curie temperature above which the phenomenon vanishes. This time, however, a flip of one atom or ion with respect to its appropriate sub-lattice is permitted but only by magnetic-dipole selection rules. The absorption is thus very weak but the other side of the same coin is that the magnetic moments only interact with each other weakly so the absorption lines are very sharp. Antiferromagnetic resonance is thus an observable phenomenon. The main problem with observing it is that antiferromagnetism does not necess-arily set in as soon as the temperature falls below the Curie temperature. Often the specimen adopts a paramagnetic configuration and only when a much lower transition temperature has been reached does it go over into its antiferromagnetic configuration. This lower temperature is often called the Néel temperature and signified by T_N. The antiferromagnetic resonance frequency starts to fall rapidly as the rising temperature approaches T_N [611] and finally becomes zero at T_N. Antiferromagnetic resonance has been observed [612] in very many transition metal systems, for example FeF_2 ($\tilde{v}_a = 57.7$ cm^{-1}, $T_N = 78$ K), NiO ($\tilde{v}_a = 36.6$ cm^{-1}), MnO ($\tilde{v}_a = 27.5$ cm^{-1}), MnF_2 ($\tilde{v}_a = 8.7$ cm^{-1}), CoF_2 ($\tilde{v}_a = 28.4$ cm^{-1}, $T_N = 38$ K) but it has also been studied in rare-earth compounds, for example $DyPO_4$ [613] and in the very interesting case of α oxygen [614] where below the α/β transition temperature (23 K) a line is found at 27 cm^{-1}.

The spin vectors of the magnetic entities in the lattice can be thought of as equivalent to the displacement vectors used in normal lattice dynamics analysis. One may thus conceive of a "magnon" in the same way that one does of a phonon. Antiferromagnetic resonance then consists of the absorption of a photon with the simultaneous creation of a $\overline{k} = 0$ magnon. The Néel temperature is then directly related to the wavenumber $\tilde{v}$ of the $\overline{k} = 0$ magnon mode by the usual relation $h\tilde{v}c = kT_N$. Two magnon processes can also take place but because of the relative weakness of the single magnon absorption process, the two phenomena are often of similar intensities [615].

Ferrimagnetic resonance can be observed in such substances as the rare-earth-iron garnets which have the general formula $5Fe_2O_3 3R_2O_3$. These have positive values of J but since there are two *unequal* sub-lattices, the selection rule mentioned above no longer applies. Nevertheless the phenomenon is one of the most difficult to observe experimentally because very low temperatures are required and the absorption intensity is very weak. Richards and his colleagues [616] have observed the resonance in ytterbium iron garnet at 3 cm^{-1} ($T \sim 1.2$ K) and Sievers and Tinkham [617] have observed those for holmium, samarium and gadolinium iron garnets.

5.7 Electric resonance spectroscopy

It was not until 1913, that is seventeen years after the discovery of the Zeeman effect, that the corresponding electric-field effect was observed. The first observations were made by Johannes Stark and, because of his pioneering work, the effect is named after him. The reason for the long delay was that first-order (i.e. E-dependent) Stark effects are very rare, unlike the Zeeman case where, when an effect is observed at all, it is usually first order, and Stark had the good fortune to hit upon one of the few known examples. Second-order (i.e. E^2-dependent) Stark effects are observed for all dipolar molecules so, provided one has available a high enough field, one can produce Stark shifts for almost any molecule. For this reason Stark-modulation is frequently employed in microwave spectrometers [618]. The explanation of the rarity of first-order Stark shifts is that the electric-dipole operator is odd under the parity operation so that for any state with a definite parity (positive or negative) the expectation value of the dipole moment will be zero. First-order Stark effects can only be observed therefore for states of *mixed* parity. In atomic spectroscopy the only common example is the hydrogen atom [619] where, because all levels with $n > 0$ are degenerate (that is $\mathscr{E}$ is independent of l), these levels are mixtures of base states of alternating parity. The overall parity is therefore mixed. The dipole moments of these mixed states can be very high, of the order 10 D. A similar situation arises for symmetric tops for states with $K \neq 0$ where the mixing of the positive K states with the equal energy negative K states provides the mixed parity. Inversion, as in ammonia, removes, of course, the degeneracy so a first-order Stark effect is not observed until the Stark splitting gets large in comparison with the inversion splitting.

The second-order Stark effect can be thought of as arising from the removal of the M degeneracy of the components of a level of a given J. As in the magnetic case, the applied field "labels" space and states of different M split away from one another. In the case of a linear molecule each level breaks up into $(2J + 1)$ distinct components. The magnitude of the splitting is obtained by diagonalising a rather complicated Hamiltonian [620] and the calculation may be taken to various levels of sophistication. For moderately low fields applied to a linear molecule, the result is

$$\Delta \mathscr{E} = [A + BM^2]\mu^2 E^2, \qquad (5.7.1)$$

where A and B are somewhat complicated functions of J. This equation is very important. It shows for example that since $\mathscr{E}$ depends on both J and M, the observed *transition* frequencies will be field dependent: in other words there will be *observable* Stark effects. Secondly it then provides an easy way of determining dipole moments in excited states. Thirdly it yields a way of making unambiguous assignments of the lowest possible value for J in a transition, merely from the pattern of the split lines. When working with a complicated spectrum this can prove very valuable.

Apart from the spectroscopic advantages of the Stark effect, it has the very attractive practical use for bringing molecular or atomic transitions into exact resonance with a fixed frequency laser line. This can be used to increase the number of available optically pumped laser lines but it also gives us the new and very powerful technique of electric resonance laser spectroscopy. This technique has been used to explore the energy levels of stable molecules such as

D_2O [621] (see Fig. 5.30) but it is particularly suitable for the study of free radicals. So far, OH [622], CH [623], HO_2 [624], NH [625], NH_2 [626], PH [627], PH_2 [628], HCO [629] and CH_3O [630] have been investigated but the scope is almost limitless. For this application, where nearly always the concentration of the radicals is very low, it is usual to put the Stark cell inside the laser cavity. Oka and his colleagues [631], for example, have studied HCO using a CO_2 laser and Johns and McKellar [632] have reported spectra of HNO and DNO obtained with a CO laser. The technique can be used anywhere in the infrared where one has a suitable *cw* laser [633] but in the far infrared there may be problems [621], because, to get high values of E, one would like to have the electrodes close together but then the cell begins to look like a cut-off waveguide.

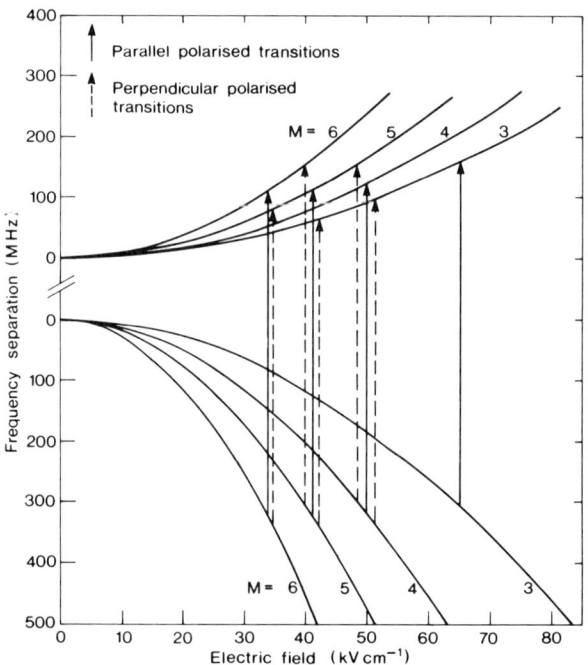

FIG. 5.30. Stark effect of the $6_{15} \rightarrow 6_{24}$ pure rotation line of D_2O: from the work of Duxbury and Jones [621].

Another very useful application of the Stark effect is to induce transitions which are not allowed under field-free conditions. Thus $\Delta M = \pm 1$ transitions can be induced by having the radiation field vector and the static electric field perpendicular to one another.

Laser Stark spectroscopy is thus a versatile and powerful technique which can give the molecular spectroscopist information which would be very difficult to get any other way. It has its experimental difficulties, breakdown of the low-pressure gas and wave-guiding cut-off for example, but since it applies to all polar molecules, that is the great majority of molecular species, it is much more generally applicable than is the experimentally easier technique of Laser Magnetic Resonance which applies only to that relatively rare class of molecule which possesses a permanent magnetic moment.

Chapter 6
Practical Infrared Equipment

6.1 Broad-band hot-body sources

The discussion of black-body or "full" radiators, given in Chapter 2, rested on the idea of a cavity connected by only small holes to the outside world. This, naturally, is only of academic interest to the practical engineer designing viable infrared sources since he will want to get finite amounts of power out of his source. Cavities, therefore, are usually reserved for those relatively rare occasions when an absolute calibration of the radiant power is required. In all other applications free-standing hot bodies are used which are either unenclosed or else surrounded by a protective wall which is transparent at the wavelengths of interest. The problem for the designer is to get the emissivity of the hot-body up as near to unity as possible and to choose the optimum operating temperature. Different solutions to this problem emerge in the different regions of the infrared.

In the near infrared, all practical sources will be operating in the curved region of Fig. 2.14, T_s will be low and the important consideration will hence be to get the temperature as high as possible. The best solution is a tungsten filament operating in an inert atmosphere where temperatures of the order 3000 K can be realised. Modern quartz–halogen lamps are ideal. The emissivity of the filament would normally be low because metals reflect strongly, but if its surface is roughened the emissivity will rise because of the formation of numerous micro-pseudocavities. Values as high as 0·7 have been achieved in this way in the near infrared. Such values are amply good enough since the departures from ideality of the emissivity can be more than compensated for by the increase in temperature. Thus suppose one is working at 5 µm where the equivalent temperature (see section 2.2.3) is 2878 K, then a 2900 K quartz–halogen lamp will have $T_s = 1$ whereas the hottest conveniently available true black-body ($T = 1330$ K) would have a T_s value of only 0·46.

From equation (2.2.49) it follows that the quartz–halogen lamp will be intrinsically 4·5 times brighter. Even if one allows for the imperfect emissivity, this ratio is only reduced to 3·2. This has been confirmed experimentally by Taylor, Rupert and Strong [634] who compared the emission, at 5 μm from a 2900 K tungsten source with that from a 1330 K globar (see later). They found a ratio of 3·9 and when allowance is made for the imperfect emissivity of the globar (0·8), the ratio is reduced to 3·12 in excellent agreement with the theory. At shorter wavelengths the effect becomes still more marked. Thus at 2 μm the expected ratio would be 14 to 1.

In the mid infrared, tungsten lamps cannot be used because the quartz envelope becomes opaque and since there is no satisfactory material to act as an alternative, one has no choice but to use unshielded hot bodies as the sources. The commonest mid-infrared source in most commercial spectrometers is some variant on the Nernst glower. This device was invented by Nernst in 1904 hopefully as a viable source of visible radiation for illumination purposes to compete with the carbon filament lamp but neither device was able to compete effectively with the more efficient devices which soon came along, such as the tungsten filament lamp. However, as an infrared source, it, and its less cumbersome descendants, still hold a central position. The original glower was a cylinder made by firing at 1700 °C a mixture of zirconia with various rare earths. The dimensions of the cylinder were, diameter 1–3 mm, length 30–40 mm, and a favourite recipe for the mixture was zirconia 85 % and yttria 15 % with perhaps a small proportion of thoria. Electrical connections were made at each end through platinum wires wound tightly round the cylinder. The connections were finished off by immersing the junctions in some more of the mixture. The chief difficulty with the Nernst glower was its large negative coefficient of resistivity. Thus at room temperature the resistance of the cylinder might be in excess of 10 000 Ω, yet at the operating temperature (1900 °C) it might only be a few Ω. Ballast resistors are therefore essential but in addition the glower has to be preheated so that the resistance can be made low enough for a self-sustaining current to flow. In the early days of the art this preheating was done with a gas flame, but in modern versions a thin heating wire running through or else wound tightly round the outside serves the dual role of a preheater and an electrical connector. In the warm-up phase the current is carried by the rhodium wire and in the operating stage mostly by the, now hot, refractory mixture.

The Nernst glower has the advantages that it can be operated in air so no windows are required, that it is long and thin so it easily couples optically to a dispersive spectrometer and that its temperature is readily controllable. It has the disadvantages that it readily suffers brittle fracture and the electrical connections often go open circuit. To overcome these difficulties it was customary in prior years to leave the glower switched on all the time and therefore running continuously. The introduction of the heating wire has done much to relieve the electrical difficulties and most of the remaining drawbacks have been eliminated by some ingenious design modifications introduced by the Perkin–Elmer Corporation. In this redesign there is a hollow cylinder of sintered alumina typically having an inner diameter of 2 mm and an outside diameter of 3 mm. This hollow tube is packed with a mixture of zirconium silicate and aluminium oxide and there is a rhodium heating wire passing down the centre. This form of glower operates on a much reduced applied voltage (28 V) and consumes much less power (30 W) yet achieves a surface temperature of 1500 K. This is, of course, lower than that achieved by the original design but in the mid-infrared temperature is no longer so overriding a consideration and the new form compensates for this by being usable over a much wider spectral range, in fact over the entire 3–25 μm fingerprint region. The new design has a very small variation of resistivity with temperature and can therefore be run off a simple power supply. It can be switched off and on repeatedly without affecting seriously its operating lifetime and altogether is a much more attractive commercial proposition than the original Nernst glower.

Another type of commercial source can be regarded as a hybrid of a metallic ribbon type with a Nernst glower type. It is essentially a nichrome strip immersed in a fused silica sheath. The nichrome radiates at the very short wavelengths (< 3 μm) where the fused silica is transparent and the sheath radiates for the remainder of the infrared out to quite long wavelengths (500 μm) where silica starts to transmit again. Because of the development of these convenient commercial sources, the globar which was much used in the past is now only of historical interest. The globar was made from a rod of synthetic carborundum (i.e. silicon carbide) and operated at a temperature of approximately 1400 K. It was relatively large (6 mm diameter) and consumed rather a large amount of power (200 W) but it was cheap and robust and could be used over a wide range of the infrared [435]. In operation it was essentially a grey body with an average emissivity of ~ 0.8.

When one comes to consider far-infrared sources, temperature might be thought to be even less of a consideration because one is now in the Rayleigh–Jeans region where the radiant power is proportional merely to the first power of the temperature. However, by the same arguments the total far infrared power radiated is very low indeed and quite unlike the mid infrared one can soon find oneself in the situation of being source power limited. One needs therefore to work at as high a temperature as one can but if one does one will be generating unavoidably much larger quantities of mid- and near-infrared radiation and this can cause serious problems by leaking round the experimental system and being misidentified as the wanted far-infrared radiation. This leakage of short-wave radiation is usually referred to, for historical reasons, as "stray light" and to avoid it, it is necessary to introduce suitable short-wave filters. Unfortunately, due to materials and constructional problems, filters transparent at far-infrared wavelengths yet totally opaque in the mid and near infrared do not exist, so source design is always governed by balancing the amount of absorption of far-infrared radiation that can be tolerated in the filter against the degree of rejection of near-infrared radiation that must be achieved. The answer to this balancing decision is different for different instruments; thus for a grating instrument where "stray light" is a major hazard, the near infrared rejection has to be high even at the cost of significant far-infrared attenuation, whereas for an interferometer where the only hazard introduced by short wavelength radiation is the possible non-linearity induced in the detector by overloading the filtering need not be so extreme.

The best available compromise, as a far-infrared source, the medium pressure mercury in quartz arc, was introduced by Rubens in the early years of this century. Despite its venerable status, there have been only minor changes in design over the intervening seventy years or so. The Philips HPK 125 is typical of modern versions of this arc lamp. It is manufactured primarily as an ultraviolet source for, e.g. health lamps, which is fortunate because if it were only sold on the small far-infrared market it is doubtful whether the manufacturers would consider it worthwhile continuing in production. The cylindrical quartz envelope of the lamp is about 30 mm long and 12 mm in diameter. When running, the quartz envelope is at a temperature of about 1000 K and does all of the radiating at wavelengths between 3 and 100 μm. At longer wavelengths, however, the envelope becomes transparent, ceases to radiate effectively and the plasma inside, which behaves roughly like a black body at 10^4 K, takes over. For this reason, high-pressure mercury arc lamps can be used out to at least 5000 μm (i.e. 60 GHz or $2 \, \text{cm}^{-1}$).

The optimum shape for the source depends on the shapes of the limiting stops in the subsequent optical system. Thus a grating spectrometer which features tall thin slits is best matched by a similarly shaped source. Since grating instruments have dominated the infrared market throughout most of its existence it is not surprising to find that most sources which are available have just this shape. Interferometers, on the other hand, have circular limiting stops so would be best matched by a circular source. Perfectly circular sources have yet to be developed, but the short stubby HPK 125 lamp is a reasonable approximation and this is an extra commendation for its use in far-infrared interferometry. On the other hand, with careful design, it is possible to produce a source which crams the very maximum of radiant power through the restricting slits of a grating instrument. Such a source would be made of metal in the form of a thin-walled hollow cylinder. Parallel to the axis of the cylinder is cut a thin slot in the wall. The radiation emerging from the slot will be a much closer approximation to black body radiation than that emitted by the outside walls and one will have gone some way towards recovering part of the luminosity advantage that interferometers enjoy over dispersive spectrometers.

6.2 Infrared lasers

6.2.1 *Noble gas lasers*

The electronic transitions in neutral atoms which can be inverted to provide the bases of laser action lie mostly at the longer infrared wavelengths. The reason for this is that the myriad excited states of the atom all have to be squeezed in between the ground state and the first ionisation limit and the transition frequencies between adjoining connected states tend to be low. In fact the only visible region neutral atom laser line of any significance is that at 0·6328 μm produced by the helium/neon laser. The production of shorter wavelength lasers, e.g. in the blue region, requires the use of ionised atoms which, because the second ionisation potential is always larger than the first, have their energy levels rather more spaced out. The number of infrared laser lines which are known from neutral atoms is large [636] but they are nearly always very weak and therefore of little technical importance. Lines from neon, for example, are known which span virtually the whole infrared but the power levels available are nearly always fractions of a milliwatt. This is of interest to the atomic spectroscopist who finds power levels of this order adequate for absolute frequency determinations but only the line group near 1·15 μm and the 3·39 μm line from He/Ne are of interest to the infrared technologist who needs much more power. Some use has been made in the past of the 95·8 μm ($3^1P \rightarrow 3^1D$) and the 216·3 μm ($4^1P \rightarrow 4^1D$) lines from pure helium but lines from optically pumped lasers (section 6.2.5) have now replaced these, so present interest in infrared laser emission from neutral atoms is concentrated exclusively on the near infrared lines of He/Ne.

He/Ne lasers for infrared operation are often quite long—sometimes several metres—but are usually of quite narrow bore ($\sim$ 1 cm). The Brewster windows are made from Spectrosil (fused quartz) and the cavity is defined by silicon substrate multilayer dielectric mirrors. These mirrors have to be made specially (see section 3.8) for each application since the reflectivity is sharply peaked spectrally. The laser can be driven either by RF induction (27–40 MHz), by longitudinal DC excitation with a conventional hollow cathode or else transversely with a slot-type hollow cathode. This latter is, however, rare despite its advantage of low running voltages. The RF method has the advantages that no metal electrodes are required to pass through the glass walls and that the discharge produced is particularly "quiet". Commercial diathermy equipment is quite satisfactory. The RF frequency is not critical provided it is greater than the collisional relaxation frequency. Much higher frequency microwave excitation is also possible using a magnetron and this is particularly appropriate if pulsed operation is required. Longitudinal DC excitation involves a high striking voltage (several tens of kilovolts) but the running voltage is much less so a ballast resistor is needed. The operating current is of the order 10 mA. The gas mixture of 1 torr of helium with 0·1 torr of neon is sometimes circulated through a buffer volume especially if high-power operation is required. Sealed off operation is possible but it requires particularly careful assembly so most users simply top up the mixture from time to time through a stopcock. Continuous-wave lasers operating at 3·39 μm can give power levels of the order 100 mW in single mode operation. The output frequency can be stabilised by locking to the methane line (see section 5.2.3.2) at 2947·912 cm^{-1} and such a stabilised laser is an attractive standard of length. Higher power levels than this require pulsed operation and then it is normal to use much wider tubes since in this way collisional de-excitation at the wall during the brief time of the pulse can be avoided. It is possible in this way to achieve peak power levels of the order of several hundreds of watts but in truth it has to be admitted that there are not many applications for high power pulsed He/Ne lasers since there are more attractive higher power alternatives, such as Nd/YAG, available and for most applications the exact wavelength is not critical.

6.2.2 The carbon dioxide laser

6.2.2.1 Basic properties and theory of the laser mechanism
The carbon dioxide laser which operates in two bands near 9·4 μm and 10·6 μm is without doubt the most important source of coherent infrared radiation. [637]. There are many reasons for this pre-eminence but some of them are:

a. It radiates in the technically important 10 μm "fingerprint" region where the atmosphere is also transparent.

b. It is easy to construct and operate and for those not wanting to bother with making their own, several commercial models are available.

c. It can run continuous wave, pulsed, or Q-switched and the power levels available can easily run into kilowatts continuous wave and megawatts pulsed.

d. In the *T*ransversely *E*xcited *A*tmospheric (TEA) form it can provide intense very brief pulses from a compact and inexpensive assembly.

e. Mode control is easy giving the experimentalist considerable flexibility.

f. The laser can operate on either of two entire vibration/rotation bands and, in addition, it is possible to make it work on the higher analogues—that is the "sequence" bands. Consequently the number of possible output frequencies is very large.

g. The laser will work with a wide range of "fuels"—all the isotopic forms of CO_2 plus the isoelectronic N_2O and its isotopic variants—thus still futher increasing the number of possible output frequencies.

h. By increasing the gas pressure, the individual vibration/rotation lines can be broadened and at a sufficient pressure it becomes possible to get a continuously variable output frequency.

i. The power levels available from the CO_2 laser are so high that it is easy to use it as a pump for secondary "stimulated fluorescence" lasers and in this way extend the lasing activity all the way to the millimetre waveband.

The lasing gas is a mixture of CO_2, N_2 and He and when an electric discharge is passed through this mixture a high population is built up of N_2 molecules in their first vibrationally excited state. This state is metastable because, in a homonuclear diatomic molecule, it must have the same parity as the ground state, and it may only be depopulated by collision, either with the walls or with other molecules in the gas. Vibrational energy transfer (see Fig. 6.1) may then take place to the CO_2 molecules which are thereby transferred in large numbers from the ground state to the state where the antisymmetric stretching mode v_3 is singly excited. This is a very efficient process since the two states are virtually resonant. Some CO_2 molecules, it is true, are directly excited by electron collisions but this is a much less efficient mechanism for populating the upper level of the laser.

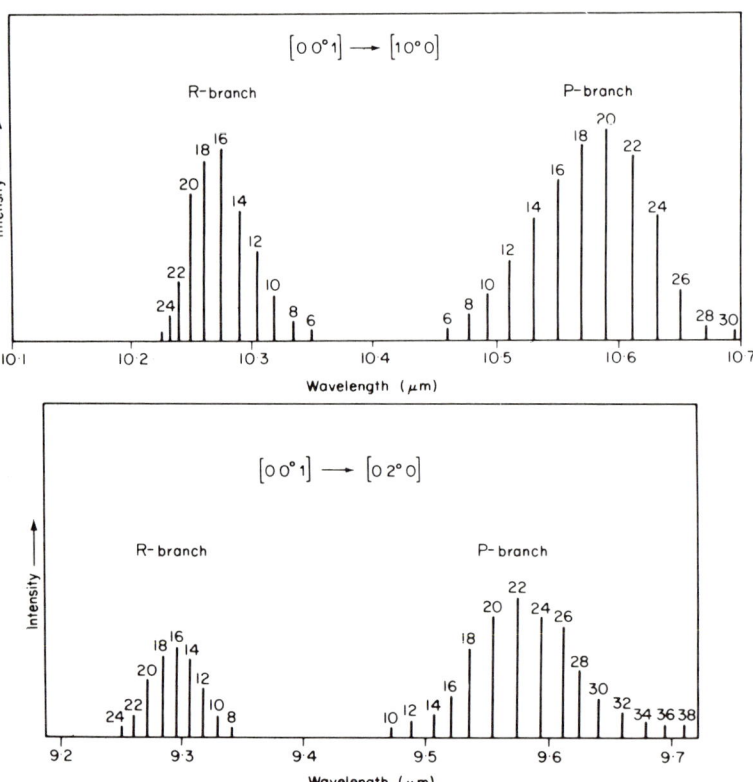

FIG. 6.1. Energy levels and observed emission spectrum of the CO_2 laser. The state labels (10⁰0) and (02⁰0) have only schematic meaning because of the strong Fermi resonance.

The spectroscopy of the CO_2 molecule has been briefly touched on in Chapters 2 and 5. The vibrational state of the molecule can be specified by the code $(v_1 v_2^l v_3)$ where the subscripts refer to the symmetric stretching mode v_1, the bending mode v_2 and the antisymmetric stretching mode v_3 respectively. The superscript l on v_2 refers to the vibrational angular momentum induced by the Lissajous motion arising from the simultaneous in-phase excitation of the degenerate bending mode in two planes at right angles. The quantum number l is thus formally equivalent to the similar quantity which arises in atomic spectroscopy and for this reason it is usual to refer to states with $l = 0, 1, 2$, etc. as Σ, Π, Δ, etc. Each of these classes can be further divided into two subclasses, those which are even $(+)$ and those which are odd $(-)$ with respect to inversion at the centre of symmetry. The nuclear statistics for a molecule such as CO_2 with the two oxygen atoms identical forbid the existence of J odd levels in a Σ^+ state and of J even levels in a Σ^- state. The most dramatic consequence of the resulting missing levels is that the spacings of the lines in the P- and R-branches will be twice what one might expect them to be. When the oxygen atoms are different as in $^{16}O \ ^{12}C \ ^{18}O$, the missing rotational lines reappear. The two states (10^00), i.e. the symmetric stretching mode and (02^00), one component of the overtone of the bending mode, are to the harmonic approximation coincident in energy but, because of a strong Fermi resonance, they become widely separated. The other component of (020), namely (02^20), is of different symmetry (Δ) and, being unaffected by the resonance, occurs at an energy which is almost exactly that which one would expect for either (10^00) or (02^00) in the absence of resonance. The state (02^20) thus retains its identity but (10^00) and (02^00) become completely mixed and it is improper, in a strict sense, to use either label to refer to either of the final states. Nevertheless it has become conventional to label the upper level as (10^00) and the lower as (02^00). These two states are relatively unpopulated under the discharge conditions because both are connected strongly, via radiative cascades, to the ground state. The connection of the state, loosely labelled as (10^00), to the ground state is a consequence of the resonance since the true state (10^00), would be, of course, metastable.

There are several selection rules which govern the infrared activity: thus there are the "strong" ones, namely $(+) \leftrightarrow (-)$ and $\Delta l = 0, \pm 1$, and the weaker ones such as "only one quantum number to change in an allowed transition". Transitions from (00^01) to either of the resonating levels satisfy the strong rules but violate the less strong. One might thus expect these difference bands to occur in the absorption spectrum but quite weakly. In practice, however, both are found to be reasonably intense and readily observable under normal laboratory conditions. The band at 9·4 μm, notionally $(02^00) \rightarrow (00^01)$, is, in fact, often used as an example of a parallel band of a linear molecule in student demonstrations! We thus have the situation that a population inversion exists between two states which are connected by a non-zero ($A = 0·21 \ s^{-1}$) transition moment and lasing therefore results.

Because the CO_2 laser is so important, its emission frequencies have been intensively studied and absolute values both for the line frequencies and for the line difference frequencies have been determined by several groups [638]. Thus Bridges and Chang [639] measured many difference frequencies for normal $^{12}C^{16}O_2$ and subsequently other authors investigated the isotopically substituted versions. As early as 1968, Siddoway [640], using a grating spectrometer, measured the lines from $^{14}C^{16}O_2$, whilst in 1974, Freed, Ross and O'Donnell [641], using the difference frequency technique, determined very precise constants for $^{12}C^{18}O_2$, $^{13}C^{16}O_2$ and $^{13}C^{18}O_2$. Evenson and his colleagues [642] at NBS and the group at NPL [643] have played a prominent role in the measurement of absolute frequencies, as part of the international comparison programme aimed at determining the true time frequency of the ^{86}Kr wavelength standard. From their work and from that of their colleagues [644] has come very accurate spectral information about the common form of CO_2. Thus the centrifugal distortion constants D, H, and L, are all known but the latter two are so small, for this very rigid molecule, that frequencies, accurate enough for most purposes, can be calculated invoking only D. The line frequencies can therefore be written,

$$P(2m) = v_0 - (B' + B'')(2m) + (B' - B'')(2m)^2 + 2(D' + D'')(2m)^3$$
$$- (D' - D'')(2m)^2[(2m)^2 + 1],$$

$$R(2m) = v_0 + (B' + B'')(2m + 1) + (B' - B'')(2m + 1)^2$$
$$- 2(D' + D'')(2m + 1)^3 - (D' - D'')(2m + 1)^2[(2m + 1)^2 + 1],$$
$$\cdots \ (6.2.1)$$

where $m (= 1, 2, 3,$ etc. in the P-branch and $0, 1, 2, 3,$ etc. in the R-branch) is numerically $(J''/2)$. The R-branch frequencies can be found from the P-branch formula by replacing $(2m)$ by $-(2m + 1)$. Mathematically therefore, the P- and R-branch lines can be regarded as the observed components of a larger set which is generated by a single running integer. The line parameters and the resulting line frequencies for $^{12}C^{16}O_2$, are listed in Tables 6.1a and 6.1b. Data for the two commoner isotopic variants are listed (in units of cm^{-1}) in Table 6.2. Line positions can readily be calculated from the data in Table 6.2, using equation (6.2.1).

6.2.2.2 Longitudinally excited glow-discharge CO_2 lasers

The CO_2 laser is so useful a device and so flexible in its operation that many different versions have been evolved to meet the differring needs of spectroscopists, metrologists, technologists etc. We can begin a survey by considering the moderate power, continuous wave, longitudinally excited variety which has become a routine "workhorse" in many laboratories. A schematic of its construction is given in Fig. 6.2. The gas mixture ($0·8\ CO_2$: $1N_2$: $7He$) is flowed fairly rapidly through the discharge tube with an effective ambient pressure of about 10 torr. The power supply delivers 10–15 kV and the tube current is restricted to about 100 mA by means of a suitable ballast resistor. The optimum gas pressure, composition and tube current are found to vary slowly and in a rather complex way with tube diameter but it is found that diameters in the range 2–5 cm give fairly uncritical operation. It is very important that the lasing gas be kept as cool as possible so that self-absorption of the laser radiation by thermally excited molecules is suppressed. Either flowing the gas, from a reservoir, into the tube and then out to an exhaust or else circulating it in a closed-system through a heat exchanger is vital to effective operation.

TABLE 6.1a
CO_2 Laser line frequencies and wavelengths

Band	Designation	Frequency (THz)	Wavelength (μm)
	P(32)	27·969476	10·718559
$00^01 \rightarrow 10^00$	P(30)	28·027458	10·696384
$v_3 \rightarrow v_1$	P(28)	28·084696	10·674585
	P(26)	28·141192	10·653154
Parameters (THz)	P(24)	28·196948	10·632089
$v_0 = 28·8088385$	P(22)	28·251967	10·611384
$B' = 1·1606180 \times 10^{-2}$	P(20)	28·306251	10·591034
$B'' = 1·1697538 \times 10^{-2}$	P(18)	28·359800	10·571036
	P(16)	28·412616	10·551385
$D' = 3·9728 \times 10^{-9}$	P(14)	28·464700	10·532079
$D'' = 3·4222 \times 10^{-9}$	P(12)	28·516052	10·513112
	P(10)	28·566675	10·494481
$H' = 0·51 \times 10^{-15}$	P(8)	28·616567	10·476185
$H'' = 5·74 \times 10^{-15}$	P(6)	28·665730	10·458218
	P(4)	28·714163	10·440578
	P(2)	28·761866	10·423262
	v_0	28·8088385	10·406267
	R(0)	28·832051	10·397889
	R(2)	28·877927	10·381370
	R(4)	28·923071	10·365167
	R(6)	28·967482	10·349276
	R(8)	29·011158	10·333695
	R(10)	29·054097	10·318423
	R(12)	29·096299	10·303457
	R(14)	29·137761	10·288795
	R(16)	29·178480	10·274437
	R(18)	29·218455	10·260380
	R(20)	29·257683	10·246623
	R(22)	29·296161	10·233165
	R(24)	29.333886	10·220005
	R(26)	29·370855	10·207141
	R(28)	29·407063	10·194573
	R(30)	29·442509	10·182300
	R(32)	29·477187	10·170321

Water cooling, by means of a jacket surrounding the laser tube, can also be very helpful. CO_2 is not indefinitely stable under the discharge conditions and it slowly breaks up into CO and O_2. This does not give any difficulties in the flow through "throw-away" regimes but it is a serious problem for sealed systems. The best solution here is to have a large gas reservoir, at ambient temperature, in which is disposed a suitable heterogeneous catalyst [645] which can bring about the recombination of the CO and O_2. The helium gas, in the mixture, permits stable discharges to be maintained in higher CO_2 partial

TABLE 6.1b

CO_2 Laser line frequencies and wavelengths

Band	Designation	Frequency (THz)	Wavelength (μm)
	P(32)	31·0427197	9·657416
$00^01 \to 02^00$	P(30)	31·1014934	9·639166
$v_3 \to 2v_2$	P(28)	31·1595093	9·621219
	P(26)	31·2167624	9·603573
Parameters (THz)	P(24)	31·2732483	9·586227
	P(22)	31·3289628	9·569179
$v_0 = 31\cdot88996029$	P(20)	31·3839017	9·552428
	P(18)	31·4380615	9·535971
$B' = 1\cdot1606180 \times 10^{-2}$	P(16)	31·4914387	9·519808
$B'' = 1\cdot1706333 \times 10^{-2}$	P(14)	31·5440301	9·503936
	P(12)	31·5958329	9·488354
	P(10)	31·6468443	9·473060
$D' = 3\cdot9728 \times 10^{-9}$	P(8)	31·6970622	9·458052
$D'' = 4\cdot6868 \times 10^{-9}$	P(6)	31·7464844	9·443328
	P(4)	31·7951091	9·428886
	P(2)	31·8429348	9·414724
$H' = 0\cdot51 \times 10^{-15}$	v_0	31·8899603	9·400841
$H'' = 6\cdot96 \times 10^{-15}$	R(0)	31·9131726	9·394004
	R(2)	31·9589960	9·380534
	R(4)	32·0040173	9·367338
	R(6)	32·0482362	9·354414
	R(8)	32·0916526	9·341758
	R(10)	32·1342669	9·329370
	R(12)	32·1760795	9·317246
	R(14)	32·2170913	9·305386
	R(16)	32·2573034	9·293786
	R(18)	32·2967172	9·282444
	R(20)	32·3353342	9·271358
	R(22)	32·3731564	9·260526
	R(24)	32·4101860	9·249946
	R(26)	32·4464254	9·239614
	R(28)	32·4818774	9·229530
	R(30)	32·5165449	9·219690
	R(32)	32·5504313	9·210092

pressures, helps to keep the mixture cool by increasing the thermal conductivity and collisionally depopulates the lower level of the transition and thus assists in the maintenance of the population inversion. A well-cooled flowing gas laser can achieve a power conversion efficiency of the order of 20% which compares well with the theoretical maximum of 40%. The power output is found to be linearly proportional to the tube length and under ideal conditions one can get figures of the order of $80 \, W \, m^{-1}$. The power is, however, only weakly dependent on the tube diameter so one can choose the smallest value which will permit the propagation of a good TEM_{00} mode without serious diffraction loss. The diameter will then increase as the square-root of the tube length (equation 2.1.6). The tube-lengths used in ordinary laboratory applications are at most a few metres but versions have been made which were several hundred metres long and which gave out tens of kilowatts of continuous-wave power! Such lasers have been used for heavy machining (see

TABLE 6.2

Parameter	$^{13}C^{16}O_2$	$^{12}C^{18}O_2$
v_3	2283·47	
$2v_2$	1265·81	
v_1	1370·05	
v_3-2v_2	1017·6593	1083·7229
v_3-v_1	913·4249	966·9554
B_{001}	0·3872777	0·3440939
B_{10^00}	0·3897215	0·3465276
B_{02^00}	0·3909203	0·3474072
D_{001}	$1·346·10^{-7}$	$1·091·10^{-7}$
D_{10^00}	$1·215·10^{-7}$	$0·964·10^{-7}$
D_{02^00}	$1·599·10^{-7}$	$1·206·10^{-7}$

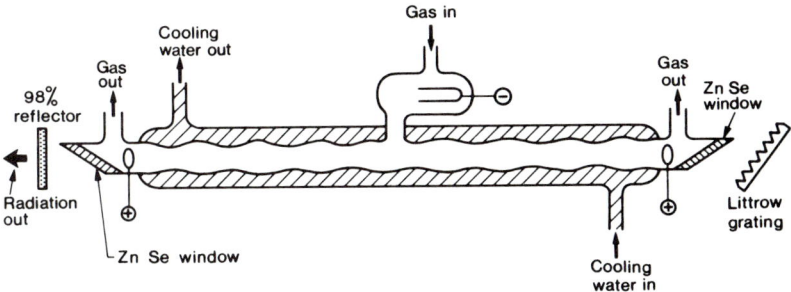

FIG. 6.2. Schematic of the construction of a longitudinally excited glow-discharge CO_2 laser.

section 7.4.2) but in a crowded workshop had to be assembled with several folds, rather like a concertina, in order to economize on space.

At the usual power levels in the laboratory (~ 10 W), the choice of window material does not present any difficulties. There are several possible candidates but the best are the II-VI compounds such as CdTe, ZnS or ZnSe which combine low loss ($< 10^{-2}$ cm^{-1}) with good thermal conductivity and low thermal expansivity and which are, moreover, completely non-hygroscopic. The competing alkali halides, e.g. KBr, are cheaper and even less absorbing but the water pick-up problem, characteristic of all the alkali halide materials, has prevented their finding general applications as the windows of CO_2 lasers. With medium to high power lasers, however, one does encounter problems with the windows since the increasing absorption with temperature character- istic of semiconducting materials becomes a major design consideration. This phenomenon is due to thermal promotion of electrons across the forbidden gap and into the conduction band and low-gap materials, such as Ge, Si, or CdTe, are no longer suitable since they show "thermal runaway" effects [646] for window tempertures as low as 100 °C and this limits the intensity at the window to values of 100 W cm^{-2} or less. ZnSe wich is a high gap ($\mathscr{E}_g = 3$ eV) material is much better and it does not show runaway effects until the window has reached temperatures of the order 500 °C. Since such temperatures would not be encountered in practice this means never! ZnSe is indeed so useful for general purpose work in the 10 μm region that it is being manufactured in large quantities and is now available in single crystal form as slabs many inches in diameter. Antireflection coatings for ZnSe can made using ThF$_4$ and simple two-layer coats can readily give reflectivities of less than 0·1 % per surface [647]. Thermal runaway is usually very unwelcome but in a less drastic form, i.e. power-dependent transmissivity, it can be used to produce passive mode- locking of the pulsed CO_2 output: p-germanium [648] is very effective in this application. Damage to the cavity mirrors again presents no problems at low powers where multi-layer dielectric coatings on a germanium substrate are ideal, but at higher powers damage problems start to obtrude [649]. Coated metallic mirrors can always be used, the best being silver freshly deposited on a copper or molybdenum substrate and protected from corrosion by a coating of ThF$_4$ or Y$_2$O$_3$, since these have high damage thresholds (~ 50 J cm^{-2}) [650] but unless one is prepared to use an intracavity beam-splitter, one has to have at least one non-metallic mirror. Damage problems become acute when one is working with high-power *pulsed* lasers since then the power densities can become enormous. A vast amount of technical research has gone into attempts to alleviate the damage problem and although great progress has been made, for example the discovery that damage often grows out from a micro- inhomogeniety [651], it still remains true that damage threshold rather than laser technology is often the limiting factor in the design of a high-power pulsed laser.

Metrological uses of CO_2 lasers usually call for some degree of frequency stabilisation. Coarse stabilisation depends on restricting the laser output to a single line by means of an intracavity etalon but fine stabilization is also required. This is usually achieved by splitting off part of the beam and taking it to a fast infrared detector. The signal from this detector is fed via a Rowley-Shotton [652] control loop to a piezo-electric element on one of the cavity mirrors. In this way the cavity length is servo-controlled so that the cavity is always exactly resonant at the top of the gain curve. An alternative is to use control of the current passing through the laser. If the laser wanders off resonance, its output will fall which is equivalent to its electrical impedance rising [653]. The resulting fall in current can be sensed by picking off a voltage across a suitable series resistor and this again used to control the cavity length via a piezo electric element. In the first set-up one is optimising the power *produced* by the laser whereas in the second one is optimising the power *consumed*. In practice, however, the two approaches give equivalent results.

The relaxation times inside the CO_2 active medium (> 1 μs) are such that Q-switching is very easy. Rotating mirrors can be used [654] but there is also a wide range of absorbing gases, e.g. SF_6 [655], which can act as saturable absorbers and hence provide passive Q-switches. Q-switching is a useful trick for anyone who has a basically continuous wave system but occasionally needs high-power pulses. If, however, brief intense pulses are needed on a regular basis it is better to go in for a TEA laser as described in the next section.

6.2.2.3 *Variant designs of electrically excited CO_2 lasers*
When one comes to consider scaling up the conventional CO_2 laser design, so that it will work in the high-power regime, four major problems are encountered. These are:

a. Low gas pressures are required in order that a glow-discharge may be sustained and the gain per metre is therefore correspondingly low.
b. Very high striking and running voltages are necessary—of the order of a megavolt—and producing these in the ordinary laboratory is not easy quite apart from the health hazard. Laboratories which are equipped for high-voltage work normally use a combination of the Marx bank (in which a set of capacitors is charged in parallel and then, using thyristor switches, is discharged in series) together with a pulse-shaping transmission line of the Blumlein type [656]. This is not only expensive but also calls for very specialised resources and very skilled technicians.
c. The electron energies necessary to sustain the discharge are very different from the optimum values for exciting the molecules.
d. The best electrical and mode conditions lead to a situation where there is a relatively slow rate of de-excitation of the lower level and again the gain per metre figure is correspondingly low.

It is not, unfortunately, possible to find a design which overcomes all these problems, so several variants have emerged which address themselves to particular experimental situations.

For high-power, very-brief-pulse work, the gas pressure and operating voltage problems can both be overcome, very elegantly, by going over to transverse excitation. For experimental convenience the gas pressure in the laser is chosen to be near atmospheric, so that absolutely gas-tight construction is unnecessary, and one arrives at the very popular and very versatile, *T*ransversely *E*xcited *A*tmospheric (or TEA for short) laser [657]. The "tube" in this type of laser is frequently just a rectangular box made from pieces of perspex screwed and glued together. TEA lasers give out intense pulses at the megawatt level which are exceedingly brief (~ 100 ns) and since these are produced from a laser which can be only 100 cm long and which requires only a simple power supply, the device is very attractive indeed to a variety of users [658]. Thus simple sealed-off versions have been used for military applications, rather more sophisticated versions have been used for optical pumping and to drive the Spin Flip Raman laser (see later) and state-of-the-art versions have been used as the initial oscillators in the ultra-high-power systems used in fusion research (section 7.4.4). The efficiency of TEA lasers is very high ($\sim 30\%$), mostly because of the high gas pressure, and these lasers therefore are well to the fore in the efficiency league for turning electrical power directly into monochromatic radiant power.

The main problem with transverse excitation is to get a uniform discharge over a large volume of the gas—the difficulty stemming from the tendency of high-pressure glow discharges to break up into a series of arcs. The original solution was to have a series of cathodes in the form of pins each fed via a series resistor. The anode was a continuous rod. The idea behind this "pin-rod" design [659] is that arcs take a finite time to develop and that if one has a running discharge over a large number of gaps, each individual discharge will be over before it can degenerate into an arc. The next solution to emerge rested on the early work of Rogowski [660] in classical electrostatics. This approach uses special electrodes [661] which are mechined to a mathematically chosen ("Rogowski") profile which ensures a uniform field not only between the electrodes but at their edges as well. It is shown diagrammatically in Fig. 6.3(a). Electrodes of this type were originally made from castings taken from a master mould but nowadays they are usually manufactured on a computer controlled automatic mill. There are some alternative designs, or improvements, to the Rogowski profile but the differences in practice are rather small. It is possible to run the laser with just a single high-voltage pulsed power supply but this so-called "single discharge" method has hardly ever been used in practice. The preferred technique uses a trigger wire—as shown in Fig 6.3(a)—to pre-ionise the gas so that there are plenty of current carriers present when the main pulse starts to flow [662]. The ionisation is caused both by electron bombardment and also by the presence of photo-electrons produced by the hard UV irradiation of the cathode—the UV coming from the trigger wire to cathode preliminary discharge. One thus has what is formally a double discharge system. Double discharge Rogowski TEA lasers have occupied a predominant position in the field and continue to do so but work continues on the development of the pin-rod design and some new versions (see Fig 6.3(b)) which feature a running spark are proving rather promising. The wires are encapsulated in glass and when the HV pulse is applied, sparks run along the gaps and these ionise the gas [663]. It is not really clear, however, just exactly how the gas is ionised. CO_2 is, apart from a slight window at 1200 A°, quite opaque throughout the vacuum UV. Radiation cannot therefore penetrate to any significant depth and it is unlikely that either CO_2, N_2 or He can be ionised in this way. It is conceivable that efficient ionisation depends on the presence of impurities (pump oil?) which have low ionisation potentials.

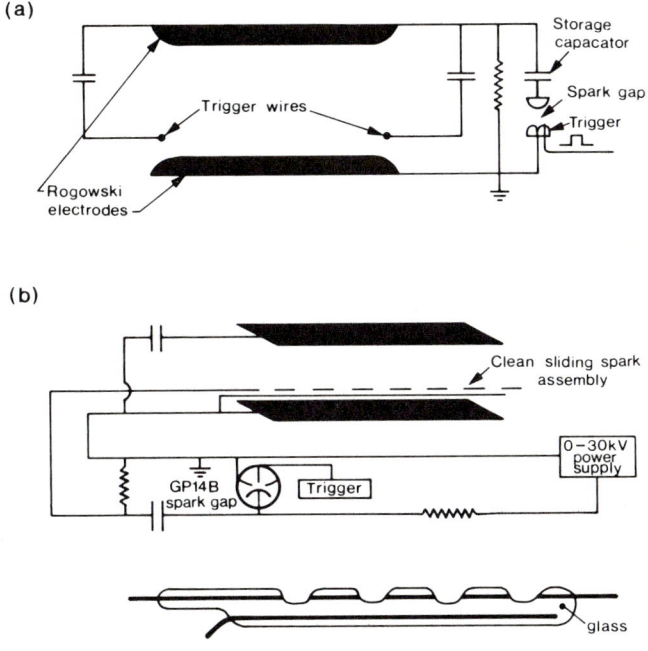

FIG. 6.3. Two forms for the discharge electrodes in TEA lasers.

With the TEA laser, the problem of decomposition of the CO_2 by the discharge becomes even more acute and considerable effort has been expended towards overcoming it. The principal motivation came from military users who need to have sealed-off lasers for battle-field applications. The basic decomposition equation is

$$2CO_2 \rightleftharpoons 2CO + O_2 \qquad (6.2.2)$$

so one approach is to invoke the Law of Mass Action and put in some carbon monoxide into the original mixture. This will have the effect of shifting the equilibrium in (6.2.2) to the left, but even so it is found that sealed-off lasers using such a mixture can only work at low repetition rates ($\sim$ 2 per second). The presence of oxygen produced by the decomposition is most undesirable since oxygen atoms can readily absorb electrons and thus impede the efficient operation of the pre-ionising discharge or spark. A small amount of hydrogen in the gas mixture can be helpful, since it tends to mop up surplus oxygen. An alternative approach is to use a catalyst, such as heated platinum, which helps the gas mixture to attain its true equilibrium more rapidly.

It is very desirable for TEA lasers to be made from a material transparent in the visible region since one can then readily observe the quality of the discharge. Non-sealed off lasers, using either gas circulation or else a reservoir are therefore usually made, as remarked above, from perspex. Such an assembly can easily be made sufficiently gas tight and, as a constructional refinement, it is normal to have the cavity mirrors also inside the box so only one infrared transmitting window is required. With sealed-off lasers there is no practical way of avoiding glass/metal seals unless one uses RF induction excitation and one usually then has the mirrors outside the "bottle" so that repairs to the mirrors, in the event of damage, can be more readily achieved.

When one is seeking high-power *continuous-wave* or *near continuous-wave* operation one turns instead to designs where the points (c) and (d) of the list are addressed. The two forms of the laser which have shown the most promise in this connection are the *ioniser–sustainer* type and the *fast-gas-transport* type. The ioniser sustainer is a device for getting round the difficulty that the average kinetic energy of the electrons in the discharge is about 10 eV whilst the optimum excitation energy is only 2 eV. If one drops the voltage across the tube to reduce the electron energy to a few eV or else achieves the same result by increasing the pressure so that collisions are more frequent, the arc will just go out. If, however, one injects a high-energy (100–200 keV) electron beam into the gas through a thin window (e.g. aluminium) the arc will start again and keep going just as long as the electron beam is being injected [664]. This type of laser only works in a pulsed mode because the electron-beam is pulsed but the pulse length is relatively long ($\sim$ 100 μs) so quasi-continuous wave experiments are possible. As an example, lasers have been made [665] with a total gas volume of the order of 40 litres, a gas mixture in the proportions $1CO_2$: $2N_2$: $3He$ at a total pressure close to one atmosphere and which with a 130 keV electron beam sustainer gave out 1200 J per pulse in fundamental mode and rather more in multimode operation. This is equivalent to a peak power of 24 MW! True continuous-wave operation at high powers requires the efficient depopulation of the laser lower level; in other words one needs a way round the thermal bottleneck. The most effective way of doing this is to flow the gas round the loop (or out into the open air) not gently but at near supersonic speeds. Flow at these speeds down a long tube is hardly practicable so transverse gas flow is used and the necessary rate of gas volume transfer is achieved by the use of a high capacity blower [666] and one has the usual design of fast gas-transport laser. This version has been shown to saturate at gas flow rates of the order of $2 \, m^3 \, s^{-1}$. Small fast gas-flow lasers, which are of the order of a few cubic metres in size, can give out 1 kW continuous wave at an efficiency of about 10%. Large ones can give up to 30 kW with an electrical power-in/radiant-power-out, efficiency of about 17% [667].

Almost the opposite pole to the fast-flow laser is the sequence band laser [668] where the lasing gas is deliberately allowed to get hot so that the laser can no longer work on its main transitions. Of course under these conditions no line will lase but if one separates the functions of generating the radiation and absorbing it, a very useful device results. All that is necessary to achieve this is to run the laser normally but to include within the cavity a cell filled with heated CO_2 gas. The CO_2 molecules in the cell will be transferred thermally to the state $(10^0 0)$ and $(02^0 0)$ and will hence strongly absorb the primary laser lines. However the CO_2 molecules, in the laser tube, in the $(00^0 1)$ state which now cannot decay by stimulated emission can be further excited by collision with N_2 or else with the electrons and be transferred to the $(00^0 2)$ state from which they *can* decay via the transitions $(00^0 2) \rightarrow (20^0 0)$ and $(00^0 2) \rightarrow (04^0 0)$ since the anharmonicity will slightly shift these "hot" band lines away from the fundamental frequencies where the absorption is high. One can get therefore an entirely new set of frequencies corresponding to the "hot" or "sequence" bands [669]. Sequence band lasers are particularly useful in double resonance spectroscopy where one seeks to get as close a match as possible of the pumping frequency to the transition to be pumped. If there is no suitable line for CO_2, N_2O or their isotopic varieties a change to one of the sequence bands may well provide a suitable candidate. The intracavity cell is fitted with ZnSe Brewster windows and is usually heated with an electrothermal mantle.

The waveguide principle, outlined in section 3.6.2, has also been applied to the construction of CO_2 lasers [670] since it provides a way of avoiding some of the difficulties encountered with the conventional design. Principal amongst these is that of ensuring a clean TEM_{00} mode output. The large diameter Fabry–Perot resonator has high gain and an almost unavoidable strong radial variation of temperature when the laser is running. Both of these factors encourage the generation of radiation in the transverse modes of the resonator. This can be suppressed to a certain extent by the use of carefully designed fore-optics which ensure that a TEM_{00} mode just fits into the laser tube. If, however, one wants, as one frequently does, to be able to change the CO_2 line on which the laser is operating, then one will want to have a grating as one of the Fabry–Perot mirrors and then this method of off-axis suppression is no longer available. Instead, one has to incorporate carefully designed and carefully placed limiting apertures (i.e. circular iris stops) into the cavity and whilst this works well for a single line, one runs into problems when one tries to find a combination of aperture size and aperture placing which gives good performance over a wide range of possible lasing lines. These problems arise because the mode diameter near the grating must be much less than that of the laser tube and because the mode volume is a function of the laser frequency [671]. Wave-guide lasers behave very differently because they have very narrow bore diameters and with a simpler mode structure, it is much easier to get single mode output. The basic low-loss modes of the waveguide (section 3.6.2) comprise TE(transverse circular electric), TM(transverse circular magnetic) and hybrid EH types. The CO_2 waveguide laser is so important that considerable attention has been paid to the analysis of these modes and to the combinations of the cavity dimensions, within the geometry of Fig. 3.22, which give effective coupling to the resonator modes and hence low overall loss [672]. If the radius of the waveguide is a, the radius of curvature of the mirrors R and the distance of the mirror from the waveguide exit aperture D, then, assuming that the mirrors are large enough to intercept virtually all of the diffraction pattern of the aperture, three low-loss regimes emerge [673]. The first two are not surprising since they have obvious interpretations within geometrical optics. These are (a) $D/R = 0$, that is plane mirrors butted tight up against the waveguide aperture, and (b) $D/R = 1$, that is curved mirrors set at their radius of curvature from the aperture. The third configuration, (c) $D/R = 0.5$ and $2\pi a^2/\lambda R = 2.415 \ldots$ is however not interpretable in elementary terms. The solution is unique to the EH_{11} mode which is very similar, in its amplitude and phase variations, to the lowest order Gaussian beam mode. Coupling of this resonator mode to the EH_{11} waveguide mode is optimal [674] when its beam-waist is given by $w_0 = 0.6435a$. The coupling can then be as high as 98%. From equation (2.1.16), it follows that

$$2\pi w_0^2/\lambda = R,$$

i.e.

$$2\pi a^2/\lambda R = (0.6435)^{-2} = 2.415 \ldots$$

as remarked above.

All three configurations have been used in practical waveguide lasers but the present practice seem to be polarising towards preferring (a) when high power from compact lasers is required [675] and the mode properties are only of secondary importance and to prefer (c) when a clean Gaussian TEM_{00} beam is required and the extra resonator length can be tolerated. Type (c) lasers certainly give excellent beam quality and moreover they have the advantage that this good performance is independent of output frequency.

Waveguide lasers have several other advantages over conventional designs. Thus the narrow bore facilitates cooling and the efficiency of the laser can therefore be increased. Indeed by using a dry ice/acetone coolant it is possible to increase the gain per metre by more than 50%. The narrow bore also funnels down the discharge and a stable current can be made to flow through higher gas pressures [676] than can be used with conventional laser tubes. Pressures as high as 350 torr have been used and this not only gives higher efficiency and a much wider tuning range (a few GHz) it also permits operation in a mode-locked regime where one can obtain pulse-trains with individual pulse-lengths of the order of nanoseconds. The ideal material for the narrow tubes is beryllia, but at the diameters required (~ 1 mm) this is difficult to form and finish to an acceptable accuracy. Alumina and boron nitride ceramics will both take a good polish and both have been used to make laser tubes but the present practice seems simply to prefer the use of wider (~ 4 mm) pyrex tubes. Waveguide lasers, because of their mode of operation, cannot be excited readily in the conventional longitudinal fashion unless Brewster windows are used (Fig. 6.4(a)). Several alternative solutions have therefore emerged. Thus a square-section waveguide with two opposite faces made from metal can be excited transversely [677]. Electron-beam ioniser sustainer operation is possible especially for the higher power applications, but, for the ordinary laboratory "workhorse" laser, radio-frequency excitation [678] via a coil wound round the waveguide is both clean and effective (Fig. 6.4(b)). This is used in the laser offerred commercially by Edinburgh Instruments. Because of their high-gain, waveguide lasers giving acceptable power levels can be physically small. Indeed lasers only a few cm long have been made and operated at the 2 W level. Such highly portable devices have many potential applications in industry, com-

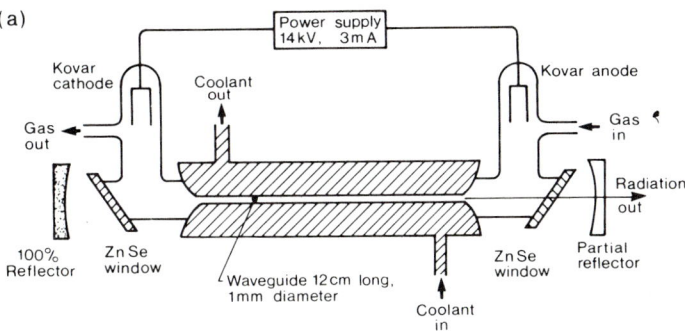

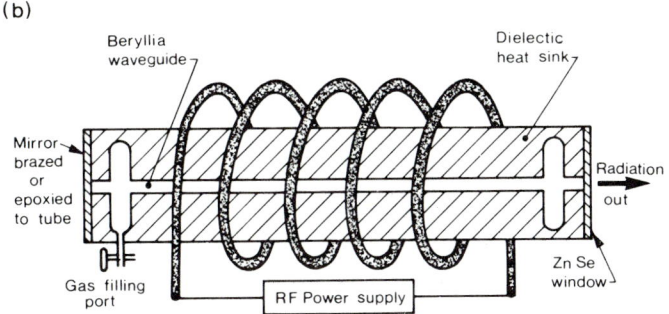

Fig. 6.4. Schematic of two possible designs for waveguide CO_2 lasers.

munications and convert surveillance, but the possible widespread availability of such potentially dangerous devices must give some cause for concern.

6.2.2.4 *Specialised designs of CO_2 laser*

The CO_2 lasers described so far are of enormous technical importance providing as they do highly flexible sources of coherent radiation for an important atmospheric window. Nevertheless there are still two drawbacks which limit the possible range of application of the laser. Firstly the emission is in the form of a set of discrete frequencies spaced by about 60 GHz and the tuning range for each is very small—some tens of MHz for conventional lasers and only of the order of a GHz or so even for waveguide lasers. Secondly, even with all the design modifications sketched out above, the available power from the laser is, for some applications, still not high enough.

The first drawback can be overcome by working at pressures of the order of 40–50 atmospheres where pressure-broadening is sufficient to bring about overlap of the lines (see Fig. 6.5). The output frequency of the laser can then be selected coarsely by the use of the intracavity etalon and finely by scanning the cavity length. One has then, of course, the same problem mentioned above, that in the high pressure required one cannot have a self-sustaining arc discharge but the solution is the same, external ionisation via a high-energy

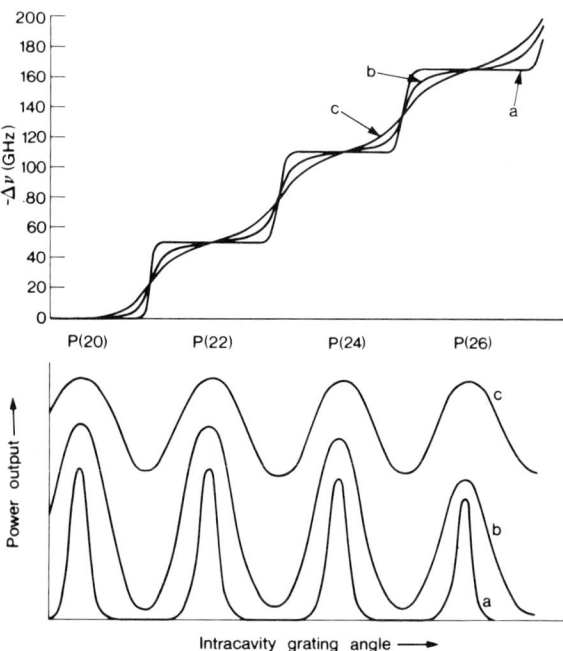

Fig. 6.5. Frequency (upper inset) and power output (lower inset) as functions of intracavity grating angle for the CO_2 laser at low (a), medium (b) and high (c) pressure.

electron beam. The beam ionises the gases and then a non-self-sustaining discharge from a subsidiary high-voltage power supply can flow [679]. The gas mixture, which is used, usually omits the helium, since this has an adverse effect on medium to high-pressure discharges.

The scanning range of this type of laser includes many fundamental absorption bands of a large number of molecules of interest and, because of this, the laser might be expected to be an attractive proposition to the infrared spectroscopist and the infrared analyst. Unfortunately the lasers which can be built or bought at the moment are too cumbersome and costly to compete in ease of operation and cost-effectiveness with conventional delay-type instruments, and the resolution which they can offer (~ 0.003 cm^{-1}) does not compare with that available from diode lasers. Nevertheless there are some important applications where a combination of tunability, coherence and high power is indispensable and for these applications the medium-pressure CO_2 laser fits the bill admirably [680]. An excellent example is provided by remote long-range monitoring of atmospheric composition [681]. The laser, being tunable, can be switched back and forth from a frequency where the atmosphere is relatively transparent to one where it displays a characteristic absorption (see section 7.2.1). Then using either a retroreflector, or more usually relying on particulate back-scatter, one can determine the amount of a given atmospheric constituent or pollutant along the line of sight. The back-scattering method is preferable since it gives both range and concentration information, but of its nature the phenomenon is weak and one needs a very powerful initial laser pulse to get an acceptable S/N for the returning pulse. The power of the medium-pressure laser is hence most welcome, but even so one really needs to use techniques such as heterodyne detection before one gets usable S/N ratios. The tunability of the laser is now a disadvantage since there is a difficulty of providing a suitable local oscillator. Some promising results have been obtained using a diode laser as the local oscillator and the combination of the two lasers could well make a major impact on tropospheric pollution monitoring [682].

When one comes to consider scaling up the CO_2 laser for really high-power operation, the problem one is faced with is that large area discharges tend to break up into a series of highly localized arcs and the efficiency falls catastrophically. For these applications one must therefore find a suitable non-electrical method of excitation. Interestingly one encounters a fortunate, if unexpected, phenomenon, namely that although *all* transitions can be inverted if one is prepared to work hard enough, some are so easy to invert that almost any effort will succeed. Happily the transitions involved in the operation of the CO_2 laser come into this "easy" class, and CO_2 lasers can be made to work with several types of non-electrical excitation. As early as 1963 [683] laser theoreticians were considering rapid gas expansion under adiabatic conditions as a means of bringing about population inversions: the idea being that supersonic adiabatic expansion is one of the most efficient ways known of suddenly chilling a gas and there might well occur differential rates of cooling for the various excited states, leading to the possibility of population inversions. Lasers based on this principle are called gas-dynamic lasers [684]. Several types have so far been made to work, but the CO_2 gas-dynamic laser is still the most important example and the one most widely used. The basic operation of the laser (see Fig. 6.6) is that the gas mixture of typically 7·5% CO_2, 91·3% N_2 and 1·2% H_2O (with perhaps some helium) is first heated to temperatures of the order of 1500 K at pressures of the order of 17 atmospheres in a reservoir or "combustion chamber". It is then very rapidly expanded via a set of specially designed nozzles or "venturi" at supersonic or hypersonic velocities down to room temperature and pressure. The population of the upper states is "frozen" at its 1500 K value for times given by the reciprocal of the vibrational relaxation time, but the lower state is much more rapidly equilibrated by the resonant collisions with H_2O molecules and a significant population inversion results. The H_2O therefore plays a crucial role in the operation of the laser. Basically what it does is to "quench" or "catalytically cool" the lower state of the lasing transition so in one way of looking at it a gas dynamic laser works by under populating its lower level rather than as the bulk of lasers do by overpopulating the upper states. The inverted gas flows transversely across the Fabry–Perot cavity and intense stimulated emission occurs on the CO_2 laser lines. The exhaust gases can either be allowed to flow out into the atmosphere, as in Fig. 6.6, or else be recirculated by means of a compressor. Both types of laser have been built and have worked very well [685]. The operation being purely thermal there are no *a priori* scaling problems and lasers of arbitrary size and power are, in principle, possible.

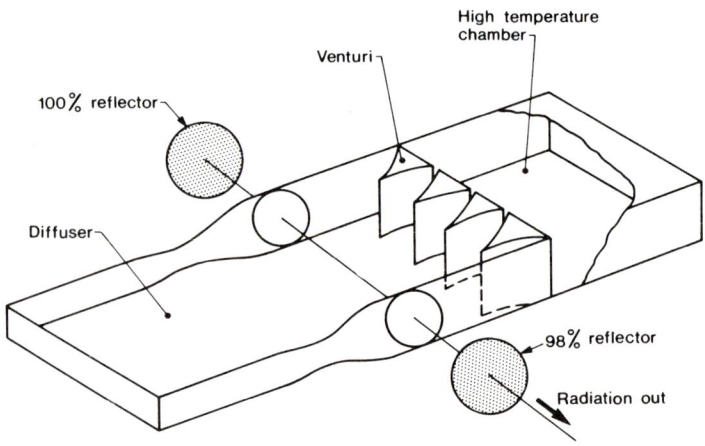

Fig. 6.6. Schematic of an open-exhaust gas-dynamic laser.

The gas dynamic laser converts thermal power directly into radiant power—it is essentially a heat engine—and it can therefore be analysed from a thermodynamic point of view. The essential results which emerge are: (a) that the laser operation will be restricted to transitions whose upper level has an absolute equivalent temperature roughly equal to that of the reservoir, and (b), provided this is true, the efficiency of the laser can be very high (about 50 % of the possible lasing energy can be extracted). The nearly resonant CO_2/N_2 levels ($\tilde{v}_0 = 2349 \, cm^{-1}$) (Fig. 6.1) have an equivalent temperature of 3·3 $\times 10^3$ K, so immediately one can see that even for a favourable case like CO_2 it is going to be difficult to get optimum reservoir temperatures without thermally decomposing the gas or causing rapid erosion of the metallic parts of the reservoir.

It is these points which indicate a practical compromise of 1500 K and which also, of course, restrict gas-dynamic lasers to the infrared. There are few, if any, primary lasers whose upper states have equivalent temperatures of 1500 K or less, so all gas-dynamic lasers operate at less than optimum performance, but even so the power levels that have been achieved are most impressive. In the mid-1960s the prototype models gave values of the order of tens of kilowatts, and by 1970 the more engineered versions were delivering 60 kW for times of the order of several milliseconds [686]. By 1973, quasi-continuous-wave operation at the 400 kW level had been reported [687], and still more powerful lasers were under construction to serve as possible drivers for laser fusion systems (see section 7.4.4). It is now abundantly clear that gas-dynamic CO_2 lasers have a dominant role to play in all situations where very high power coherent infrared radiation is required.

6.2.3 *The carbon monoxide laser*

The CO laser, which gives out a rich array of lines spanning the 5–8 μm region, has been known since 1964 [688] but its subsequent development has been rather sporadic and for much of its history it has been overshadowed by the more popular, much more easy to use and far more rapidly developing CO_2 laser. Nevertheless the need for a good coherent source, for wavelengths shorter than those supplied by the CO_2 laser, spurred on the CO technology and now this laser can be had in all the usual formats (i.e. continuous electric discharge, ioniser-sustainer pulsed discharge, TEA discharge, gas dynamic expansion, etc.) and commercial versions are starting to appear. A particularly attractive feature of the CO laser is its high-energy conversion efficiency ($> 20\%$) and its high output (powers of the order of hundreds of kilowatts continuous-wave, energies of the order of several kilojoules pulsed) [689].

The original layout [688] of the CO laser was the conventional glass tube enclosing a low pressure of CO gas through which was passed a pulsed or continuous-wave electrical discharge. It was found difficult to achieve satisfactory continuous-wave operation with this arrangement, however, and research diverted to mixing techniques [690] in which electronically excited (that is "active") nitrogen atoms, produced in a subsidiary discharge, were mixed with "cold" CO in the laser tube. These experiments revealed very clearly the importance of the gas temperature in determining which lines would oscillate. To get a wide spectral range it was found necessary to be able to operate with the laser tube temperature lying anywhere between liquid nitrogen and ambient temperature [691]. Two further versions of the laser which were subsequently developed were the CO chemical laser and the CO gas-dynamic laser. The CO chemical laser [692] usually works on a fuel of CS_2 and O_2 which upon flash photolysis gives

$$O_2 + h\nu \rightarrow O + O$$
$$CS_2 + O \rightarrow CS + SO$$
$$CS + O \rightarrow CO^* + S + 75 \text{ Kcal mol}^{-1} \qquad (6.2.3)$$

This form of the laser has been developed to the stage where high powers are available on very many lines but the overall *electrical* efficiency is low. The gas dynamic version involves, as does the corresponding CO_2 form, an initial very high stagnation pressure $(6 \cdot 8 \cdot 10^6$ Pa, i.e. $1000 \text{ lb in}^{-2})$ and temperature (2000 K). The gas is then allowed to expand at supersonic speeds through a nozzle with a high (~ 3000) area ratio. GDLs with CO as the working gas have reached power levels of the order 100 kW in "macropulses" lasting some milliseconds [693]. High power output can also be obtained from high pressure lasers in which the discharge is sustained by means of an externally injected electron beam [694]. This form operates at low temperatures in a pulsed mode which is not very convenient but the increased durability is welcome and the very high power (kilojoules per pulse) is invaluable in applications where all that is wanted is just brute power.

Useful though these variants are in their specialised applications, there is no doubt that for ordinary laboratory work, a simple electrically excited laser is very attractive. In the late 1960s, therefore, great interest was aroused by the discovery that the "thermal bottleneck" which made continuous-wave operation difficult could be bypassed by the use of small-bore laser tubes cooled to liquid nitrogen temperature and filled with special gas mixtures [695]. The subsequent introduction of xenon as a component improved matters still further and by diligent engineering research it gradually proved possible to raise the operating temperature of the laser tube to much more convenient values in the vicinity of room temperature. This form of laser is now very popular and in its commercial forms (for example Edinburgh Instruments PL3) is widely used for spectroscopy.

The observed transitions from the CO laser all occur in the ground $X^1\Sigma$ ground state, but they involve a large number of vibrationally excited states each of which contributes a few rotational transitions to the overall manifold. The result is a set of lines stretching from 4·9 to 8·3 μm [696]. Some indication of the range is given in Table 6.3. Nearly all the observed lines are P-branch transitions. The only exceptions occur in some chemical CO lasers where a few R-branch lines are observed. The inversion mechanism has its origins in the anharmonicity of the CO potential function and in the slow rate of vibrational/translational relaxation. Because of the anharmonicity (see section 5.2.3) the vibrational levels get closer and closer together as the vibrational quantum number increases. Under conditions of strong stimulation, the rapid translational/rotational relaxation leads to cool translational temperatures but the slow vibrational relaxation results in high vibrational temperatures. This sort of behaviour is not uncommon but in the case of CO and some other molecules, the detailed balancing of radiative and relaxational processes leads to a highly non-Boltzmann distribution amongst the vibrational states [697]. Treanor and his colleagues have shown [698] that the quasi-equilibrium

TABLE 6.3

Observed lines from the carbon monoxide laser

The subscript on P denotes the v″ value, the figure in brackets the J″ value

Line group	Wavenumber range (cm^{-1})	Line group	Wavenumber range (cm^{-1})	Line group	Wavenumber range (cm^{-1})
$P_3(13-15)$	2012·733 ⎱ 2004·339 ⎰	$P_{15}(05-13)$	1736·777 ⎱ 1707·877 ⎰	$P_{27}(05-11)$	1440·141 ⎱ 1421·385 ⎰
$P_4(10-18)$	1999·143 ⎱ 1965·835 ⎰	$P_{16}(06-14)$	1708·166 ⎱ 1679·289 ⎰	$P_{28}(06-11)$	1413·084 ⎱ 1397·382 ⎰
$P_5(10-17)$	1973·298 ⎱ 1944·527 ⎰	$P_{17}(07-13)$	1679·674 ⎱ 1658·212 ⎰	$P_{29}(06-10)$	1388·954 ⎱ 1376·596 ⎰
$P_6(09-17)$	1951·455 ⎱ 1981·986 ⎰	$P_{18}(08-13)$	1651·303 ⎱ 1633·530 ⎰	$P_{30}(06-10)$	1364·889 ⎱ 1352·674 ⎰
$P_7(09-16)$	1925·711 ⎱ 1897·663 ⎰	$P_{19}(06-13)$	1633·320 ⎱ 1608·909 ⎰	$P_{31}(06-11)$	1340·877 ⎱ 1325·697 ⎰
$P_8(08-16)$	1903·877 ⎱ 1872·231 ⎰	$P_{20}(06-12)$	1608·531 ⎱ 1587·917 ⎰	$P_{32}(07-11)$	1313·967 ⎱ 1301·891 ⎰
$P_9(08-16)$	1878·258 ⎱ 1846·918 ⎰	$P_{21}(05-11)$	1587·106 ⎱ 1566·902 ⎰	$P_{33}(06-11)$	1292·958 ⎱ 1278·132 ⎰
$P_{10}(08-15)$	1852·711 ⎱ 1825·620 ⎰	$P_{22}(05-10)$	1562·459 ⎱ 1545·881 ⎰	$P_{34}(06-10)$	1269·050 ⎱ 1257·385 ⎰
$P_{11}(08-15)$	1827·240 ⎱ 1800·402 ⎰	$P_{23}(05-13)$	1537·886 ⎱ 1511·230 ⎰	$P_{35}(06-10)$	1245·167 ⎱ 1233·642 ⎰
$P_{12}(07-15)$	1805·509 ⎱ 1775·255 ⎰	$P_{24}(05-12)$	1513·388 ⎱ 1490·430 ⎰	$P_{36}(07-10)$	1218·489 ⎱ 1209·892 ⎰
$P_{13}(07-15)$	1780·179 ⎱ 1750·204 ⎰	$P_{25}(04-13)$	1492·080 ⎱ 1462·874 ⎰		
$P_{14}(06-14)$	1758·494 ⎱ 1729·053 ⎰	$P_{26}(06-12)$	1461·510 ⎱ 1442·141 ⎰		

population distribution in an anharmonic oscillator is given by

$$N(v+1) = N(v)\exp[2v\omega_e x_e hc/kT - (\omega_e - 2\omega_e x_e)hc/kT_{\text{vib}}], \quad (6.2.4)$$

where ω_e and $\omega_e x_e$ are the usual molecular constants (see section 5.2.3) and T_{vib} is the vibrational temperature characterising the population of the $v = 1$ level. When $\omega_e x_e$ is zero, this reduces to the conventional form but when $\omega_e x_e$ is not zero, it follows that beyond a certain value of v, the vibrational population will inevitably become inverted. This conclusion depends on the assumption of lack of efficient energy deactivation processes which cannot, of course, be true. But for a significantly anharmonic oscillator under conditions of strong excitation we will expect to get the population pumped higher and higher up the rungs of the vibrational ladder [699]. One would not in the real case expect to get a total inversion so the whole band will not lase but over limited ranges of J there may be inversion for P-branch transitions and these lines *will* lase. Polanyi [700] has shown that the condition for inversion on a P-branch line is,

$$T_{\text{rot}} < 2(J+1)(B_e/\omega_e)T_{\text{vib}}. \quad (6.2.5)$$

This can be confirmed qualitatively by noting that in the extreme case where T_{rot} is zero but T_{vib} is greater than zero, all the molecules must be in $J = 0$ states and hence that the P(1) lines must all be inverted. By the same argument it follows that R-branch lines can only be inverted when there is an overall vibrational inversion and this, as mentioned above, will only happen rarely.

The availability of so many transitions covering nearly the whole range of vibration/rotation states within the ground electronic state has given spectroscopists a splendid opportunity for testing the validity of the anharmonic oscillator/non-rigid rotor approach to be interpretation of molecular spectra. Thus one can take any of the P-branches and fit them to the theoretical expression

$$\tilde{v}_{\text{obs}} = \tilde{v}_0 - (B' + B'')J'' + (B' - B'' - D' + D'')(J'')^2 + 2(D' + D'')(J'')^3$$
$$- (D' - D'')(J'')^4 + \ldots \quad (6.2.6)$$

If one does this for say the $v = 4 \to v = 5$ P-branch, the result is

$$\tilde{v}_{\text{obs}} = 2037 \cdot 7482 - 3 \cdot 6876J'' - 0 \cdot 0175(J'')^2 + (12 \cdot 244 \times 10^{-6})(J'')^3$$
$$+ \{0\}(J'')^4 \quad (6.2.7)$$

from which one has $B'' = 1 \cdot 8526 \text{ cm}^{-1}$, $B' = 1 \cdot 8351 \text{ cm}^{-1}$, $D' \sim D'' = 6.12 \times 10^{-6} \text{ cm}^{-1}$. Roh and Rao [701] have carried out analyses of this nature for all the observed bands and have applied a computer program to give the best overall fit. Their result for the overall energy function,

$$T(v, J) = \omega_v(v + \tfrac{1}{2}) + B_v J(J+1) - D_v J^2(J+1)^2 + H_v J^3(J+1)^3 - \ldots$$
$$(6.2.8)$$

where

$$\omega_v = \omega_e - \omega_e x_e (v + \tfrac{1}{2}) + \omega_e y_e (v + \tfrac{1}{2})^2 - \omega_e z_e (v + \tfrac{1}{2})^3$$
$$+ \omega_e a_e (v + \tfrac{1}{2})^4 - \omega_e b_e (v + \tfrac{1}{2})^5 + \ldots$$
$$B_v = B_e - \alpha_e (v + \tfrac{1}{2}) + \gamma_e (v + \tfrac{1}{2})^2 - \ldots$$
$$D_v = D_e - \beta_e (v + \tfrac{1}{2}) + \varepsilon_e (v + \tfrac{1}{2})^2 - \ldots$$
$$H_v = H_e - \delta_e (v + \tfrac{1}{2}) + \ldots \tag{6.2.9}$$

leads to the numerical constants

$$\omega_e = 2169 \cdot 8173, \qquad \omega_e x_e = 13 \cdot 289545, \qquad \omega_e y_e = 1 \cdot 06521 \times 10^{-2},$$
$$\omega_e z_e = 5 \cdot 37 \times 10^{-10}, \qquad \omega_e a_e = 1 \cdot 1486 \times 10^{-6}, \qquad \omega_e b_e = 3 \cdot 3124 \times 10^{-8},$$
$$B_e = 1 \cdot 931284, \qquad \alpha_e = 1 \cdot 7510 \times 10^{-2}, \qquad \gamma_e = 1 \cdot 72 \times 10^{-6},$$
$$D_e = 6 \cdot 1258 \times 10^{-6}, \qquad \beta_e = 2 \cdot 9 \times 10^{-9}, \qquad \varepsilon_e = 4 \cdot 3 \times 10^{-10},$$
$$H_e = 7 \cdot 7 \times 10^{-12}, \qquad \delta_e = 3 \cdot 4 \times 10^{-13}, \tag{6.2.10}$$

all in units of cm^{-1}.

Some results for some of the lower-lying levels of CO are given in Table 6.4. These constants agree well with those derived in the earlier work of Mould, Price and Wilkinson [702] who studied the spontaneous emission from a radio-frequency discharge through CO. Either from the constants in equation (6.2.10) or else from inspection of the smooth trends in Table 6.4 it will be seen that the traditional approach gives an excellent account of the observed behaviour and the fact that one can account for virtually all the levels of the ground state with relatively few constants underlines this. The CO laser is

TABLE 6.4

Vibration/rotation constants for the first six levels of the carbon monoxide molecule

v	T_v	$\tilde{v}_0(v)$	B_v	D_v	H_v
0	1 081·587 621		1·922 529	$6 \cdot 1244 \times 10^{-6}$	$7 \cdot 53 \times 10^{-12}$
		2143·273 127			
1	3 224·860 748		1·905 022	$6 \cdot 1224 \times 10^{-6}$	$7 \cdot 19 \times 10^{-12}$
		2 116·791 359			
2	5 341·652 107		1·887 519	$6 \cdot 1212 \times 10^{-6}$	$6 \cdot 85 \times 10^{-12}$
		2 090·375 884			
3	7 432·027 991		1·870 020	$6 \cdot 1208 \times 10^{-6}$	$6 \cdot 51 \times 10^{-12}$
		2 064·027 832			
4	9 496·055 823		1·852 523	$6 \cdot 1213 \times 10^{-6}$	$6 \cdot 17 \times 10^{-12}$
		2 037·748 257			
5	11 533·804 080		1·835 031	$6 \cdot 1227 \times 10^{-6}$	$5 \cdot 83 \times 10^{-12}$
		2 011·538 141			
6	13 545·342 220		1·817 541	$6 \cdot 1250 \times 10^{-6}$	$5 \cdot 49 \times 10^{-12}$

therefore not only a most useful source for the mid infrared, it provides us also with perhaps our best example of a fully analysed molecular vib/rotor.

The possible uses for CO lasers are rapidly increasing. One very important area is in the machining of metals where a short wavelength is desirable and a high-power CO laser would therefore be preferred to a similar power CO_2 laser. Spectroscopic applications are not so plentiful as for the CO_2 case because molecular fundamentals in the 5–6 μm region are relatively scarce. However, one good practical application is in the monitoring of NO and N_2O in the atmosphere [703]. The CO laser bands do overlap the 6.3 μm band of water vapour so some of the lines are strongly absorbed in the atmosphere but the laser can be constrained to give out only the non-attenuated lines by including within the cavity a cell filled with water vapour. The power of the CO laser can be turned to advantage in applications where CO radiation is used to pump a secondary laser. Its use as the pump for a spin-flip Raman laser will be mentioned later.

6.2.4 *Glow-discharge lasers*

Several molecular species of which the best known examples are H_2O, D_2O, HCN, DCN, NH_3, OCS, H_2S, and SO_2, when enclosed at a low pressue in a Fabry–Perot cavity and subjected to a glow discharge, give out stimulated emission in the long-wave infrared [704]. The mechanism for the laser action depends on vibrational/rotational perturbation as discussed in section 2.5. These molecular species are seldom stable under the discharge conditions, where the average energy of the electrons (~ 10 eV) is considerably more than the binding energy of the molecule, so the lasers have to be operated in a continuous flow mode. Indeed the electrical decomposition is so serious that the H_2S (main lines at 225·3, 162·4, 103·3, 87·47, 80·50, 61·50, 60·29 and 33·64 μm), the OCS (lines at 132 and 123 μm), and the SO_2 (strong lines at 140·85 and 192·67 μm) lasers can only be operated pulsed and then with a very low duty cycle (~ 2 Hz). For these reasons they have not found many technical uses. The main technical glow-discharge lasers are the hydrogen cyanide, the water vapour and to a lesser extent the ammonia laser.

The HCN laser which gives out lines at 337 μm (890·7607 GHz), 335 μm (894·4142 GHz), 327 μm (964·3134 GHz) and 310 μm (967·7658 GHz) is widely used as a far-infrared source, mostly because of the simplicity of the laser construction, the convenience of direct electrical excitation and the relative transparency of the atmosphere at its operating wavelengths. It has found applications in dielectric measurements [705] in metrology [706], in fundamental standards work [707], in plasma diagnostics [708] and in high-resolution spectroscopy [709]. The DCN laser gives out two strong lines at 189·95 and 194·70 μm whose origin is thought to lie [710] in perturbations between the (22^00) and (09^10) vibrational levels near $J = 21$. The corresponding sequence transitions at 190·01 and 194·76 μm are also radiated under strong excitation conditions. The DCN laser, however, is restricted to specialised uses, when compared with the HCN version, because its emission lines are absorbed by the atmosphere.

The original HCN and DCN lasers [711] were relatively low-power devices (up to 10 mW maximum) but modern versions (Fig. 6.7) can give out up to 250 mW [712]. Waveguide versions can reach similar power levels. The gas mixture used is usually $N_2 + CH_4$ (or CD_4) + He in the volume ratio $1:3:12$. Under the discharge conditions HCN (or DCN) is formed by a variety of reactions, for example

$$N_2^* + CH_4 \rightarrow HCN^* + NH_3$$

or else

$$2N^* + 2CH_4 \rightarrow 2HCN^* + 3H_2 \qquad (6.2.11a)$$

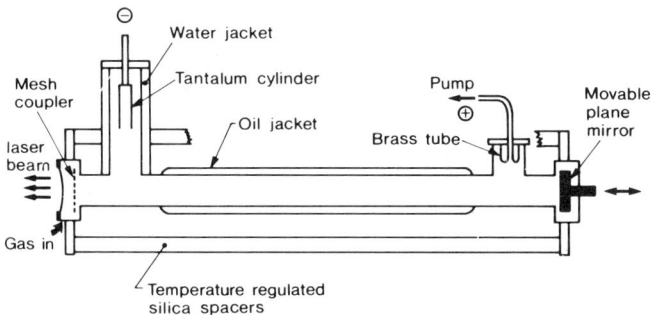

Fig. 6.7. Constructional details of a modern high-power HCN laser.

Either way the highly excited HCN molecule relaxes back to the ground state down the various vibrational ladders and if it passes through the perturbed levels it will radiate in stimulated emission. Clearly this is not a very favourable process since the majority of paths back to the ground state do not pass through these few states. Some recycling of the energy must take place or else there must be resonances with the electrons in the discharge if the moderately high small-signal gain (~ 0.1 m^{-1}) and the correspondingly fair to medium observed power from so low pressure a gas ($\sim$ few torr) is to be adequately explained. The recycling cannot go on for very long because HCN is itself unstable in an electric discharge. One of the products of its decomposition is a rather mysterious "polymer", which is coloured between yellow and dark brown depending on the conditions of its formation and which tends to get deposited heavily on the relatively cool glass tube. It is also formed to a significant extent on the laser mirrors. Its presence on both causes substantial losses and eventually if the layer gets too thick the laser will not work at all. The polymer can be removed mechanically but that involves the dismantling of the laser. It can also be removed by running the laser tube hot ($\sim 160°C$) but that is obviously not acceptable if rubber 0-rings are being used. A good solution seems to be to every now and again run the discharge with an oxygen plus water vapour "fuel". The oxygen and hydrogen atoms produced in this way gradually convert the polymer to gaseous products. The HCN laser can be run in either a pulsed mode or else continuous-wave though nowadays pulsed operation is becoming rare. The discharge current lies between 1 and 2 A and the voltage across the tube is between 2 and 4 kV. This presents an obvious safety hazard since were one to touch either of the metal end-pieces one might be electrocuted as the discharge would switch to a path to earth through the end-piece and through its toucher! Hot cathodes (usually tantalum) in side-arms are often used to increase the electron flow and to reduce the striking potential but this is not essential. For reasons spelled out in Chapter 2 it is necessary to be able to adjust the length of a far-infrared laser cavity so as to ensure a match of the resonator with the molecular resonance. This is usually achieved by mounting one of the mirrors on a micrometer spindle. Cavity-length stabilisation is also sought by the use of invar or else temperature regulated silica connecting rods.

The H_2O laser gives out a large number (~ 60) of possible lines but its two most widely used ones arc at 28 μm (10·718 073 THz) and at 119 μm (2·527 953 THz). The origin of these lines [82] has been shown earlier in Fig. 2.38. The basic process stems from perturbation of rotational levels of the states (100) and (001) by the overtone of the bending fundamental (010), i.e. (020). The reason for the strong interaction is that the bending frequency is almost exactly half the frequencies of either the symmetric or the antisymmetric OH stretches which are themselves virtually degenerate. The potential energy function is markedly anharmonic so strong Fermi resonance results but there is also Coriolis interaction. The Fermi interaction accounts for the (020), (100) interaction whilst the Coriolis interaction accounts for that btween (020) and (001). The H_2O laser is very convenient to operate since there are no deposition problems and the rate of decomposition of the H_2O molecules is slow. The D_2O version gives out far fewer lines and only four (at 171·67, 107·731, 84·2791 and 71·944 μm) can be had continuous-wave [713]. Nevertheless these lines do lie in regions where other sources are not available so the D_2O laser certainly has its uses. Thus Uehara and his colleagues [714] have used the 171·67 μm line to study the lower frequency component of the $J = 2 \to J = 3$ pure rotation line of NH_3.

Ammonia itself gives out several lines in glow-discharge operation [715]. The principal lines are at 32·13, 31·951, 25·12, 24·918, 23·86, 23·675, 22·71, 22·563, 15·47 and 14·78 μm. Some of these lines have been shown [716] to be members of the P-branch of the transition $v_2 = 3^s \to v_2 = 2^a$ (see Fig. 5.9). The upper level ($v_2 = 3^s$) has a total energy of 2384 cm^{-1} which makes it nearly resonant with N_2 so there is a strong possibility that the laser operates by collisional energy transfer as in CO_2. The N_2 molecules would be produced by the decomposition of the ammonia molecules by the electronically catalysed inverse of the Haber process, i.e.

$$NH_3 + e \rightarrow NH_3^* + e$$

$$NH_3^* + NH_3 \rightarrow N_2^* + 3H_2 \qquad (6.2.11b)$$

The remaining lines probably arise from collisional cascades in which levels remote from those initially inverted are reached and overpopulated by means of a series of collisions. The technical value of the NH_3 laser is that its lines nicely span the gap between the H_2O lines and the CO_2 lines but nowadays with so many optically pumped lines available in the same region the ammonia laser is tending to dwindle in importance.

6.2.5 Infrared optically pumped lasers

6.2.5.1 Introduction—continuous-wave far-infrared lasers

An attractive and in practice very efficient way of producing a population inversion is to take a three-level system $\mathscr{E}_1 < \mathscr{E}_2 < \mathscr{E}_3$ where all three levels are connected by allowed transitions and apply to it intense monochromatic radiation tuned to the transition frequency $v_{13} = (\mathscr{E}_3 - \mathscr{E}_1)/h$. This transition will then become partly saturated and the population of level $\mathscr{E}_3$ will become highly suprathermal. The intermediate state $\mathscr{E}_2$ will be unaffected by the pumping and will have its normal ambient population so an inversion will result between $\mathscr{E}_3$ and $\mathscr{E}_2$. Laser action on the allowed transition $v_{32} = (\mathscr{E}_3 - \mathscr{E}_2)/h$ then becomes possible provided the pumped medium is enclosed within a suitable Fabry–Perot resonator. Lasers based on this mode of operation are referred to as "optically pumped" to distinguish them from those primary lasers which are pumped by broad-band radiation from a flashtube and from those where the "pumping" is by collisional transfer, electron bombardment or chemical reaction. Any laser can be used for optical pumping provided it is powerful enough and provided it has a coincidence with an absorption line of a suitable molecule but only the CO_2 laser, its isotopic variants and its isoelectronic forms (e.g. N_2O) are ever used in practice. Likewise any absorptive system could be pumped but only gases in fact are used because pressure variations provide a practical possibility for adjusting the relaxation rates, so that steady-state inversions can be maintained. For these reasons nearly all present-day optically pumped lasers work in the long-wave infrared region and involve low-pressure gases pumped by the CO_2 laser. The population inversion mehanism for a typical far-infrared optically pumped laser has been outlined previously in section 2.4.3.

Because of the dominant role of the CO_2 laser, optical-pumping practitioners have evolved simple codes to designate uniquely which CO_2 line is being used as the pump. The 9.4 or 10.4 μm bands are denoted by just a 9 or a 10 and the individual line in the band is specified by the conventional spectroscopic shorthand. Thus 10 P(18) means the transition from $J'' = 18$ to $J' = 17$ in the 10·4 μm band. From Table 6.1 its wavelength would be 10·571 036 μm. The same sort of notation is used to denote the isotopic species: thus $^{17}O^{13}C^{18}O$ could be written (17 13 18) or, deleting the redundant digits, (738). The common species is then (626). More than 1300 CO_2/N_2O optically pumped laser lines are now known [85, 717]. Some of these which have proved particularly valuable in practice are listed in Table 6.5.

TABLE 6.5

A listing of some of the stronger continuous-wave optically pumped laser lines in the far-infrared and millimeter-wave regions

Nominal wavelength μm	Emitting molecule	Measured frequency THz	Pump line	Power available mW
37·5	CH_3OH	8·0	CO_2 9P32	~10
41	CD_3OD	7·3	CO_2 10R18	~60
42·2	CH_3OH	7·109 810	CO_2 9P32	~10
46·7	CH_3OD	6·42	CO_2 9R8	~10
57·0	CH_3OD	5·24	CO_2 9R8	~10
64·0	CH_3OH	4·68	CO_2 9R18	⪢20
67	NH_3	4·47		~10
65·6	CH_3OH	4·60	CO_2 9P34	~10
70·5	CH_3OH	4·251 673	CO_2 9P34	~100
80·6	CH_3OH	3·72	CO_2 (sequence)	~10
81·5	NH_3	3·678	N_2O 10P13	~100
96·5	CH_3OH	3·105 937	CO_2 9R10	>10
103·1	CH_3OD	2·907 089	CO_2 9P30	~10
112·6	D_2O	2·662	CO_2 (sequence)	~10
117·2	CH_3OD	2·557 365	CO_2 9P26	>10
118·8	CH_3OH	2·522 782	CO_2 9P36	>100
121·7	CH_2F_2	2·463	CO_2 9R22	~10
147·8	CH_3NH_2	2·027 753	CO_2 9P24	~10
152·9	$^{15}NH_3$	1·961	$^{13}CO_2$ 10R18	~200
163·03	CH_3OH	1·838 839 3	CO_2 10R38	~10
164·8	CH_3OH	1·819 315	CO_2 9R10	~10
165·9	CH_2F_2	1·807	CO_2 9R20	< 100
170·6	CH_3OH	1·757 526 3	CO_2 9P36	~10
186·11	CH_3OH	1·610 80	CO_2 9R18	~10
189·80	CD_2F_2	1·579 250	CO_2 10R34	~10
192·78	CH_3F	1·555 1	CO_2 10R32	~10
198·0	CH_3NH_2	1·514 0	CO_2 9P38	~10
198·80	CH_3OH	1·508 0	CO_2 9P38	~10
206	CD_3F	1·46	?	~10
214·7	CH_2F_2	1·396		~10
218·0	CH_3NH_2	1·375 0	CO_2 9P24	~10

TABLE 6.5 (*Contd.*)

Nominal wavelength μm	Emitting molecule	Measured frequency THz	Pump line	Power available mW
229·10	CH_3OD	1·309·0	CO_2 9P6	~ 10
235·5	CH_2F_2	1·273	CO_2 9R6	~ 10
237·60	CH_3OH	1·261 80	CO_2 9P34	~ 10
247	CD_3F	1·21		~ 10
255	CD_3OD	1·18	CO_2 10R36	~ 20
293·6	CD_3Cl	1·020 925	CO_2 9P24	~ 10
297·0	CD_3OH	1·009 0	CO_2 9R18 10R36	~ 10
304·1	DCOOD	0·985 889 7	CO_2 10R24	< 10
326	$^{13}CD_3F$	0·920		~ 10
349·3	CH_3Cl	0·858 2	CO_2 10R18	~ 10
372·7	CH_3F	0·804 4	CO_2 9P50	~ 10
375	$^{15}NH_3$	0·799	CO_2 10R42	~ 20
375·54	CF_2CH_2	0·798 287 0	CO_2 10P12	~ 10
380·6	DCOOD	0·787 7	CO_2 10R12	~ 10
383·3	CD_3Cl	0·782 167	CO_2 9R34	~ 10
392·07	CH_3OH	0·764 642 7	CO_2 9P36	< 10
393·6	HCOOH	0·761 607 7	CO_2 9R18	< 10
394·20	HCOOH	0·761 0	CO_2 9R16	< 10
405·58	HCOOH	0·739 161 0	CO_2 9R10	< 10
418·61	HCOOH	0·716 256 8	CO_2 9R22	> 1
418·7	CD_3OH	0·715 987 6	CO_2 10R36	< 10
432·63	HCOOH	0·692 951 4	CO_2 9R20	> 1
432·67	HCOOH	0·692 895 0	CO_2 9R20	> 1
433·2	CDF_3	0·692 ·0		~ 10
447·14	CH_3I	0·670 463	CO_2 10P18	> 10
449·8	CD_3Cl	0·666 502	CO_2 10R20	~ 10
458·00	CHF_2CH_3	0·651 0	CO_2 10P20	~ 10
460·6	CD_3I	0·650 928	CO_2 9R12	~ 10
461·3	HCOOD	0·649 941	CO_2 10P16	~ 10
461·4	$^{13}CH_3OH$	0·649 766 7	CO_2 9P12	< 1
464·8	CD_3Cl	0·645 052	CO_2 10R20	~ 10
469·02	CH_3OH	0·639 184 6	CO_2 10R38	~ 10
470	$^{13}CD_3F$	0·638		> 1
488·2	CDF_3	0·614 1	CO_2 10R38	> 1
490·4	CD_3I	0·611 334	CO_2 9R22	~ 10
494·6	CH_3CN	0·606 074 7	CO_2 9P6	< 10
496·07	CH_3F	0·604 333 0	CO_2 9P20	> 1
496·10	CH_3F	0·604 373	CO_2 9P20	> 1
500·60	CD_2F_2	0·598 9	CO_2 10R24	> 5
508·37	CH_3I	0·589 7	CO_2 9P34	~ 10
508·37	CH_3I	0·589		
			CO_2 9P34	~ 10
513·02	HCOOH	0·584 388 2	CO_2 9R28	< 10
529·3	CH_3I	0·566 40	CO_2 10P36	< 10
531·1	CH_3Br	0·564 5	CO_2 10P24	< 10

TABLE 6.5 (Contd.)

Nominal wavelength μm	Emitting molecule	Measured frequency THz	Pump line	Power available mW
533·0	CHF_2CH_3	0·560 0	CO_2 10P20	~10
545·2	$CH_3{}^{81}Br$	0·549 87	CO_2 10P38	~10
545·4	CH_3Br	0·549 68	CO_2 10R32	~10
548·7	CD_2F_2	0·546 4	CO_2 10R28	~3
553·0	CD_3OH	0·542	CO_2 10R08	<10
561·3	DCOOD	0·534 109 6	CO_2 10P20	>1
567·9	DCOOD	0·527 926 0	CO_2 10R26	<10
570·67	CH_3OH	0·525 427 5	CO_2 9P16	~40
599·5	CD_3I	0·500 029 0	CO_2 10R22	~10
614·1	CD_3I	0·488 174	CO_2 10R22	~10
635·4	C_2H_3Br	0·471 850 5	CO_2 10R26	~1
647·9	CH_3CCH	0·462 720	CO_2 10P14	~1
658·5	CDF_3	0·455 3	CO_2 10R10	>3
662·82	CH_2CF_2	0·452 301 5	CO_2 10P24	~3
699·42	CH_3OH	0·428 628 5	CO_2 9P34	<10
747	CH_3Br	0·401	?	~10
764·1	CH_2CF_2	0·392 00	CO_2 10P10	~10
871·6	CD_3OH	0·343 962 4	CO_2 10R18	<10
890·11	CH_2CF_2	0·336 800	CO_2 10P22	~10
926·2	HCOOD	0·323 677	CO_2 10R14	~10
990·00	CH_2CF_2	0·302 800	CO_2 10P22	~10
1020·00	CH_2CF_2	0·294 000	CO_2 10P14	~10
1063·29	CH_3I	0·281 950	CO_2 10P38	<1C
1221·8	$^{13}CH_3F$	0·245 372	CO_2 9P32	~1·4
1223·66	CH_3OH	0·244 996 9	CO_2 9P16	~10
1253·74	CH_3I	0·239 118 9	CO_2 10P32	~1
1572·6	$CH_3{}^{79}Br$	0·190 63	CO_2 10P04	>1
1886·9	CH_3Cl	0·158 883	CO_2 9P26	>1
2650	$CH_3{}^{81}Br$	0·113 1	CO_2 10P10	~0·1

The operation of continuous-wave optically pumped lasers can be analysed in terms of the balance between the radiative and the relaxational processes taking place in the active medium. To full rigour this requires the use of a density-matrix formalism [718] but under the conditions of practical oper- ation where the gain in the pumped laser is high and the pumping power is likewise high, a very much simpler formalism, almost identically equivalent to that used for analysing chemical kinetics, may be used. Because of the similarity it is usually referred to as the "rate-equation" approach [719]. The treatment starts with the equation

$$\left(\frac{P_0}{V}\right) = \left(\frac{T_0}{A_0 + T_0}\right)\alpha_0 I_0^s \qquad (6.2.12)$$

given by Hodges [720] for the volumetric power generation in a laser operating well above threshold. In this equation, the subscript 0 refers to the output wavelength, V is the volume, T the output mirror transmission, A the cavity losses, α_0 the small signal gain and I_0^s the saturation parameter. For a typical laser where $T/(A + T)$ might be 0·5, α_0 might be between 0·1 and 1·0 per metre and I_0^s might lie between 1 and 100 mW cm^{-2}; this equation indicates an output power lying between 0·5 and 500 mW l^{-1}. Clearly, therefore, one wants rather large volumes to get significant power out but, as will be seen later, this requirement does tend to conflict with other design criteria. The next step is to use the rate equations to calculate the form of the two parameters in (6.2.12) so that their dependence on the operating conditions of the laser can be investigated. The result given by Hodges [720] is

$$\left(\frac{P_0}{P_p}\right) = \frac{1}{(1 + g_i/g_k)}\left(\frac{v_0}{v_p}\right)\left(\frac{T_0}{A_0 + T_0}\right)\left(\frac{A_p}{A_p + \alpha_p}\right)\left[1 - \frac{hv_0}{kT}f_j\frac{\tau_v}{\tau_J}\right] \quad (6.2.13)$$

where the subscript p refers to the pump, g_i and g_k are the degeneracies of the upper and lower radiating levels, α_p and A_p are respectively the absorption coefficient and the cavity loss at the pump frequency and the factor in square brackets represents the effects of the molecular dynamics. The first two terms are Manley–Rowe factors which are obviously constant for any chosen system whilst the second two represent cavity efficiency factors.

The second of these factors highlights the problem in getting effective absorption of the pump power, a difficulty stemming from the non-tunability of infrared pumps. The microwave spectroscopist has intense *tunable* pumps so microwave–microwave double resonance is a routine procedure [144]. In the infrared, the pumping lasers are fixed frequency devices so coincidences with pumpable absorption lines can only arise fortuitously. Fortunately there are several favourable factors which greatly increase the chance of an accidental coincidence. These are (a) the magnitudes of the observed force constants and of the effective masses of the vibrating entities in the molecules ensure that nearly all small organic molecules will have at least one fundamental in the 10 μm region; (b) the discrete absorption in this region for polyatomic molecules is greatly enhanced by the presence of sum, difference, combination and hot bands all with measurable intensity; (c) all of these bands have a rich rotational structure and this can become even more densely packed in the case of non-rigid molecules such as methanol which have internal rotational degrees of freedom; (d) the CO_2 laser itself has many lines and this number can be increased greatly by the use of either isotopically varied forms or else by sequence band operation; (e) one can also use N_2O as the fuel and here the absence of nuclear statistic selection rules leads to a line-spacing only half that observed with the CO_2 laser. These factors combine to produce a situation where even for small rigid molecules it is very rare to find no close coincidences [721]. However, the coincidences which are observed are seldom exact so one has to investigate the experimental consequences of off-resonant pumping. There are two cases to consider: firstly continuous-wave operation where the maximum permissible frequency offset is of the order of the line width and secondly pulsed operation where, for reasons to be given later, much larger offsets can be tolerated. The continuous-wave lasers work at very low pressures (see later), of the order 300 mtorr, so pressure broadening is negligible and the observed line width will be due almost entirely to Doppler broadening. This for a small molecule in the 10 μm region will be about 50 MHz and, since the average offset will be of this order or larger, the typical pump line will lie in the wings of the line to be pumped. The absorption coefficient α_p will be correspondingly small and the low number density of molecules in the low-pressure gas will make it, of course, still smaller. The best one can hope for is $\alpha_p \sim 1 \text{ m}^{-1}$ but observed values are often smaller so to get reasonable absorption one must use multiple reflection techniques in order that the effective length d which appears in equation (6.2.13) will be larger than the physical length of the cavity. In the far-infrared optically pumped lasers this is a fairly straightforward affair since the gas does not significantly absorb its own emission line (i.e. A_0 is small) but for mid-infrared optically pumped lasers where $\mathscr{E}_2$ lies in the ground rotational manifold and the emission line is significantly absorbed, different considerations apply and it may prove necessary to work in the saturation regime. The rate–equation approach indicates that in many cases the efficiency of optically pumped lasers will be high. Typical values for the efficiency within the cavity would be 10% but in favourable cases this can range as high as 36%. The external efficiency will be less because of the windows but values as high as 20% are nevertheless possible. When one bears in mind the high power of the CO_2 pump, it will be realised that optically pumped lasers are by far the brightest sources available in the far infrared. It is hardly surprising therefore that they hold so dominant a position in long-wave science and technology.

Two possible designs for far-infrared optically pumped laser cavities are shown in Fig. 6.8. In the first [722], a quasi-parallel beam from the CO_2 pump laser is focussed through a perforated metal reflector onto the circular aperture of a hole-coupled FIR cavity which contains the lasing gas. The walls of the glass or metal partial waveguide and the remote mirror then serve to provide an effective absorbing path length for the pump beam which is several times the physical length of the cavity. It is quite easy to transmute this basic design to yield a fully wave-guided laser [723]. The second design shown in Fig. 6.8(b) gives a still larger absorbing path length. It is sometimes called the "zig-zag" laser [724] and it has been especially used for applications where very high power pulses are required. In these one can use a master oscillator followed by a similarly pumped synchronous amplifier. The best known example is provided by the use of pulsed optically pumped lasers containing D_2O or CH_3F [725] to provide FIR lines for Thomson scattering measurements on thermonuclear plasmas (see section 7.4.4.).

All these efforts to improve cavity performance will be, of course, in vain unless the final term in equation (6.2.13) is positive so this factor has to be carefully considered. The quantities in it not so far defined are f_J which is the Boltzmann factor for the upper level, τ_{vib} the vibrational relaxation time and τ_J

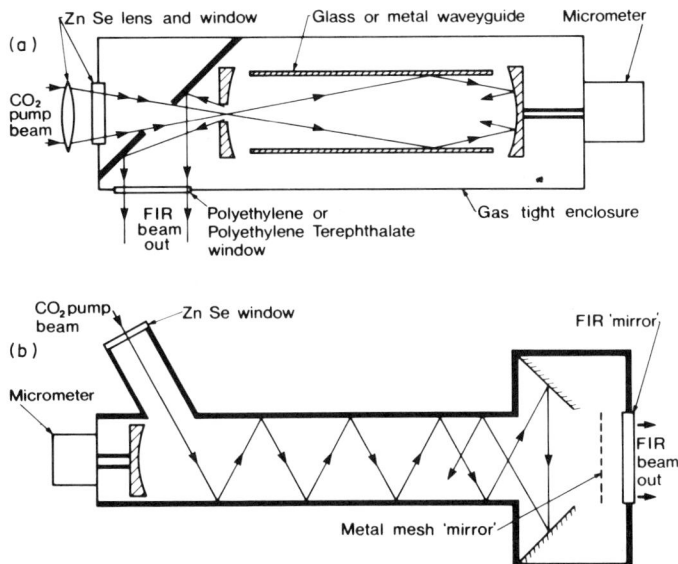

FIG. 6.8. Two designs for efficient far-infrared optically pumped cavities.

the rotational collision relaxation time. This latter is directly related to the homogeneous line width Δv_h and hence the pressure by [720]

$$\tau_J = (\Delta v_h)^{-1} = (\sigma p)^{-1}, \tag{6.2.14}$$

where σ is of the order 40 MHz torr^{-1} for the highly polar molecules used in optically pumped lasers. The physical significance of the ratio of the two relaxation times is that one wants a long rotational relaxation time to slow down thermalisation of the upper state rotational levels and a short vibrational time to relieve the thermal "bottleneck" by depopulating the lower level. Vibrational relaxation times are unfortunately usually rather long and that which is significant in the present context is substantially the time for the molecules to diffuse out of the beam region and be de-excited at the walls. In this case one has

$$\tau_{vib} \propto pr^2, \tag{6.2.15}$$

where r is the laser tube diameter. It follows that the positive gain condition sets an upper limit to the pressure which is of the form

$$p_c \propto r^{-1}. \tag{6.2.16}$$

For practical systems this pressure limit lies between 30 and 300 mtorr. It is obviously very desirable to use laser tubes as narrow as possible which is one of the reasons for the popularity of the waveguide configuration. The other is that hybrid output couplers [726] can be used to provide excellent mode control. They also permit the use of somewhat higher pressures but even so continuous-wave optically pumped lasers always have to work in a very low pressure regime.

Pulsed lasers can work at more conventional pressures since all that is required is that the pulse length be less than the collisional relaxation time, typically 10 ns at 1 torr. In fact pressures up to 5 torr have been used. Q-switching is not possible with optically pumped lasers because of the short rotational relaxation times which prevent meaningful energy storage. However the very brief, very intense pulses which are available from TEA CO_2 lasers provide ideal pumping and intense pulses from certain optically pumped lasers, mainly D_2O, $^{12}CH_3F$, $^{13}CH_3F$ and HCOOH, are readily available. Single oscillators can give 100 kW and master oscillator/amplifier combinations can extend this into the MW regime [727].

It was mentioned above that nearly all optically pumped lasers work with off-resonant pumping—the frequency offset typically being of the order of the line width that is some tens of MHz. It is very important to note, however, that the stimulated fluorescence will be exactly resonant at the natural line-centre frequency of the long-wave emission. This is because a homogeneously broadened transition can be truly pumped at frequencies *anywhere* within the profile and a heterogeneously broadened line can be thought of as made up of a large number of independent narrower homogeneously broadened lines. Thus pumping off-resonantly will excite only a set of molecules with an appropriate Doppler shift and this set will decay to the lower level ($\mathscr{E}_2$) without any change in their motion with respect to the laboratory frame of reference. This is a very important point indeed, for since the emission frequencies are absolutely true, their measurement provides a powerful means of doing precise high-resolution spectroscopy. As an example, from the true time frequencies of the two excited state transitions in CH_3F (Fig. 2.38) it is possible to derive B and D values for this state which are two orders of magnitude more precise than anything which could be derived from conventional spectroscopy.

6.2.5.2 Mid-infrared lasers—the ammonia laser

The v_2 band of ammonia stretches across much of the mid infrared and because of the complications introduced by the double minimum in the potential (see Fig. 5.9), it has a rich rotational fine structure. In 1970, Shimizu [728] showed that there were several close coincidences between lines in this band and lines of the CO_2 and N_2O lasers. In particular the P(13) line of the 10·78 μm band of the N_2O laser is in virtually perfect coincidence with the $a \rightarrow s$ component of the Q($J = 8$, $K = 7$) line of the ammonia band. This coincidence can even be used to provide a passive Q-switch for the N_2O laser and it has proved a very fruitful starting point for double-resonance studies [729]. Chang, Bridges and Burkhardt [730] showed that it could also be used as the basis of an optically pumped laser which gave some useful lines in the far infrared. The origin of those at 81·48 μm (122·7 cm^{-1}) and 263·4 μm (38 cm^{-1}) is shown in Fig. 6.9(a).

In 1976, Chang and McGee [731] showed that pumping of ammonia with the 9R(16) line from a CO_2 TEA laser gave rise to strong mid-infrared emission at 12·812 μm. The origin and assignment of this line are given in Fig. 6.9(b). Subsequently it was shown that pumping with 9R(30) could give rise to several laser emissions, one of which, the line at 12·08 μm, was particularly strong (Fig. 6.9(c)). The observation of several emission lines when only a single level is pumped is interpreted to mean that both radiative and collisional cascades are involved [732]. Since then, the ammonia system has been extensively studied and rate-equation modelling taken to considerable degrees of sophistication [733]. A particularly interesting example comes from the work of Pidgeon and his colleagues [734], at Heriot–Watt University, who have pumped $^{15}NH_3$ with the 9R(18) line from a $^{13}CO_2$ laser ($\tilde{v} = 927\cdot30040$ cm^{-1}). They obtained strong emission at 152·9 μm as explained

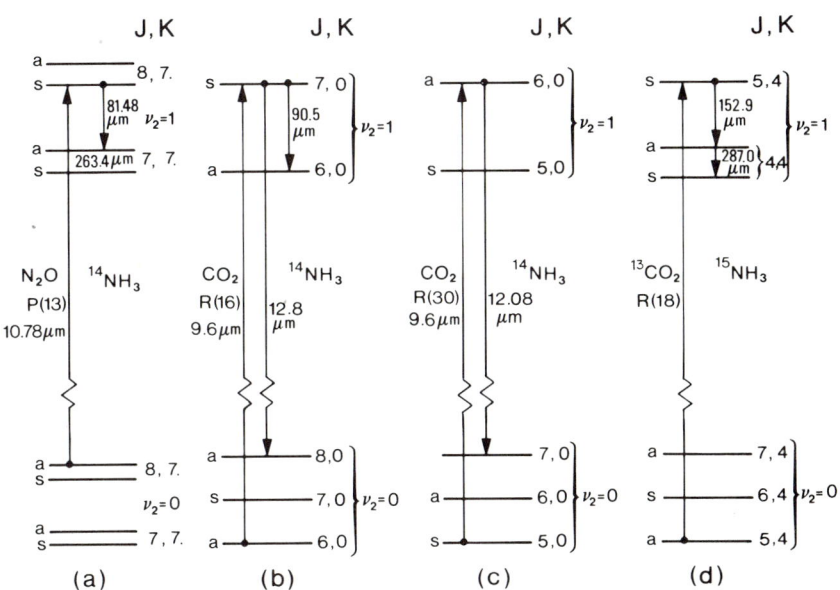

FIG. 6.9. Some optically pumped lasing transitions in ammonia. All the lines lie within the v_2 manifold, a global picture of which is given in Fig. 5.9.

in Fig. 6.9(d). The coincidence is virtually exact (< 16 MHz) and the strong, near-resonant, interaction gives rise to a series of quantum effects [735] which cannot be explained by a simple rate equation approach.

The strong mid-infrared lines from ammonia near 12 µm are technically very important, not only for their own sake, but also because they can be used to pump a spin-flip Raman laser (see section 6.2.10) which will then give tunable radiation at still longer wavelengths [736]. The problem, as always with mid-infrared optically pumped lasers, is self-absorption of the stimulated emission caused by the finite population of the lower level. It is necessary therefore to both saturate the transition and pump the entire volume of the gas. The saturation is not too difficult to achieve, at least for pulsed operation, since one has available the intense radiation from TEA lasers, but to reach the whole volume of the gas requires essentially collinear pumping and that introduces experimental difficulties since one has to simultaneously optimise two independent cavities with a single output coupler. An early solution, due to Walker [737], and illustrated in Fig. 6.10(a), used specifically coated dielectric mirrors which were transparent at the CO_2 pump wavelength but highly reflecting at the ammonia output wavelength. This design and its descendents [738] proved very successful but the dielectric-coated mirror proved sus-

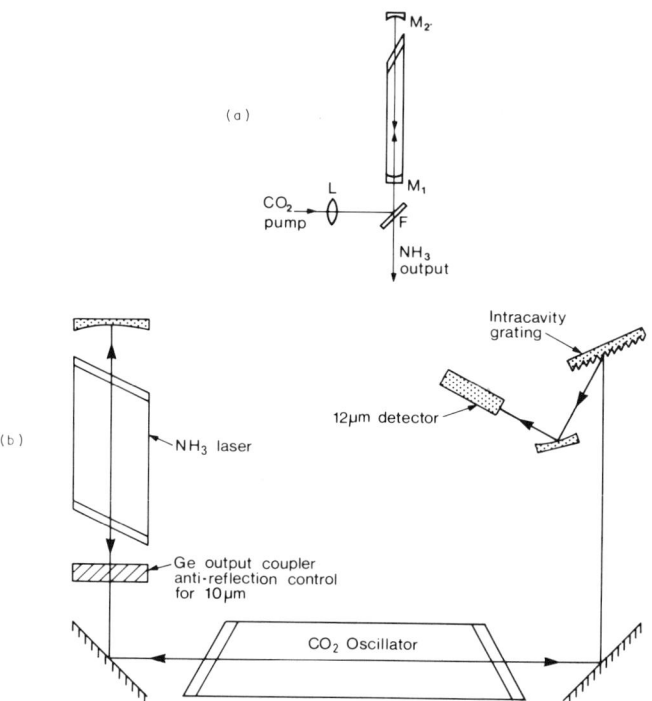

FIG. 6.10. Two designs of cavity suitable for mid-infrared optically pumped lasers.

ceptible to damage problems and the output power was thus limited. Also, the cavity was awkward to align and it was difficult to wavelength tune the combined cavities. An improvement came from the work of Harrison and his group [739] who realised firstly that TEA CO_2 lasers and NH_3 optically pumped lasers are both high-gain devices so it is not necessary to have a very high reflectivity output coupler and secondly that the CO_2 plasma is transparent at the NH_3 laser frequencies. One of their designs is shown in Fig. 6.10(b). The crucial points are the use of a common output coupler made from germanium ($T = 65\%$ at 9·3 μm, $T = 57\%$ at 12·8 μm) and the use of an intracavity grating which not only selects the operating line of the CO_2 laser but also finally separates the two beams. These lasers are very efficient, energy conversion efficiencies of close to 30 % being obtained. The power output can be of the order of 400 kW for the optimum gas pressure (7 torr) and cavity length(1 m).

6.2.5.3 *Pulsed far-infrared optically pumped lasers*

It was implied earlier that pulsed optically pumped lasers differ in two important ways from their continuous-wave counterparts. Firstly higher gas pressures can be used and secondly much higher frequency offsets can be tolerated. The pulse lengths used in practice vary considerably ranging from the order of nanoseconds all the way up to quasi-continuous-wave operation produced by mechanically chopping the output from a continuous-wave pump laser. However, it is only with the very short pulses that unusual effects appear. In this regime, frequency offsets of some GHz, more than 50 times the line width, are not uncommon and the stimulated emission often occurs as a doublet with one component at a fixed frequency whilst the other moves as the pump frequency is varied. Clearly one is not dealing with a linear situation where the output is governed by the balance of simple rate processes. The off-resonant pumping can be interpreted in terms of a dynamic or "AC" Stark effect or else in terms of a non-linear susceptibility (see section 6.3.1). The two effects are equivalent—in physical terms one can say that the intense electromagnetic field perturbs the wave function of the molecule and forces it into a virtual state resonant with the field. This should lead to a second-order very small shift in the laser frequency, the so-called Autler–Townes or "light" shift [740], but for all intents and purposes one gets a line at the correct frequency. The second, and pump-frequency dependent, line is almost certainly due to stimulated Raman emission [741]. In this one gets, in the simplest way of thinking about it, emission from the virtual state, as shown in Fig. 6.11 and a line is obtained whose frequency offset from the true line exactly equals the frequency mismatch between the laser and the molecule. Doublets of this nature are normally only observable when emission is only possible in the pulsed mode. When the emission is also available continuous-wave, the smaller frequency difference and the greater intensity difference combine to make the Raman line unobservable.

Pulsed optically pumped lasers have necessarily rather restricted applications. The only two to have seen extensive use are the CH_3F laser at 494 μm
and the D_2O laser at 66 μm. The D_2O laser is particularly important because
of the power which is available—several Megawatts having been obtained
after amplification. This line, like several others from pulsed optically pumped
lasers, has so high a gain that it is possible to obtain it in superradiant emission
without the need to employ a resonant cavity [742].

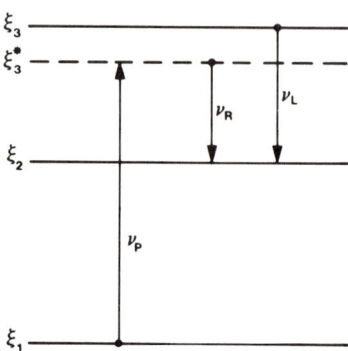

Fig. 6.11. Explanation of the doublet emission from pulsed optically pumped lasers.

6.2.5.4 *Recent developments*

The availability of a large number of lines from optically pumped lasers has
proved a powerful stimulus to the development of physical and chemical
investigations in the long-wave infrared and there is naturally intense research
going on to find new contenders and to improve old ones. A very promising
line is the use of waveguide CO_2 lasers as the pumps. These operate at higher
pressure (~ 80 torr) so they are slightly tunable and a better coincidence with
the pumped transition can be ensured [743]. Using such lasers, several lines,
previously only available at the microwatt level, have now been obtained at the
tens of milliwatts level and many new lines have been discovered. Another
promising line is to use the Stark effect to shift molecular transitions into
better coincidence with the pump [744]. The same result could, in principle, be
obtained by using an electro-optic modulator to put sidebands on the pump
line but at the moment the modulators which are available are not efficient
enough to give usable output power in the sidebands. This technique can,
however, be used perfectly well in a spectroscopic context in which one puts
sidebands on the optically pumped output and thus produces a slightly
tunable far-infrared probe [745]. Optically pumped lasers are at the moment
rather more noisy than one can comfortably tolerate. Undoubtedly some of
the noise is produced by erratic feedback of the pump radiation into the pump
laser cavity. Ways of isolating the two cavities by means of polarising elements
have given promising results but there is still room for improvement.

Because of their near-ideal performance characteristics, optically pumped lasers are attractive candidates for metrological applications. Some effort has therefore been devoted to developing automatic stabilisation systems. A promising approach uses a microprocessor controlled feedback loop which continually monitors the ratio of the output power to the pump power. If this ratio changes, the microprocessor can either alter the cavity tuning or else the electric power supplied to the pump laser. With this system it has proved possible to stabilise the amplitude to $\pm 0.5\%$ and the frequency to ± 100 kHz. This is good enough for the laser to serve as a local oscillator in an astronomical heterodyne spectrometer, but if better spectral purity is required then it will be necessary to use a frequency counting system, that is a non-linear detector simultaneously fed from a standard frequency and from the laser under monitor to automatically control the cavity length. Such a system has been described by Koepf, Fetterman and McAvoy [746] who have shown that it can be made rugged enough for the stabilised laser to be taken to the remote observatories on top of high mountains without degradation of the stabilisation.

6.2.6 Infrared chemical lasers

6.2.6.1 Introduction
A chemical laser is one where the necessary population inversion is brought about as a result of a chemical reaction. There are three basic types [747].

1. Photodissociation

$$AB + (\text{light}) \rightarrow A^* + B$$
$$A^* \rightarrow A + (\text{laser light})$$

2. Chemically pumped

$$AB + (\text{light}) \rightarrow A + B$$
$$A + CD \rightarrow AC^* + D$$
$$AC^* \rightarrow AC + (\text{laser light})$$

3. Hybrid (or energy transfer) chemosensitised

$$AC^* + M \rightarrow M^* + AC$$
$$M^* \rightarrow M + (\text{laser light})$$

Chemical reactions have, for a long time, been known to give transient highly non-Boltzmann distributions so chemical mechanisms are potential sources of inverted populations and hence of laser action. As one example [748], the reaction

$$F + HI \rightarrow HF + I$$

leads initially to a relative population distribution, 0·64, 0·79, 0·95, 1·00, 1·21, 1·61 and 1·01 in the vibrational states $v = 0$ to $v = 6$ of the HF molecule. Thermalisation within any given degree of freedom (vibrational, rotational or translational) is unfortunately very fast, for example energy exchange reactions of the type

$$HF(v = 2) + HF(v = 0) \rightarrow 2HF(v = 1)$$

rapidly equilibrate the vibrational population, so one cannot expect to simply do the reaction in a resonator and get whole bands to lase. Fortunately, however, the rates of thermalisation *between* degrees of freedom are very different: rotational to translational being fast vibrational to rotational being slow, so for relatively long periods after the reaction is over, one can get high vibrational but low rotational/translational temperatures. Discrete line lasing of the CO type is hence a marked possibility. Many chemical laser systems are in fact known [749] and their investigation has developed into a thriving branch of modern physical chemistry [750]. The point is that the stimulated emission following flash photolysis is not only easy to detect, in comparison say with spontaneous infrared fluorescence, it is also a more powerful diagnostic tool for following in detail the convoluted kinetics of these very complicated reactions. Apart from this fundamental motivation, chemical lasers are also of considerable interest to the infrared technologist. There are three main reasons: (1) they provide powerful coherent sources in spectral regions not otherwise covered very well, thus HF gives out lines between 2·6 and 3·6 µm and DF gives lines between 3·6 and 5·0 µm; (2) they provide some possible candidates for very high power laser systems; (3) they provide coherent sources which do not need an external power supply, an obviously attractive consideration for lasers which are to be used in remote or inaccessible terrain.

6.2.6.2 *Photodissociation lasers*

A very simple example of a photodissociation laser is provided by the flash-photolysis of caesium vapour. In this vapour there is present the equilibrium

$$2Cs \rightleftharpoons Cs_2$$

where the bonding energy of the caesium dimers is very low (< 2 eV). When an intense Q-switched pulse of radiation from a Nd/YAG laser is applied to caesium vapour, the Cs_2 molecules are dissociated and the result is excited caesium atoms, preferentially in the $7^2P_{1/2, 3/2}$ states (see Fig. 5.4). Lower lying states are virtually empty so strong laser emission [751] is obtained on the lines

$$7^2P_{1/2} \rightarrow 7^2S_{1/2} \qquad 3{\cdot}095 \text{ µm}$$
$$7^2P_{1/2} \rightarrow 5^2D_{3/2} \qquad 1{\cdot}376 \text{ µm}$$
$$\text{and} \qquad 7^2P_{3/2} \rightarrow 5^2D_{5/2} \qquad 1{\cdot}360 \text{ µm}$$

These are followed by the cascade lines

$$5^2D_{3/2} \rightarrow 6^2P_{1/2} \qquad 3{\cdot}010 \text{ µm}$$
$$5^2D_{3/2} \rightarrow 6^2P_{3/2} \qquad 3{\cdot}613 \text{ µm}$$
$$\text{and} \qquad 5^2D_{5/2} \rightarrow 6^2P_{3/2} \qquad 3{\cdot}489 \text{ µm}$$

In a similar way laser action is possible in rubidium and potassium vapour.

Photodissociation lasers based on all the hydrogen halides are known and characterised [752], but undoubtedly the most important is the HF (or DF) laser. It was first described by Kompa and Pimentel [100] and by Deutsch [753] in 1967 but since then has been investigated exhaustively by numerous workers [754]. The main pumping reaction, in essence, is the so-called "cold reaction",

$$F + H_2 \rightarrow HF^* + H + 35 \text{ kcal mol}^{-1},$$

but the v and J dependence of the various reaction pathways are so complex that Kompa remarks [747] that to account for the v behaviour alone calls for the use of 60 coupled differential equations! This complexity is reflected in the extreme pressure sensitivity of the laser—getting it to go on a particular line calling for very good control of the gas mixture and of the overall pressure. This laser has one of the highest gains known [755] so it is of great technical value. The problem, of course, is to provide the required fluorine atoms to get the reaction going but fortunately almost any inorganic fluoride will flash-photolyse to give fluorine atoms. Thus UF_6. WF_6, MoF_6, SbF_5, IF_5, CF_4, CF_3I, ClF_3, N_2F_4, F_2O and F_2 itself have all been used in the laser fuel. Undoubtedly, however, the most convenient are CF_4 and SF_6 which are non-reactive, non-corrosive and non-poisonous gases which can be stored conveniently in pressurised steel bottles. The hydrogen is usually supplied as hydrogen gas but almost any organic compound can be used since these usually flash photolyse to give hydrogen. During the course of the flash, fluorine molecules will form and these can contribute to the laser action via the so-called "hot reaction":

$$H + F_2 \rightarrow HF^* + F + 98 \text{ kcal mol}^{-1}.$$

Laser action can also be initiated by an electric discharge, though the starting mechanism now is very complex indeed, possibly being a combination of electron collision decomposition and ultraviolet photolysis. Electric discharge HF lasers give out a wider range of lines than does the flash photolysis set-up but the operation is more erratic. A list of all the known lines is given in Table 6.6.

When the lasers are running strongly, cascade pure rotational transitions are often observed further into the infrared. Thus in a laser fuelled by various fluorine substituted methanes (CF_4, $CClF_3$ or $CBrF_3$) and hydrogen [756] rotational lines have been observed stretching from 10 to 22 μm. These comprise $v = 0$, $J'' = 15$ to 27, $v = 1$, $J'' = 11$ to 22, $v = 2$, $J'' = 12, 19, 20, 21, 28$ and 29, $v = 3$, $J'' = 12, 13, 14$. The observed wave numbers can be computed from the relation

$$\tilde{v}(v, J) = 2B(v)(J + 1) - 4D(v)(J + 1)^3 + 2H(v)(J + 1)^3(3J^2 + 6J + 4), \quad (6.2.17)$$

using the parameters, for HF, given originally by Mann et al. [757], which are listed in Table 6.7. It is interesting to note how large the centrifugal effects are for the hydrogen halides as compared, say, with CO or CO_2.

TABLE 6.6
Output wavelengths from the HF/DF laser

HF transition $v \to v - 1 \;\; J' \to J' + 1$		Wavelength μm	Relative power[a]
$1 \to 0$	P(3)	2·608 34	—
	P(4)	2·639 62	0·6
	P(5)	2·672 60	—
	P(6)	2·707 36	0·65
	P(7)	2·743 95	1·8
	P(8)	2·782 45	6·6
	P(9)	2·822 94	6·8
	P(10)	2·865 49	—
	P(11)	2·910 18	5·6
	P(12)	2·957 13	12·2
	P(13)	3·006 41	5·0
	P(14)	3·058 16	1·0
	P(15)	3·112 47	5·8
	P(16)	3·169 47	5·4
	P(17)	3·229 30	0·5
$2 \to 1$	P(1)[b]	2.666 77	—
	P(2)[b]	2.696 27	0·2
	P(3)	2·727 47	0·6
	P(4)	2.760 44	0·8
	P(5)	2·795 22	1·0
	P(6)	2·831 89	1·0
	P(7)	2·870 52	6·0
	P(8)	2·911 18	9·0
	P(9)	2·953 95	7·0
	P(10)	2·998 92	—
	P(11)	3.046 18	6·0
	P(12)	3·095 83	4·5
	P(13)	3·147 97	7·0
	P(14)	3·202 72	—
	P(15)	3·260 19	—
	P(16)	3·320 53	0·5
$3 \to 2$	P(1)	2·790 21	0·3
	P(2)	2·821 30	0·4
	P(3)	2·854 20	0·8
	P(4)	2·889 00	0·65
	P(5)	2·925 72	0·8
	P(6)	2·964 47	1·2
	P(7)	3·005 30	1·4
	P(8)	3·048 31	0·6
$4 \to 3$	P(1)	2·922 10	—
	P(2)	2·954 88	—
	P(3)	2·989 61	—

TABLE 6.6 (*Contd.*)

DF transition $v \to v-1 \; J' \to J'+1$		Wavelength μm	Relative power[a]
1 → 0	P(8)	3·680 00	—
	P(9)	3·716 00	—
	P(10)	3·752 00	—
	P(11)	3·790 00	—
	P(12)	3·830 00	—
2 → 1	P(8)	3·800 00	—
	P(9)	3·838 00	—
	P(10)	3·876 00	—
	P(11)	3·916 00	—
	P(12)	3·957 00	—
3 → 2	P(8)	3·927 00	—
	P(9)	3·965 00	—
	P(10)	4·005 00	—
	P(11)	4·046 00	—

[a] For the $CF_4 + H_2$ flash photolysis laser.
[b] Only observed for $H_2 + F_2$ flash photolysis.

TABLE 6.7
Spectroscopic parameters for HF in units of cm^{-1}

v	$T(v)$	$B(v)$	$D(v) \times 10^3$	$H(v) \times 10^7$
0	0	20·5590	2·119	1·61
1	3 961·60	19·7883	2·073	1·60
2	7 751·02	19·0355	2·019	1·68
3	11 373·04	18·2988	1·954	1·43
4	14 831·83	17·5837	1·919	1·23

The line-centre vacuum wavelengths, for the vibrational lines, which are listed in Table 6.6, can also be computed from these parameters. The best way of doing this is to take equation (6.2.1) to the next level of approximation and to rewrite it in the alternative notation

$$\tilde{v} = \tilde{v}_0 - \Sigma BJ + (\Delta B - \Delta D)J^2 + (2\Sigma D - \Sigma H)J^3 - (\Delta D - 3\Delta H)J^4$$
$$- 3\Sigma HJ^5 + \Delta HJ^6 - \dots \qquad (6.2.18)$$

where $\Sigma B = B' + B''$ and $\Delta B = B' - B''$ etc. and $\tilde{v}_0 = T' - T'''$. All the observed lines can be accounted for within the experimental error at this level of approximation.

All the other hydrogen halides, as mentioned above, can be induced to lase but the only one used much technically is the HCl laser [758], which gives a series of lines between 3·5 and 4·9 μm, though occasional use is made of the

HBr laser which gives out lines in the same region [759]. The bromine $(4^4 P_{1/2})$ laser activated by flash photolyses of IBr [760] or of CF_3Br [761] is of some theoretical interest and also of practical value since the formation reaction is reversible and the laser can therefore be operated sealed off. The $NO(X^2 \Pi_{1/2, 3/2} v > 0)$ laser worked by flash photolysis of NOCl [762] is also of interest but all of these pale into insignificance when compared with the iodine laser.

Laser action on the $^2P_{1/2} \rightarrow ^2P_{3/2}$ magnetic dipole transition in atomic iodine was first reported by Kasper and Pimentel [763] in 1964 who obtained it following the flash photolysis $(\lambda \approx 2700 \text{ Å})$ of CF_3I or of $i.C_3F_7I$. The mechanism of the laser was subsequently explored in detail by Pimentel and his colleagues and by Polanyi [75, 76]. The iodine laser is so important because the inversion is virtually complete, the power absorption from the flash large yet the laser gain small because of the weak transition moment for a magnetic dipole line. Energy storage can thus be high and when the laser is finally allowed to operate one will get an intense pulse with no significant loss due to amplified spontaneous emission. Reported performance of the laser has indicated pulse energies of 65 J and peak powers of 1 gigawatt. Systems comprising a mode-locked master oscillator plus amplifying chain can yield very high powers indeed and the iodine laser is one of the few candidates under active investigation as the driver of a possible laser fusion system [764]. Because of this possibility, very detailed modellings of all the possible reactions and energy exchange mechanisms following the flash photolysis have been carried out [765]. The laser wavelength, 1·315 µm, lies in an interesting spectral region so there may well be humbler but still very useful applications of the laser in machining and microfabrication.

6.2.6.3 Chemically pumped lasers

The CO_2 and CO lasers are both technically chemically pumped lasers, in that the excitation is due to energy transfer, but it is better to retain the term only for these lasers where a chemical reaction is essential for the laser mechanism. The CO_2 laser has been pumped [766] by nitrogen molecules produced by the flash photolysis of hydrazoic acid (HN_3) but the best chemical way of pumping it is to use a subsidiary DF chain [767]. There are three steps

$$F + D_2 \rightarrow DF(v) + D$$
$$D + F_2 \rightarrow DF(v) + F$$
$$DF(v) + CO_2 \rightarrow DF(v - 1) + CO_2(001)$$

This form of pumping is useful for intense pulse operation of the CO_2 laser. It has the merit of not requiring mains electricity since nearly all the output energy comes from the fluorine/deuterium reaction. The flash is just necessary to start the reaction going.

6.2.6.4 *Continuous-wave operation of chemical lasers*

Pulsed chemical lasers are very useful for all sorts of applications but there is no doubt that continuous-wave lasers are necessary for many other applications. The first continuous-wave chemical lasers appeared in 1968 [768]. The principal line of investigation has been the development of flame lasers. Many systems are now known but indubitably the HF and CO flame lasers are the most important. Apart from their value technically, these laser systems are providing the physical chemist with a new tool to explore the immensely complicated world of chemical reactions and the energy exchanges which go on in flames.

The experimental arrangement for a continuous-wave laser [769] consists of a high-pressure region, or plenum, in which fluorine atoms are produced by mixing arc-heated nitrogen gas with cool SF_6. The high-pressure gas containing free fluorine atoms is then expanded, at supersonic speeds, through a "throat" or venturi and past a set of fine nozzles which serve to diffuse hydrogen into the very rapidly expanding gas. The mixture, in which the $H_2 + F \rightarrow HF^* + H$ reaction is occurring, then flows transversely across a conventional Fabry–Perot cavity and finally flows out to a suitable disposal unit. Infrared radiation on the HF laser lines is thus produced in the cavity and it can be extracted in any of the usual ways.

6.2.7 *Near-infrared tunable dye and colour-centre lasers*

In the visible and very near (< 1·2 μm) infrared regions, one of the most useful sources of coherent radiation is the dye-laser [770]. This not only gives out moderate power, narrow-band (< 10 MHz) radiation, it also has the very attractive feature that the centre frequency can be tuned over a wide (~ 50 nm) spectral range. In a dye laser, the operating medium is a solution of a fluorescent dye (e.g. Rhodamine 6G) in a non-quenching solvent. The widths of the spectral fluorescence curves observed for many dyes is quite large (see Fig. 6.12) and so if this radiation could be obtained in stimulated emission we would have a large potential range of tunability. The fluorescence mechanism for many dyes is a four-stage one. The absorption of short-wave radiation takes the molecule to an excited state which has the same configuration as the ground state—this by the Frank–Condon effect. This state then by non-radiative transitions relaxes rapidly with a characteristic time τ'_{rel}, to the stable form of the electronically excited state. This state has a relatively long lifetime τ_{lum} but finally decays with the emission of a photon to an unstable configuration of the ground state, again due to the Frank–Condon effect. Finally this state relaxes rapidly by non-radiative transitions (relaxation time $= \tau''_{rel}$) to the true ground state. One thus has a four-level system, the preferred one for laser action since the inversion is not with respect to either the ground or the primary excited state, and, moreover, it is in principle possible to invert the luminescing states at *any* level of pumping power. When the dye solution is

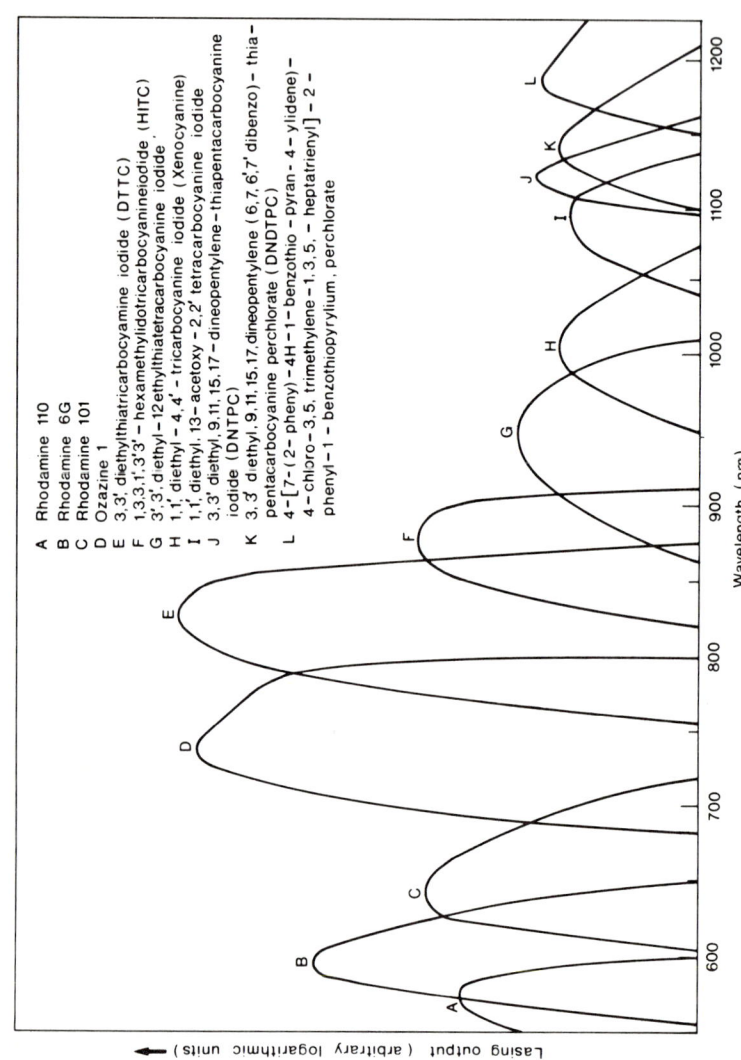

Fig. 6.12. Spectral fluorescence curves for some dyes used in red and near-infrared dye-lasers.

therefore included in a resonator and pumped with reasonable power in one of its absorption bands stimulated emission will be observed. If the resonant cavity contains a frequency selective element, e.g. a grating or etalon, the output frequency will be tunable over the entire range of the fluorescence curve where there is gain and at high enough pumping power this can be, as remarked above, considerable. The actual form of the resonator is commonly the three-mirror form, extensively discussed by Kogelnik and his colleagues [771] and illustrated in Fig. 6.13. The cell containing the dye is set such that its windows intercept the wavefront inside the cavity at the Brewster angle. There is thus no loss for the favoured direction of polarisation. However, since there is a finite angle between the cell windows and the beam, a measure of astigmatism will be introduced but this can be almost perfectly cancelled out by that of the opposite sense introduced by the off-axis spherical mirror M_2. The two spherical mirrors are spaced at the nearly confocal separation so the beam waist at the dye cell will be very small, indeed at these near-infrared wavelengths it is more usual to describe it as a diffraction-limited spot. The pumping, at short wavelengths, is usually done using discharge lamps, but in the visible and near infrared primary lasers, such as argon ion, krypton ion or Nd/YAG are attractive candidates and this pumping radiation can be focussed down to similarly small spot-sizes thus giving an excellent match. It is often possible to choose the coating on the dielectric mirror M_1 such that it strongly transmits the pump radiation whilst strongly reflecting the longer wavelength laser radiation. In this case collinear pumping (as in Fig. 6.13) can be employed. Because the pumped and lasing volume can be so small, it is often possible to dispense with the dye cell and just use a small jet of the solution. In both cases the problem of the heating caused by the pumping is solved by flowing the dye solution round a closed loop containing a heat exchanging cooler. In the second arm of the cavity $(M_2 - M_3)$ the radiation is quasi-parallel, being brought gradually to a second beam waist at the surface of M_3. Because of this very gentle convergence, tuning elements are very conveniently inserted in this arm.

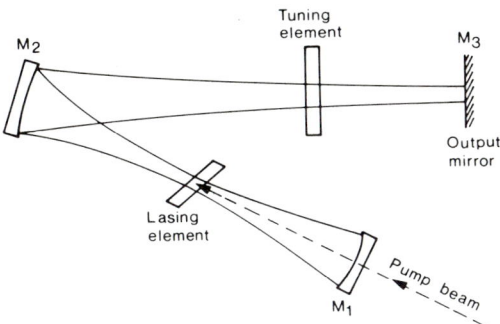

FIG. 6.13. Three-mirror cavity used for near-infrared tunable lasers.

As remarked above, dye lasers can only be used, at the moment, at wavelengths shorter than 1·2 μm and only then with the use of exotic dyes [772]. There does not seem any real prospect that this wavelength limit can be lengthened by the use of still more exotic dyes since the fluorescence efficiency tends to fall off and the dyes become photochemically unstable. Fortunately a new type of fluorescent material is now available which can be incorporated directly into the dye laser resonator configurations without any significant modification and which can extend the operation of tunable lasers out as far as 3·3 μm. This class of material is that of the alkali halides having the rock-salt structure (face-centred cubic) and which contain colour centres. Several groups have made significant contributions to colour-centre laser development but here we follow closely the elegantly lucid account of Mollenauer and Olson [773].

It has been known for a long time that alkali halide crystals can develop visible colouration when subjected to ionising radiation [774], to exposure to alkali metal vapours [775], to electron beams or else even merely to mechanical working. The nature of the colour centres has only been deduced in recent years [776], so, in previous times, a neutral alphabetical code was used for them, one had F centres, U centres etc. The F centre (from the German word *farbe* for colour) is now known to be an anion vacancy at which an electron is trapped. The U centre (which is only coloured in the vacuum ultraviolet!) is an anion vacancy containing an H^- ion. The quantum-mechanical treatment of the F centre involves the solution of the problem of an electron moving in a cubic potential. Because of the high symmetry (T_d) the stationary states are found to be very similar to those of the hydrogen atom and the near ultraviolet absorption, which gives the yellow/red visible colouration, corresponds to the $1s \rightarrow 2p$ resonance absorption of the centre. F centres have a four-level absorption/emission/relaxation scheme similar to that of dyes but unfortunately the crystals absorb at the lasing frequency due to photoionisation of the electron into the conduction band and the oscillator strength (or f-number) of the transition is inherently small. F centres themselves are thus unsuitable for lasers. However there exist several centres closely related to the F centre and several of these are eminently suitable for the construction of tunable lasers.

The first of these F-like centres is the so-called F_A centre. This arises when one of the cations surrounding the anion vacancy is a foreign ion, for example Li^+ in NaCl. F_A centres can be divided into the common F_A (I) class which so far as relaxation is concerned is very similar to the basic F-type. Also, since the configurations will be similar, the luminescing transition will be likewise weak: essentially this arises from poor overlap of the wave functions involved. The rarer F_A (II) centres, however, are quite different. Here the relaxed state of the electronically excited state involves a double well due to the appearance of a symmetric double vacancy. In atomic terms one would say that the relaxed upper state was similar to the H_2^+ molecular ion. The luminescing transition is now strong and Mollenauer and Olson show that absorption of this radiation by photoionisation does not occur at appreciable amplitude. As an additional benefit, the directional discrimination brought about by the presence of the foreign ion causes the p_z orbital to split away from the (p_x, p_y) pair. This causes the appearance of a separate longer wavelength absorbing band which is convenient because most of the useful pumps lie in the visible region. So far only two F_A (II) colour centres have been shown to give laser emission—these are KCl : Li and RbCl : Li. The quantum efficiency is strongly temperature dependent and falls to zero at about room temperature—the crystal has therefore to be maintained in a cryostat at a temperature less than 200 K. The tuning range available is 2·5–2·8 μm with KCl : Li and 2·6–3·3 μm for RbCl : Li. Details of the preparation of the crystals have been given by Mollenauer and Olson but basically one starts with a crystal containing about 1 in 3000 impurity LiH molecules. These will therefore contribute U centres and this is very convenient since U centres are very stable whereas F-like centres are not, at least at ambient temperature. On irradiation with X-rays a fraction ($\sim 10\%$) of the U centres are converted to ordinary F centres and these after prolonged optical pumping are converted into F_A (II) centres. The crystal is then ready for use.

When one has *two* impurity ions around the vacancy, one has an F_B centre and again there are two varieties F_B (I) and F_B (II). F_B (II) lasers have been constructed [777] in KCl : Na and RbCl : Na and these have given a tuning range of $2·25 \leqslant \lambda \leqslant 2·9$ μm. However it should prove possible to extend this to longer wavelengths by the use of RbBr and RbI crystals.

The third type of centre in which lasing action has been demonstrated is the $F_2{}^+$ centre. This consists of a single electron shared between two adjacent vacancies and is modelled therefore rather well by the $H_2{}^+$ formalism. $F_2{}^+$ lasers in KCl and KBr have been demonstrated to have a tunability from 0·8 μm to 2·0 μm.

Colour-centre lasers thus extend the tunability of dye-like lasers well into the near infrared. They give out moderate power (~ 18 mW) and are therefore very suitable for spectroscopy in this spectral region. The width of the laser line is considerably less than 0·05 cm^{-1} so high resolution spectroscopy is possible and it has already been demonstrated, in a pollution monitoring context, by a study of the $v = 2 \rightarrow v = 0$ overtone of NO at 2·68 μm [778]. With computer control it is possible to operate the laser as just described, i.e. broad spectral range reasonable resolving power, or else in an alternative mode in which very high resolution ($\Delta v \sim 1$ MHz) is obtained over a narrow range ($\Delta \tilde{v} \sim 1$ cm^{-1}). Litfin and his colleagues [779] have for example demonstrated the Λ doubling ($\Delta v \sim 1$ GHz) in NO but the Doppler broadening was of the same order so the laser was not able to do itself justice. Nevertheless colour-centre lasers seem set to play as important a role in the near infrared that dye-lasers already have in the visible. This is especially true now that commercial versions (Burleigh Instruments Inc.) are available.

6.2.8 *Solid-state infrared lasers*

6.2.8.1 *The neodymium ion, Nd^{3+}, laser*

The inversion mechanism of the neodymium laser was outlined in section
2.2.4.3 where it was remarked that laser action could be obtained between the
metastable $^4F_{3/2}$ level and the ^4I ground manifold of the Nd^{3+} ion. In fact four
laser lines are possible, viz:

$$^4F_{3/2} \rightarrow {}^4I_{15/2}, \quad 1{\cdot}80 \ \mu m$$
$$^4F_{3/2} \rightarrow {}^4I_{13/2}, \quad 1{\cdot}35 \ \mu m$$
$$^4F_{3/2} \rightarrow {}^4I_{11/2}, \quad 1{\cdot}06 \ \mu m$$
$$^4F_{3/2} \rightarrow {}^4I_{9/2}, \quad 0{\cdot}88 \ \mu m$$

However the $1{\cdot}06 \ \mu m$ line is much the most important technically and by far
the easiest to get to lase and only this particular line was indicated in Fig. 2.35.
The lower level of the $1{\cdot}06 \ \mu m$ line is not the true ground state of the Nd^{3+} ion
and one has, at least formally, a four-level laser. Other things being equal, this
would be a desirable situation but practical versions of the laser are found to
operate anywhere between two limiting regimes depending on the value of the
pulse-length τ in comparison with the multi-phonon de-excitation time of the
terminal level τ_T. If $\tau \ll \tau_T$, the laser operates in a quasi-three-level mode with
the $^4I_{11/2}$ lower level mimicking a true ground state, whereas in the opposite
limit, i.e. $\tau \gg \tau_T$, it operates in the expected four-level mode. This distinction is
important in high-power applications because the limiting (or saturation)
energy density (or "fluence") is increased by a factor of two in three-level
operation [780].

The active entities of the laser, i.e. the Nd^{3+} ions, have to be accomodated in
a suitable host and the choice of this presents some problems. An ideal host
would (a) be available as large pieces of good optical quality, (b) be hard
enough to give a fine polish but not too hard for convenient working, (c) be
transparent over the spectral region where the guest ions absorb the pumping
radiation, (d) have a high thermal conductivity so that the radiation which is
absorbed but converted to heat rather than to laser light can be disposed of
readily, and (e) permit the guest ion to fluoresce efficiently at reasonably high
(preferably ambient) temperatures. There are several possible candidates
which meet some or most of these requirements: for example, fluorite, CaF_2;
calcium tungstate, $CaWO_4$; cerium or lanthanum fluoride, $Ce(La)F_3$; lead
molybdate, $PbMoO_4$; yttrium aluminate, $YAlO_3$; and yttrium vanadate,
YVO_4. But the two outstanding hosts are yttrium aluminium garnet (or YAG),
$Y_3Al_5O_{12}$, which is crystalline, and glass, silicate, phosphate or fluoride, which
is, of course, amorphous.

YAG is attractive because it has a high thermal conductivity (ten times
greater than glass) and this permits high repetition rates or else even true
continuous-wave operation. Its hardness is near ideal—8.5 on the Moh scale—
so it is not easily scratched but can readily be cut and polished. Finally, because

it is crystalline, the inhomogeneous broadening, due to the Stark effect of the crystalline field, is very small, the fluorescence line is therefore sharp and the threshold for laser action (see equation 2.4.4) is low. Unfortunately the other side of the same coin is that the lines which absorb the pumping radiation are also very narrow so the absorption of flash-lamp power is relatively inefficient. To alleviate this difficulty, high-efficiency, high-brightness xenon flash-tubes are used which give out broad-band radiation stretching from the near ultraviolet to the near infrared, thus ensuring a good match with both the transparency "window" of the YAG host and the absorption "envelope" of the Nd^{3+} guests. These lamps are mounted parallel and very close to the YAG rod and both are enclosed within a brightly polished reflector of elliptical cross-section. Xenon flash-tubes now have a well-developed technology, are long lived and moreover are available commercially. Krypton flash-lamps are even better but at the moment their cost and relative unreliability rules them out for all but the most specialised work. The doping levels used with YAG can be as high as 3% Nd_2O_3 before ion–ion interactions lead to a deterioration of the fluorescence efficiency, so the gain is high.

Glass laser rods and discs are used in all those applications where power and power alone is all that matters [781]. The low thermal conductivity of glass demands low repetition rates; indeed in some of the big experimental laser-fusion machines such as NOVA [782], at the Lawrence Livermore Laboratories, the firing rate may be as low as two per hour! The advantages of glass as a host are: (a) it can readily be fabricated in optically perfect pieces of arbitrary shape and size; (b) it can stand fairly high ($> 2\%$) doping levels without serious loss of fluorescence efficiency; (c) the absorption lines of the Nd^{3+} ions are broader in a glass host because of the increased inhomogeneous (Stark) broadening and there is more effective extraction of energy from the pumping flash; (d) although the gain is less because of the line broadening, the energy storage is higher than one would get say with YAG and so in very-short-pulse operation the peak-power will be much higher; and (f) the fluorescence (that is the spontaneous) line width is large ($\sim 30\,cm^{-1}$) so one can get very good mode-locked trains running with individual pulse-lengths of the order of picoseconds. Either YAG or glass lasers can be Q-switched in any of the standard ways. Thus electro-optic elements, rotating mirrors or bleachable absorbers have all been used [783] but these forms of Q-switching are less efficient in the near infrared than they are say in the visible and simple Q-switching is less common with the neodymium laser than it is with the ruby laser. Experimentalists prefer mode-locked operation and the use of an electro-optical switch to select just one of the pulses from the train to escape from the cavity. In this way one can get, fairly readily, very intense pulses of very short duration ($\sim 1\,ps$).

The absorption and emission lines of any rare-earth ion in the infrared and visible regions arise from transitions within the incomplete 4f shell. Such f → f transitions would be forbidden for the isolated ion since they violate the

$\Delta l = \pm 1$ dipole selection rule, but when the ion is in an ionic host, it experiences a crystal-field which is not spherically symmetric and the transitions can then become weakly allowed. A similar phenomenon is the explanation of the well-known colours of transition metal crystalline and hydrated compounds except that in this case it is d → d transitions which are involved. The basic origin of the effect is that the crystal field will in general have components of both odd and even parity and these will induce mixing of the corresponding odd and even atomic orbitals into the actually observed hybrid ionic orbital. In the particular case of rare-earth ions, one would get admixture of 5d and 5g atomic orbitals into the basic 4f orbital. The observed intensity then arises from cross terms in the transition integral which now no longer cancel identically. It is easy to see that this will always happen provided the crystal field at the ion does not have inversion symmetry. This is indeed the usual situation [784] since most of the common point groups qualify. It is, in principle, possible to calculate the expected wave functions by molecular orbital methods and then to use these to estimate the expected intensities but the calculation is far from easy. One obvious difficulty is that with the guest and host ions in virtual contact, one cannot reasonably use a point-charge formalism. However, some successful calculations along these lines have been reported [785]. More usually though, a powerful approximate method first introduced by Judd and Ofeld [786] is used. This method has the great advantage that one can calculate all the relevant parameters of a given glass composition and work out whether it will do a job or not without having to either make the glass or test it in an actual cavity.

All the other rare-earth ions have similar electronic levels to those of neodymium and many of them, when immersed in a suitable host, can be induced to lase. The three which have been investigated so far [787] are Er^{3+} which gives lines at 0·851 and 1·663 μm, Tm^{3+} which gives a line at 1·861 μm and Ho^{3+} which gives lines at 2·064 and 2·123 μm. The most suitable host appears to be $YAlO_3$. The erbium laser is the most important of these three since it operates at room temperature whereas the other two require liquid nitrogen cooling. It also does not require "sensitising", that is co-doping with other rare earth ions. This latter is not, however, a very serious point and the holmium laser suitably sensitised with Er^{3+}, Tm^{3+} or Yb^{3+} can be very efficiently pumped since it has a low threshold.

6.2.8.2 *Tunable transition-metal ion lasers*

According to the "Aufbau" Principle, the electronic configuration of a given element is obtained by adding an electron to the lowest available orbital of the previous element. Usually this follows the normal order but when one reaches argon, with the configuration $1s^2 2s^2 2p^6 3s^2 3p^6$, the next electron goes, not into the 3d orbital as one might expect, but into the 4s orbital. Argon is thus a rare gas and potassium is an alkali metal. Two elements later, however, at scandium the 3d orbital starts to fill and one gets the first set of transition elements (all metallic) stretching from scandium to zinc. These elements have the configuration $1s^2 2s^2 2p^6 3s^2 3p^6 3d^n 4s^2$. Subsequent series exist where the 4d, 5d etc. shells are progressively filled up. The electrons in d shells are partly shielded by the outer electrons and absorption bands (induced by crystal-field or hydration effects) involving d electrons can be relatively sharp. The shielding is not however as good as is that for f electrons in rare earths and the absorption bands frequently display vibronic sidebands due to coupling with the phonons of the lattice.

Doubly ionised transition-metal ions have the outer configuration $3d^{n-1} 4s^1$ and these electrons couple to give final states of F character. ^3F in the case of nickel ($3d^7 4s^1$) and ^4F in the cases of cobalt ($3d^6 4s^1$) and vanadium ($3d^2 4s^1$). Spin-orbit coupling is relatively weak but, in the crystalline environment, crystal-field effects can be large leading to splittings of the order $7000 \, \text{cm}^{-1}$. Thus for Ni^{2+} doped into MgF_2, the two split states of the ^3F configuration, namely $^3F_2$ (sometimes written $^3T_2$ to avoid the confusing usage of the symbol F with two distinct meanings) and $^3A_2$ are separated by $7500 \, \text{cm}^{-1}$. Luminescence at the corresponding wavelength, i.e. $1 \cdot 33 \, \mu m$, is not observed, however, because the configuration around the ion differs markedly for the two states and the Franck–Condon factors are small. The band thus develops marked vibronic structure and phonon-broadening is very obvious. The luminescence then occurs as a broadish band with a maximum in the $1 \cdot 7 \, \mu m$ region. This maximum is temperature sensitive because the phonon frequencies are temperature dependent: it can be varied from $1 \cdot 6 \, \mu m$ at 77 K up to $1 \cdot 8 \, \mu m$ at 250 K [788].

Stimulated emission on the $^3T_2 \rightarrow \, ^3A_2$ transition of Ni^{2+} in MgF_2, MgO and MnF_2 and on the corresponding transitions in Co^{2+} and V^{2+} was reported as early as 1963 by Johnson et al [789]. They used simple lamp-pumping using either xenon flashlamps or else merely iodine-quartz incandescent lamps. Because these latter have their radiant output peaked in the near infrared (section 2.2) it can be concluded that the pumping directly populates the $^3T_2$ upper state. The mechanism is thus similar to that of a dye laser with the Franck–Condon mismatch serving to turn what looks at first sight like a simple two-level system into a *de facto* four-level one. The mismatch energy then goes into phonon creation within the host lattice. The similarity to the dye laser has recently led several groups to investigate the possibility of using laser pumping and a conventional three-mirror cavity (Fig. 6.13) to give output tunability [790]. The results have been most encouraging. Pumping with the continuous-wave 1.33 μm line from a Nd : YAG laser has given power conversion efficiencies as high as 40% with quite low thresholds (~ 20 mW). At 80 K, the $Ni^{2+}:MgF_2$ laser could be tuned from $1 \cdot 61$ to $1 \cdot 74 \, \mu m$ and the $Co^{2+}:MgF_2$ laser from $1 \cdot 63$ to $2 \cdot 08 \, \mu m$. Temperature tuning is also available but the lasers always have to be operated below room temperature to prevent phonon-induced non-radiative decay of the upper level and consequent drop in the laser output power. The upper state lifetimes at reduced temperatures are long, of the order 10 ms, so Q-switching works well. Using a LiNbO$_3$ modulator, peak powers of the order of hundreds of watts can be obtained in pulses some hundreds of nanoseconds long with repetition rates as high as 100 Hz. These lasers appear to be ideal for scaling up to high powers and will then be very attractive for non-linear optics experiments in the near infrared.

6.2.9 *Semiconductor light-emitting diodes and lasers*

When one has a pn junction formed in a direct gap semiconductor, and the resulting diode is forward biased (i.e. the n-type side of the junction is connected to the negative pole of the power supply and the p-type to the positive pole) then carriers will be injected from each side into the junction region and carrier pairs can annihilate with the production of radiation. This is the basis of the light-emitting diode or LED. As remarked earlier LED action gives radiation with a wavelength equivalent to the energy gap and hence tends to be confined to the infrared. Ingenious doping procedures are required to induce the diode to radiate in the visible but these procedures are nevertheless eminently worthwhile as the huge market for red and green LED based displays testifies. Nowadays visible region LED displays are starting to encounter very strong competition from liquid crystal displays which require much less power so the almost unchallenged hegemony of the visible region LED may well be coming to an end. In the infrared, however, LEDs are firmly established as high brightness small area incoherent sources—they are used in fibre-optic communication (section 3.6.4.2) for example.

The earliest LED was based on gallium arsenide and to this day LEDs derived from GaAs and its relatives are still predominant. Intrinsic GaAs itself emits near 0.84 μm at liquid nitrogen temperature but this wavelength varies with temperature and with degree of doping. Almost from the beginning there was intense effort to make a GaAs laser since all the theory (section 2.4.3) indicated that it should be possible to achieve a population inversion. The method which proved successful was to polish opposite pairs of faces of the crystal so as to serve as the "mirrors" of a cavity and to use very high ($\sim 5 \times 10^8$ A m^{-2}) current densities. The lasers worked well but the heat generated was so great that liquid nitrogen cooling was essential and the lasers could only be used pulsed and with a low-duty cycle. Analysis of the operation of the laser showed that the junction lying between the bulk n- and p-type material was acting as a slab waveguide and that, as the current density went up, the refractive index difference, which would normally be very small, increased significantly. The waveguiding therefore became better but in addition the high field tended to confine the carriers to the active area so there was a better chance of radiative recombination. High current density, cryogenically cooled lasers are not really very practical propositions so the homostructure type of laser was soon abandoned in favour of the heterostructure [791] type. In this (see Fig. 6.14), the active region is a thin (~ 0.1 μm) plane parallel slice of p-type GaAs which is sandwiched between two much thicker (~ 1 μm) plates of $Ga_{1-x}Al_xAs$, p-type and n-type respectively. The whole structure is grown (either by chemical vapour deposition, CVD, by liquid-phase epitaxy, or else, and most recently by molecular-beam epitaxy) on an n-type GaAs substrate. Metallic contacts, e.g. gold, are deposited by evaporation to complete the assembly. The lattice constant of $Ga_{1-x}Al_x$ As differs hardly at all from that of GaAs: in fact in going from pure GaAs to pure AlAs the lattice constant only changes by 0.2%. There is thus no difficulty in maintaining not only a good bond between the layers but also a continuous crystalline structure through the junction. Also, since the polarisability of Al atoms is less than that of Ga atoms, the refractive index of $Ga_{1-x}Al_xAs$ is less than that of GaAs so there is no difficulty in getting the slab waveguide to work well. As a further bonus, the band gap of $Ga_{1-x}Al_xAs$ is much larger than that of GaAs so the carriers in the active stripe tend to be trapped there and electroluminescence is very efficient. Double heterostructure lasers can therefore be operated with much lower (10^7 A m^{-2}) current densities and can thus be used continuous wave at room temperature. The main thrust at the moment in this field is to make somewhat longer wave (1·3–2·00 μm) lasers to act as the sources for ultra-low loss fibre optic communication systems (section 3.6.4.2). A promising line here appears to be the use of quaternary III-V compounds such as $In_{0.72}Ga_{0.28}As_{0.6}P_{0.4}$ for the active medium.

The mode output from simple stripe waveguide lasers of the type illustrated in Fig. 6.14 is often very complex and this can lead to a very structured far-field

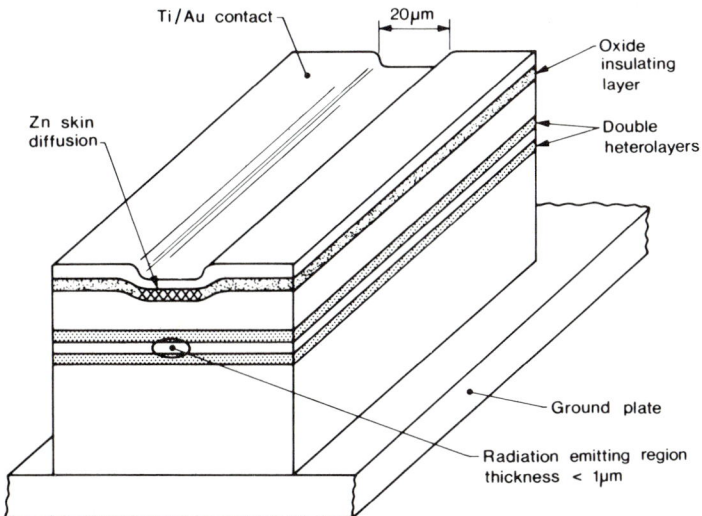

FIG. 6.14. Schematic [791] of the construction of a double heterostructure solid-state laser. The band structure through the junction is also shown.

pattern. This is obviously undesirable when the output radiation is to be coupled into a fibre-optic guide since here something much more like a Gaussian profile is required. Ways of tailoring the doping or the current gradient to give a central "spot of light" have therefore been investigated. One of the promising lines developed by STL at Harlow in England is the use of crystalline strain across the junction to concentrate the region of radiant emission [792]. Another, developed at the Bell Telephone laboratories in the USA, is the use of the so-called "buried" double-heterostructure laser [793]. These novel versions are developing rapidly and already laboratory versions have been tested which give reliable single-mode output at reasonable power levels [794].

Diode lasers are not as monochromatic as is, say, a gas laser, for a variety of reasons, not least of which is the low finesse of the "cavity" but the line widths actually observed ($\sim 10^{-4}\mathrm{cm}^{-1}$) are very suitable for spectroscopy in the mid infrared where the Doppler width is of the same order. Of particular interest, of course, is the 8–14 μm atmospheric window which coincides with the "fingerprint" region of many light organic molecules. To work in this region, the diode must be made from a semiconductor with a small band gap and extensive work has been done on lasers fabricated from lead tin telluride [795], $Pb_{1-x}Sn_x$ Te. The choice of the stoichiometry, i.e. of x, gives a coarse tuning of the laser but by altering its lattice parameter one can achieve fine tuning. This latter can be done either by altering the working temperature of the laser or else by applying hydrostatic pressure. Temperature alterations are the easier and they can usually be achieved merely by altering the drive current through the laser provided, of course, that at all times the current is above threshold. By varying the current it is possible to scan the output frequency of a diode laser by as much as 1 cm^{-1} and tuning ranges of this order have proved adequate for some very remarkable examples of high-resolution spectroscopy [796]. There is now available a large number [797] of tunable diode lasers which between them span the infrared from 2–35 μm. The $PbS_{1-x}Se_x$ laser has proved particularly valuable [798] in the near to mid infrared, 4–9 μm.

6.2.10 *Infrared spin-flip Raman lasers*

It was remarked earlier, in section 5.6.2, that, in the presence of a strong magnetic field, the continuous conduction band breaks up into a large number of discrete Landau states—the so-called "Landau ladder". The splitting between the "rungs" of the ladder is given by

$$\mathscr{E} = g\beta H \qquad (6.2.19)$$

where g is the Landé splitting Factor and β is the Bohr magneton. For electrons in atoms, g is always close to its free-space value of 2·0013. . . (section 5.6), but in semiconductors it can be very large (~ 40) and can vary with frequency, temperature and magnetic field. The splitting of the conduction band electrons in appropriately doped semiconductors has been used as the basis of a widely tunable laser—the so called "spin-flip-Raman-laser" or SFRL [799]. The almost universally used material is indium antimonide but some experiments have been reported using mercury cadmium telluride [800]. The technique is to prepare a suitably doped ($\sim 10^{23}$ m^{-3}) indium antimonide crystal in the form of a parallelipiped with two opposite faces polished to a high finish and made rigorously parallel to one another. When the crystal, at a low temperature (< 14 K), is mounted in the bore of a superconducting magnet producing a field H and irradiated with intense radiation from a CO_2 or CO laser, photons of frequency v_0 can be scattered by the Raman effect and emerge at the frequencies

$$v_s = v_0 \pm g\beta H/h, \qquad (6.2.20)$$

where the positive sign corresponds to anti-Stokes and the negative to Stokes scattering. Under normal circumstances, Raman scattering, even with a laser source, is quite incoherent and spontaneous. It is also very weak, something like one millionth of the incident power being scattered from largish interaction volumes. However, when irradiated with sufficiently intense coherent radiation, the spontaneous process goes over into a stimulated form (section 6.3) and becomes not only coherent but far more intense. Up to 10% of the incident power may be transformed into the Raman line. Stimulated Raman scattering from InSb was first demonstrated in 1970 [801] and very soon afterwards [802] tunable radiation was obtained over a wide band of the infrared. The tuning coefficient was found to be of the order of 10 cm^{-1} T^{-1} so the laser tuning range was considerable. Most of the lines of the CO_2 laser can be used as the pump, so virtually continuous coverage is possible even when the experimentalist is restricted to the lower fields available from electromagnets. With CO_2 pumping only pulsed operation is possible because the threshold is very high, so TEA lasers can be used rather conveniently as the pumps but at 5 μm, with a CO laser, continuous-wave operation is possible since the strong non-linearity in the vicinity of the band gap (i.e. the resonance Raman effect) increases the Raman scattering cross-section enormously [803]. Smith and his group at Heriot–Watt [804] and Walker and his colleagues at NPL [805] have carried out many ingenious experiments in an effort to make the SFRL a satisfactory broad-band tunable source, but their work has to a large extent been frustrated by the erratic mode-pulling and mode-hopping behaviour [806]. Attempts to overcome these problems by anti-reflection coating the parallel faces of the crystal and then using a conventional (i.e. "external") Fabry–Perot resonator have been only partly successful at 10 μm [807]. They have been more successful at 5 μm [808] but unfortunately this is a less significant advance since fewer molecules have fundamental modes in this region.

At the moment the SFRL is very valuable indeed as a fixed (but *adjustable*) frequency source but it is not easy to see how it can be used as it stands for high resolution continuous scan spectrometry [809]. The power levels available, however, of the order $100 \text{ mW} - 1 \text{ W}$ continuous-wave with a line width of the order of 10 MHz, will continue to commend it for certain specialised applications.

6.3 Infrared devices based on non-linear interactions

6.3.1 *Introduction—classical linear optics*

In the classical theory of electromagnetism, the interaction between a radiant field and a piece of matter is described by saying that the field induces oscillating voltages and currents in the material medium and that these, in their turn, generate secondary radiant fields which, superimposing on the incident field, give the final result. This full electromagnetic approach is difficult to implement quantitatively and the usual approach is to use instead an excellent approximation in which the oscillatory currents and voltages are replaced by oscillating dipoles. The classical theory is linear but it does include all possible terms, i.e. the full set of in- and out-of-phase electric and magnetic interactions. In addition, it is formulated on the assumption that the media are, in general, anisotropic and that the susceptibilities χ_{ij} will hence be second-rank tensors. The polarisation induced in a medium will, therefore, involve six tensor coefficients but, if one confines oneself to non-magnetic media, these reduce to three. Using the tensor summation convention, i.e. that sums are implied over all repeated indices, one may then write

$$P_i = \varepsilon_0 \left[\chi_{ij} E_j + \chi_{ij}^{\dagger} (\dot{E}_j/\omega) + \chi_{ij}^{\dagger\dagger} (\partial E_j/\partial z) \right], \qquad (6.3.1)$$

where i and j are x or y and the propagation is assumed to be in the z direction. The component elements of the medium, that is the molecules, will be, in general, anisotropic, but the medium itself will only be so when these elements are arranged in particular ordered patterns. In other words only crystals and some rather peculiar liquids ("liquid-crystals") can be anisotropic. The coefficients in (6.3.1) are, as mentioned above, second-rank tensors in the most general formulation but it follows that for all gases, nearly all liquids and many solids they can be taken to be scalars as was implied in Chapter 3.

Anisotropic crystals have long been of interest both theoretical and practical. Many naturally occurring crystals, for example quartz and calcite, are anisotropic and exhibit well-developed birefringence. The susceptibility, $\chi = (\varepsilon_r - 1)$, of such crystals can be thought of as an ellipsoid rather than a sphere and the directions of the principal axes of this ellipsoid relative to the natural axes of the crystal define the directions in which the crystal has to be cut to display the birefringency. In simple terms one may say that the refractive indices for light polarised along the two principal axes perpendicular to the optic axis differ. This is sometimes referred to by saying that there is an ordinary and an extraordinary index of refraction and if light polarised at an angle to one of the axes is incident on the crystal it will be split up into an ordinary and an extraordinary ray. Since these rays propagate at different speeds, the resultant, when they are finally recombined, will show a rotation of the plane-of-polarisation. In other words, crystals such as quartz are *optically active*. The same concepts can be developed in a different formalism in which one regards plane-polarised light as made up from two equal amplitude counter-rotating components. The left-hand and right-hand components will again propagate at different speeds and a rotation of the plane of polarisation will be observed which varies linearly with the distance of penetration into the medium. Of course it is not necessary that the components be ultimately recombined. If one of them can be deflected, by for example invoking total internal reflection, then one can produce plane or circularly polarised light from an unpolarised original. This is the basis of several well-known polarising devices, for example the Nicol prism. These are widely used in the visible and near-infrared regions, but further into the infrared they become less useful because of material absorption problems.

Most media, however, are essentially isotropic because either molecular rotation or else randomness of molecular orientation washes out the essential anisotropy of the molecular susceptibility tensor. It then becomes of interest to examine the properties of the coefficients in (6.3.1) from a symmetry point of view. The two relevant symmetries are that under parity and that under time-reversal [810]. The electric field vector and the polarisation vector are odd under parity but even under time-reversal: their symmetries may thus be described by $[-+]$. The linear susceptibility χ_{ij} therefore has the symmetry $[+\ +]$ and, becuase of this, it will be expected that it will be finite for *all* matter. This is so familiar an observation, essentially that the refractive index of matter differs from that of vacuum, that it hardly seems worth making, but the same analysis applied to the other terms gives some very useful results indeed. Thus the second term in (6.3.1) must have the symmetry $[-\ -]$ and can only be non-zero (or "survive") if the molecule has some in-built sense of time. That is the molecule must have going on within it some process which labels the arrow of time. For ordinary stable molecules there is no such process and all the components of $\chi_{ij}{}^{\dagger}$ are necessarily zero. The only examples amongst stable entities where $\chi_{ij}{}^{\dagger}$ would not be zero would arise say for atoms with unpaired electrons which had been separated artificially into the two kinds of spin-states. A low-pressure gas of α-spin sodium atoms might provide an example.

The susceptibility $\chi^{\dagger}$ represents a dipole-induced perpendicularly to the incident field and the resultant of the in- and out-of-phase components will thus not be parallel to the incident field: in fact it is not difficult to see that the plane of polarisation will be progressively rotated and the gas of separated spin-state atoms, mentioned above, would be strongly optically active. Because of the negative parity, the diagonal elements of $\chi^{\dagger}$ will necessarily be zero but the nature of the off-diagonal elements is undefined by symmetry arguments alone. The way to proceed then is to use first-order quantum mechanical perturbation theory to calculate, at least formally, these elements. The result is that

$$\chi_{ij} = (2N/V\varepsilon_0) \sum_r [v_r \, \text{Re} \, \langle \psi_0 \mu_i \psi_r \rangle \langle \psi_r \mu_j \psi_0 \rangle] / h [v_r^2 - v_0^2], \tag{6.3.2a}$$

$$\chi_{ij}{}^\dagger = (2N/V\varepsilon_0) \sum_r [v_r \, \text{Im} \, \langle \psi_0 \mu_i \psi_r \rangle \langle \psi_r \mu_j \psi_0 \rangle] / h [v_r^2 - v_0^2],$$

$$\tag{6.3.2b}$$

where N is the number of polarisable entities in a volume V, v_0 is the probe frequency, v_r is the transition frequency to the rth excited state and the sum is over all such states. Strictly speaking there should be damping terms in the resonance denominators to give absorptive characteristics but, in the present context, only transparent media are being considered and the susceptibilities are therefore pure real. It follows from (6.3.2b), that $\chi_{ij}{}^\dagger$ will only be non-zero when the wave functions are necessarily rather than conventionally complex. This is the same point made earlier about the need for an in-built sense of time. However it also follows from equation (6.3.2) that

$$\chi_{ij} = \chi_{ji}$$

and

$$\chi_{ij}^\dagger = -\chi_{ji}^\dagger \tag{6.3.3}$$

The quadrature susceptibility $\chi_{ij}^\dagger$ is therefore a skew symmetric tensor [811].

The third term in equation (6.3.1) introduces a coefficient with formal symmetry $[-+]$. It can therefore only be different from zero when the scattering entities are such that they cannot, by means of any physically permissible rotation or translation, be superposed on their own mirror images. The best known example arises in organic chemistry when molecules which have four different substituents on a single tetrahedrally bonded carbon atom exist in two distinct forms called dextro- and laevo-rotatory, or for shortness d and l. These two kinds of molecule are usually chemically identical reacting differently only to other molecules which have the same property of possessing an identifiable "handedness". This arises, as is well known, in biology where enzymes will often process one stereoisomer whilst ignoring the other. The most obvious difference arises, however, when a beam of linearly polarised light is incident on a solution containing the molecules when it is found that one kind rotates the plane of polarisation to the left whilst the other does the opposite. Naturally occurring sugars are optically active in this sense and the measurement of the degree of rotation over a fixed length can be used as a very convenient way of determining the concentration of the sugar. The technique is, for this reason, often called saccharimetry but it is not confined to sugar analysis. It is for all practical purposes, however, restricted to the visible region because of the use of the eye as the detector, but, since the degree of rotation tends to vary inversely as the wavelength, longer wavelengths would not be desirable anyway.

When electric or magnetic fields are applied to a material sample, new optical effects appear. Their origin lies in the field dependence of the three susceptibilities appearing in equation (6.3.1). Taking a magnetic field first, one has that H, arising as it does from rotatory currents, has formal symmetry $[+ \ -]$. The expansion of $\chi_{ij}^{\dagger}$, as a function of H, will therefore have a linear term whose coefficient will have symmetry $[+ \ +]$ and will therefore exist for all matter. This gives rise to the Faraday effect where the presence of an axial magnetic field leads always to a rotation of the plane-of-polarisation. The effect is weak and positive for diamagnetic materials and strong and negative for para- or ferro-magnetic materials. It forms the basis of several very useful microwave devices. The effect of an electric field will lead to a term with symmetry $[- \ +]$ in the expansion of the in-phase susceptibility. This term can therefore only exist for crystals which lack a centre of symmetry. It leads to the Pockels effect where initially anisotropic crystals show a marked change in their birefringence when a field is applied. Some initially non-birefringent crystals, such as hexamine, also develop a measurable birefringence in the presence of a field. Both the Faraday and the Pockels effect are first-order effects, an important point since the linear dependence means that only moderate fields are required for practical devices. Liquids, almost by definition, do not show a linear Pockel's effect but strongly polar liquids, such as nitrobenzene, show a quadratic or Kerr effect. In this, an intense static or low-frequency AC field induces a birefringence in the liquid specimen. Because of the dependence on E^2, the birefringence is independent of the sign of the field. The Kerr effect does not arise from any necessary non-linearity in the electric susceptibility, rather it arises from a partial orientation of the molecular dipoles in the applied field. The basic theory of this effect covers much of the same ground that is involved in the Debye theory of the static permittivity. However, because the molecular dipole need not be oriented along the axis of largest polarisability, the birefringence can have either sign. For nitrobenzene, for example, the Kerr constant is negative but for other liquids it is positive. Kerr cells are widely used in laser optics but because of the E^2 dependence, high fields are necessary and because of the consequent breakdown and conduction problems, they tend to be restricted mostly to use in pulsed modes.

6.3.2 *The non-linear susceptibility*

The basic assumption of the classical theory, viz. that χ is independent of the field, is obviously only a first approximation and the prospect therefore arises, for real materials, of observing non-linear optical phenomena [812]. However it was only the invention of the laser that made this a practical possibility since even with the best materials the non-linearities are very small and to get the necessary high fields and extended spatial interaction, a coherent beam is essential. Unlike the linear media just considered, where two or more fields propagating through the medium do so quite independently of one another, independent fields propagating through a non-linear medium *can* interact and the possibility therefore arises of generating new fields and of transferring energy from one field to another. This is the new topic of *non-linear optics* and many of these possibilities have indeed been demonstrated practically and are now providing the optical engineer with analogues for many of the non-linear devices used by his microwave counterpart. The rapid development of non-linear optics has thus been crucial to the emergence of optical engineering as a practical proposition and this was recognised by the award of a Nobel Prize for Physics, in 1981, to one of the pioneering giants of the art, Nicholaas Bloembergen [813].

To develop non-linear optics systematically, one should, in principle, consider the full field dependence of all the terms in (6.3.1), but for all practical cases involving the mixing of radiant fields, only the variation of χ is important. One may thus generalise (6.3.1) to read,

$$P_i = \varepsilon_0 \left(\chi_{ij}^{(1)} E_j + \chi_{ijk}^{(2)} E_j E_k + \chi_{ijkl}^{(3)} E_j E_k E_l + \text{etc} \ldots \right), \tag{6.3.4}$$

where the various non-linear susceptibilities $\chi^{(2)}$ etc. are respectively third, fourth etc. rank tensors. From (6.3.4) it follows that if two fields at frequencies v_1 and v_2 are applied to the medium, then polarisations at the harmonics $2v_1$, $2v_2$, $3v_1$, $3v_2$ etc. will be induced and also at all possible sum and difference frequencies. It is this feature of the response, completely analogous to that of a diode, which leads to the practical usefulness of non-linear optics. It is important to note that despite the use of the same symbol, the quantities involved have different dimensions, thus $\chi^{(1)}$ is a pure number, $\chi^{(2)}$ has dimensions m V^{-1}, $\chi^{(3)}$ has dimensions $m^2 V^{-2}$ etc. In the literature, these quantities are often quoted in esu, a very unhelpful practice since they are mostly used to calculate radiant powers in practical, i.e. SI units! The conversion to SI units is, however, readily achieved by multiplying by 3×10^{-4} raised to the appropriate power. The non-linear susceptibilities, like the linear one $\chi^{(1)}$, have in general both real and imaginary parts and these are related by appropriate Kramers–Kronig equations. So far only the real part of $\chi^{(2)}$ has proved of any consequence in both theory (non-linear reflection and refraction) and practice (sum/difference frequency generation), but for $\chi^{(3)}$ both parts are important. Thus the imaginary part of $\chi^{(3)}$ is responsible for all the spectroscopic applications of non-linear optics, two-photon absorption, saturation and the stimulated Raman effect, whilst the real part is responsible for the intensity-dependent refractive index discussed in section 3.4.7. The dispersion of the higher χs, like that of $\chi^{(1)}$ becomes intense in the neighbourhood of the electronic absorption bands of the medium because of the presence of resonance denominators (cf. equation 6.3.2). For most ionic crystals these electronic resonances lie in the ultraviolet but for semiconductors such as germanium, tellurium etc, they lie in the infrared, in fact at the band-gap frequency. Semiconductors are thus extremely useful non-linear materials for the mid-infrared but they cannot be used so well in the near infrared because of the strong absorption. At these shorter wavelengths the non-linear ionic crystals such as lithium niobate, $LiNbO_3$, are widely used. These crystals show a minimum of dispersion in $\chi^{(1)}$ in the near infrared and similar behaviour is observed for all the higher non-linear susceptibilities. One can thus, without serious error, treat all the χs as pure real. This "low frequency" approximation, which incidentally holds true throughout the infrared all the way down to the vicinity of the reststrahl absorption, makes a discussion of the symmetry of the tensors easier since normal permutation symmetry is involved rather than the complex (or Hilbertian) symmetry which would apply were the χs to be complex. $\chi^{(2)}$ for example is invariant to any permutation of the three indices [814]. This has the interesting consequence that $\chi^{(2)}$ must vanish if the system in question has a centre of symmetry. $\chi^{(1)}$, $\chi^{(3)}$, etc, on the other hand, will be finite no matter what the symmetry. It follows then that a centrosymmetric crystal cannot be used to generate second harmonics ($2v$) but can, in general, be used to generate third harmonics ($3v$). The same result follows at once by considering the case $j = k$ when $\chi^{(2)}$, now essentially a scalar, has the symmetry $[- +]$ and can thus only exist for a non-centrosymmetric crystal.

Technically interesting non-linear optics nearly always involves the use of single crystals and, in this situation, it is conventional to replace the tensor $\chi^{(2)}_{ijk}$ by a virtually identical† tensor d_{ijk} and, in addition, to use the contracted notation $d_{i(jk)} = d_{im}$, where m runs from 1 to 6 with the correspondence: 1 = (11), 2 = (22), 3 = (33), 4 = (23) or (32), 5 = (13) or (31) and 6 = (12) or (21). The existence of a given d_{im} and its equality, or not, with other members of the set, is determined by the crystallographic system, that is the crystal class plus point group symmetry to which the crystal belongs. Full details have been given by Byer [815]. Thus at one extreme, for C_i of the triclinic system, all eighteen components are, in principle, finite, whereas at the other, the group O_h of the cubic system, all the elements vanish identically. This latter holds, of course, as mentioned above, for all 11 of the 32 possible classes which possess a centre of symmetry. Of the remaining 20 classes, one or more of the d_{im} will be finite, but Kleinman's permutation of indices conjecture [814], which as far as is known holds exactly, reduces the number of independent components still further. Thus for the useful 42m (or D_{2d}) point group of the tetragonal system, whose classic example is zinc blende and which also includes KH_2PO_4 (KDP) and the chalcopyrites typified by $AgGaSe_2$, the two components d_{14} and d_{16}, which survive on symmetry grounds, become equal on electromagnetic

† Some authors, however, define d_{ijk} without the multiplying factor ε_0.

grounds and one therefore needs to quote merely an unsubscripted non-linear coefficient d, when referring to these crystals.

In non-linear optics, the frequencies of the induced and radiating dipoles need not be the same as that of the incident field and one has to consider the presence of several different fields. This is usually done by referring to the number of waves (i.e. fields) present to describe the situation. Linear optics is thus a "two-wave" process. The non-linear susceptibilities $\chi^{(2)}$, $\chi^{(3)}$ etc. then refer to three-wave, four-wave mixing processes etc. It is convenient within this formalism to include fields at zero frequency (i.e. DC electric and static magnetic fields) and then such phenomena as the Pockel's effect and the Kerr effect are taken formally out of the realm of linear optics and included amongst their relatives in the domain of non-linear optics. The number of possible types of interaction increases rapidly with the order of the non-linearity but only a small number of low-order non-linear phenomena has so far been investigated and characterised [816]. Thus for $\chi^{(2)}$ one has for example:

$v_1 + v_1 = 2v_1$	Second-harmonic generation
$v_1 - v_1 = 0$	Optical rectification
$v_1 = 0 + v_1$	Pockel's or linear electro-optic effect
$v_1 \pm v_2 = v_3$	Sum/difference frequency generation
$v_1 = v_2 + v_p$	Parametric oscillation/amplification

and for $\chi^{(3)}$ one has

$v_1 + v_1 + v_1 = 3v_1$	Third harmonic generation
$v_1 = 0 + 0 + v_1$	Kerr effect
$v_1 + v_2 = (v_3 + v_4)$	Coherent anti-Stokes Raman Scattering (CARS)

etc. $\chi^{(3)'}$, as mentioned earlier, is also responsible for the quadratic field dependence of the refractive index leading to phenomena such as self focussing etc. For all known media, the magnitudes of the successive non-linear susceptibilities fall off very rapidly with increasing order. This is the reason why non-linear effects due to radiant fields could not be detected prior to the invention of the laser—the electric fields which could be produced by focussing incoherent radiation were just too small. The classical examples of non-linearity depended either on a "built-in" molecular field (optical rotation) or else on the application of intense static fields (the Pockel's or linear electro-optic effect, the Kerr or quadratic electro-optic effect and the Cotton–Mouton effect). To give the discussion a quantitative aspect one notes that the components of $\chi^{(2)}$ are of the order 10^{-12} m V^{-1} so fields of the order 10^{12} V m^{-1} would be required to produce a non-linear polarisation equal to the linear one. Even with a pulsed laser this would be a tall order corresponding via equation (1.4.11) to an intensity at the focus of 10^{21} W m^{-2} or, if one can assume focussing down to a diameter of about λ, a laser pulse peak power of about 10^9 W at a wavelength of 1 μm. Of course one would not seek to achieve the whole result in a sphere of diameter one wavelength; what one seeks to do is to build up the effect by coherent interaction over very many wavelengths just as one does with a Pockel's cell or a Kerr cell and then one will hope to see the non-linear effects at very much lower power levels.

Essentially, one has a polarisation, that is a "forced" wave, whose frequency is different from that of the applied field but which propagates through the medium at the same velocity. This polarisation will, in its turn, produce, in the medium, a radiant, or "free" wave at its own frequency but, in general, because of the dispersion of the real part of $\chi^{(1)}$, the velocity of propagation will not be the same and the familiar concepts based on the application of Huyghen's construction will no longer apply. Thus, whereas in the absence of dispersion one would expect the free-wave to cancel in all directions save the forward one and in that direction for it to steadily increase in intensity as power is transferred to it from the forcing field, in the dispersive case the free-wave will eventually get out of phase with the forced wave and its power will rapidly fall as interference essentially transfers all the power back to the forcing field! Detailed calculations show that the non-linear power, say for second-harmonic generation (or SHG), will be related to the crystal length l, by [817]

$$P(2v) = 2P(v)^2 \, (\mu_0/\varepsilon_0)^{3/2} \, (\omega^2 \, d^2 \, l^2/n^3 \, A) \, \text{sinc}^2 \, (\Delta k l/2), \qquad (6.3.5)$$

where n is a geometric mean index, i.e. $(n_1 \, n_1 \, n_2)^{1/3}$, A is the area of the beam, d is the appropriate non-linear coefficient (see later) and

$$\Delta k = k_2 - 2k_1$$
$$= 2\pi\tilde{v} \, (n_2 - n_1). \qquad (6.3.6)$$

Clearly this power can grow without limit if $n_2 = n_1$, but if there is dispersion, then there will be an optimum or "coherence" length, given by

$$l_c = \lambda/2 \, (n_2 - n_1). \qquad (6.3.7)$$

This quantity can be quite large, several millimetres, for some non-linear materials, but for most it is less than 100 μm. Thus early attempts to achieve SHG, using the then conveniently available materials such as crystal quartz, always gave disappointingly low powers. The solution to this dilemma is to use a birefringent crystal and to incline it at such an angle to the beam that the refractive index, say for the ordinary ray, at the fundamental, exactly equals that for the extraordinary ray at the harmonic. In this way, it is possible to compensate perfectly for the dispersion and achieve "phase" or "index" matching. This technique is vital to modern non-linear optics since it permits interaction over the macroscopic length of the crystal. With its help, dramatic improvements in efficiency are possible. Thus for the case of Gaussian beams optimally focussed, the power output from a phase-matched crystal at the second harmonic can be written [818]

$$P_2 = 6{\cdot}35 \times 10^4 \, P_1^2 \, d^2 \, l/n^2 \, \lambda_L^3, \qquad (6.3.8)$$

where P_1 is the incident laser power, λ_L its wavelength, n is the effective refractive index of the phase-matched crystal, and l its length. It is important to note that in the Gaussian case, unlike the ideal parallel beam case considered earlier, the power output merely increases linearly with depth of penetration into the phase-matched crystal. This is due, of course, to the z-dependence of the beam-waist (equation 2.1.11). Putting some numbers now into (6.3.8), if we had 1 W of power at 1 μm with $n^2 \sim 4$ and $d = 0.63 \times 10^{-12}$ m V^{-1} (the value for potassium dihydrogen phosphate or KDP), one derives $P_2 \approx 10^{-4}$ W for a crystal 1 cm long. This is an excellent conversion ratio. With more efficient non-linear materials such as lithium niobate or barium sodium niobate still lower input powers will give acceptable output. It is even possible to include the non-linear crystal, finished off at the Brewster angle, into the primary laser cavity and achieve 100% conversion! Geusic and his colleagues [819] have in this way produced a very useful green laser at 0.53 μm, by doubling the primary radiation from a pulsed Nd/YAG laser.

Non-linear optics is so useful, potentially and actually, that considerable attention has been paid to the genesis of the material non-linearities. A very promising approach, which has been lucidly expounded by Garrett [820] is based on a model in which the molecules in the crystal are treated as anharmonic vibronic oscillators. That is, vibrational/electronic interactions are included in the Hamiltonian first-order and the anharmonicity then produces an electronic non-linearity. This is basically the same model that has been successfully applied to the interpretation of Raman intensities [821]. It has given a theoretical justification for the empirical rule discovered by Miller [822], namely that the quantity

$$\varepsilon_0^2 \delta^{(2v)} = \chi^{(2)}(2v, v, v)/\chi^{(1)}(2v)[\chi^{(1)}(v)]^2 \tag{6.3.10}$$

is essentially a universal constant, varying by no more than a factor of two or so over a range of materials for which d varies by three orders of magnitude. Taking KDP as an example, δ^{2v} has the value 3×10^9. Miller's rule has proved very useful in the search for new non-linear materials.

The treatment begins by writing down an anharmonic equivalent of equation (5.3.8), viz.

$$\ddot{x} + \gamma \dot{x} + \omega_0^2 x + Dx^2 = (e/m) E_0 \exp(-i\omega t) \tag{6.3.11}$$

where x is the electronic coordinate and D is an anharmonic coefficient. One now looks for a solution of the form

$$x = (1/2)[q_1 \exp(-i\omega t) + q_2 \exp(-2i\omega t) + c.c]. \tag{6.3.12}$$

The solution is found by substituting (6.3.12) into (6.3.11) and equating equivalent terms. The solution for q_1, namely

$$q_1 = (eE_0/m)[(\omega_0^2 - \omega^2) - i\omega\gamma]^{-1}, \tag{6.3.13}$$

gives merely the linear susceptibility, viz.

$$\chi_L = Ne^2/m\varepsilon_0[(\omega_0^2 - \omega^2) - i\omega\gamma]. \tag{6.3.14}$$

The anharmonic term is, however, more interesting. The solution for q_2, is [817]

$$q_2 = -De^2\,E_0{}^2/2m^2\,[\omega_0^2 - \omega^2) - i\omega\gamma]^2\,[\omega_0^2 - 4\omega^2 - 2i\omega\gamma]. \tag{6.3.15}$$

Defining the non-linear susceptibility, d_{NL} $(2v)$, by

$$P^{(2v)}(t) = (1/2)Neq_2\exp(-2i\omega t) = (1/2)\varepsilon_0\,d_{NL}\,E_0{}^2\exp(-2i\omega t) \tag{6.3.16}$$

gives the final result

$$d_{NL} = -\frac{D}{2}\left(\frac{e^3\,N}{m^2\,\varepsilon_0}\right)\frac{1}{(\omega_0^2 - \omega^2 - i\omega\gamma)(\omega_0^2 - 4\omega^2 - 2i\omega\gamma)} \tag{6.3.17a}$$

which can also be written

$$d_{NL} = mD[\chi_L(v)]^2\,\chi_L(2v)\varepsilon_0^3/2N^2\,e^3. \tag{6.3.17b}$$

To get an estimate for D requires the use of a particular model. Yariv [817] discusses a simple one in which a non-centrosymmetric crystal is produced from a centrosymmetric one by merely adding an extra charge to one of the lattice points surrounding the electron. This gives

$$D = -39\cdot9e^2/4\pi\varepsilon_0\,mr_0^4. \tag{6.3.18}$$

Combining (6.3.18) with (6.3.17) gives Miller's rule at once. To give some idea of the orders of magnitude of the quantities involved, one could select a set of internally consistent numbers which, whilst not actually applying to any particular non-linear crystal, would nevertheless be typical of ones such as KDP. This set would be: $N = 6 \times 10^{28}\ m^{-3}$, $\omega_0 = 7\cdot96 \times 10^{15}$ rad Hz, $\chi_L = 3$, $r_0 = 1 \times 10^{-9}$ m, $D = 7\cdot57 \times 10^{39}$, $d_{NL} = 0\cdot5 \times 10^{-12}$.

6.3.3 Infrared non-linear devices

The major value of non-linear interactions, so far as the infrared worker is concerned, is that they can provide tunable outputs from otherwise fixed frequency laser inputs and that they can be used to shift the frequency of a given phenomenon from an awkward region into a more convenient one. Undoubtedly there will be a steadily increasing list of applications as time goes by, but at the moment the most important are:

1. Two-photon absorption followed by visible region fluorescence
2. Second and third harmonic generation
3. Difference frequency generation
4. Parametric oscillation

5. Up-conversion
6. Stimulated Raman emission
7. Four-wave mixing processes

Some practical applications of these non-linear processes will now be briefly outlined.

6.3.3.1 *Sum/difference generation and parametric processes*

Two-photon absorption/fluorescence is widely used to provide visible monitor beams from near-infrared originals. Eu^{2+} doped into CaF_2 is widely used in this context, especially for doubling Nd/YAG and He/Ne 1.15 μm, but there are several others [823], such as $NaYF_4:Yb$, Er.

Apart from this very useful application, these "phosphors" are very valuable as laser intensity monitors. Thus the observed intensity of the first harmonic is proportional to the square of the laser intensity, that at the second harmonic to the cube etc. This still remains one of the best ways of determining the peak-power of mode-locked pulses from near infrared lasers.

Second harmonic generation requires, as mentioned above, phase-matching if realistic power-levels are to be produced. Essentially, both energy and momentum have to be conserved in the process. Energy conservation is automatic but the momentum condition

$$\bar{k}_2 = \bar{k}_1 + \bar{k}_1 \qquad (6.3.19)$$

usually requires the use of a birefringent crystal. Since the interaction between the three waves then takes place over relatively long distances, stringent demands are made on the optical quality of the crystal and since the power levels are high there are considerable demands on the damage resistance [824]. So important, however, have been the applications of non-linear optics that materials, many of which were hardly known before 1960, are now available as large single crystal of virtually perfect optical quality [815]. The usual materials used for second and third harmonic generation from near-infrared lasers are ammonium dihydrogen phosphate (ADP) and lithium niobate. Further into the infrared, frequency doubling of the TEA CO_2 laser is important in providing intense pulses for the 5 μm regions [825]. The usual non-linear material for this purpose is tellurium [826] which can be phase-matched but which, having a long coherence length, can also be used just as it is. Doubled pulses from CO_2 lasers are useful for pumping spin-flip lasers [827] at 5 μm, whenever a CO laser is unavailable. Tripling of the CO_2 laser to give powerful coherent radiation in the 3 μm region is beginning to be of importance. The normal non-linear material used here is a mixture of liquid CO and O_2 to which some SF_6 has been added [828]. The SF_6 has a positive dispersion between v_1 and $3v_1$ whereas the mixture of CO and O_2 has a negative one. By a judicious admixture it is possible to achieve perfect phase-matching. In the 10 μm region, several new materials, e.g. the chalcopyrite semiconductors [829] such as $CdGeAs_2$, the thiogallates such as $AgGaS_2$ [830] and the selenogallates such as $AgGaSe_2$ [831], have proved very useful. These materials can also be used for sum generation in the infrared but this tends to be a rare usage because in most cases it is better to achieve the same end by difference frequency generation from powerful visible region lasers.

Difference frequency generation was originally introduced as a means for generating far-infrared radiation from visible or near-infrared primary laser sources. Thus Zernike and Berman in 1965 [832] mixed the 1·059 and 1·073 μm lines from a Nd/glass laser in quartz under phase-matched conditions and observed coherent far-infrared generation near $100 \, cm^{-1}$. Later in 1968, Yajima and Inoue [833] succeeded in beating together the R_1 and R_2 lines from a Q-switched ruby laser in zinc telluride to produce radiation at $29 \, cm^{-1}$. Zinc telluride crystallises in the cubic system so the phase-matching equation for difference frequency generation, i.e.

$$\overline{k}_3 = \overline{k}_1 - \overline{k}_2.$$

that is

$$n_3 \tilde{v}_3 = n_1 \tilde{v}_1 - n_2 \tilde{v}_2 \tag{6.3.20}$$

cannot be satisfied in the conventional way but the coherence length defined by

$$l_c = [2(n_1 \tilde{v}_1 - n_2 \tilde{v}_2 - n_3 \tilde{v}_3)]^{-1} \tag{6.3.21}$$

is nevertheless quite large (~ 0.6 mm), so effective interaction is still possible. The non-linear coefficients are much larger for ZnTe than they are for quartz (in fact by a factor of 500!) so the twin effects lead to an efficient rate of conversion. Microwave radiation in the 60 GHz region can be produced by beating adjacent lines of the CO_2 laser against one another. Bridges and Chang [639] used this method, in GaAs, as part of their programme to measure accurate absolute frequencies for the CO_2 laser. Their work was an extension of that of Chang, Van Tran and Patel [834] who were able to satisfy the index-matching requirement by using the GaAs crystal to fill a waveguide. The phase-velocity at the difference frequency will be determined by the twin influences of the refractive index of the filler and of the physical size of the waveguide. By an appropriate choice of waveguide size, a long coherence length can be ensured. Van Tran and Patel [835] used n-InSb to mix lines from the 10·6 and the 9·6 μm bands of the CO_2 laser and thus generate coherent radiation near $100 \, cm^{-1}$. The phase-matching is not achievable conventionally in this cubic crystal but by invoking magnetoplasma effects in the impure material they could nevertheless get the refractive index at the difference frequency to equal that at the CO_2 laser frequencies. Using this technique the other way round provides a rather powerful form of magnetic resonance spectroscopy.

The development of a host of optically pumped far-infrared lasers has tended to reduce the interest in far-infrared difference frequency generation but this has been compensated for by the development of intense interest in *tunable* near-infrared generation by difference mixing of dye-laser beams. An early but nevertheless elegant example came from the work of Dewey and Hocker [836] who generated radiation in the 3–4 μm region by beating the output of a DTTCI dye-laser against its own Q-switched ruby laser pump radiation in a phase-matched $LiNbO_3$ crystal. The two laser beams were necessarily locked in time so there was no problem in synchronizing the two sets of pulses. Dewey and Hocker were only able to achieve a line width of about 10 cm^{-1} so the resolution was rather limited but modern systems can achieve much better. Thus, using continuous-wave argon ion pumping from a stabilised laser, Pine [837] has obtained line widths of better than 15 MHz with a tunability ranging from 2 to 4 μm. Using this system he has produced some excellent examples of Doppler-limited high-resolution spectroscopy. An example, the $v_3 P(7)$ line of methane, is shown in Fig. 6.15. This system has been used in particular to investigate the spectrum of SF_6 in the 10.6 μm region [838] and, by determining the anharmonic parameters to high accuracy, it proved possible to produce a good model for the initial stages of the multiphoton dissociation of this molecule (see section 7.4.1). At longer wavelengths it is possible to get similar performance by mixing in other crystals; thus proustite (Ag_3AsS_3) is usable to 12 μm [839] and CdSe [840] and Te [841] to about 25 μm. Apart from their value as tunable coherent spectroscopic probes, these lasers have some promise as drivers for uranium isotope separation machines based on selective photochemistry in the v_3 band at 16 μm.

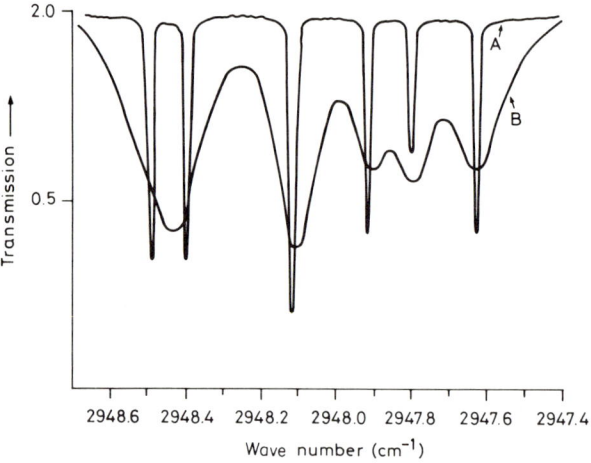

FIG. 6.15. The $v_3 P(7)$ line of methane, after Pine [837]. Curve A is for a gas pressure of 10 torr; curve B is for the same pressure of methane but with the lines broadened by the presence of 1 atmosphere of air. Compare this spectrum with those given in Figs 5.15 and 5.16.

Closely related to difference-frequency generation is the topic of parametric amplification/oscillation. In fact if one simply rearranges the basic equation to read

$$v_1 = v_2 + v_3, \qquad (6.3.22)$$

then it follows that a medium pumped by intense radiation at a high frequency v_1 will show gain at two lower frequencies, v_2 and v_3, provided the system is resonant at these frequencies and provided the phase-matching condition is satisfied. Optical parametric oscillators or OPOs [842] are conceptually close to their microwave counterparts and it is usual therefore to describe the three fields using microwave terminology. One has thus a pump v_p, a signal v_s and an idler v_i. The necessary resonance condition is achieved by mounting the non-linear crystal between two wavelength-selective mirrors which form the required Fabry–Perot cavity. If the refractive indices of the crystal are varied in any way, then the combination of signal and idler which satisfies the phase-matching condition will change and the output frequency will accordingly alter. The tuning is most readily achieved by rotating the crystal but pressure and electric field tuning are also occasionally used. In operation, the pump-beam enters the OPO through one of the mirrors which is transparent at the pump frequency; it then interacts with radiant noise fields in the crystal leading to a build up of and eventual oscillation on the signal and idler frequencies. There are two basic kinds of OPO, the doubly-resonant in which both signal and idler are fed back into the cavity and the singly resonant in which only the signal is so fed back. The doubly resonant OPO has a lower threshold and because of this it was prominent in the early days of the art but it gives unacceptably large variations of output frequency so nowadays only the singly resonant type is ever used in earnest.

OPOs can be used with both continuous-wave and pulsed pumping, but since the continuous-wave version is unstable, difficult to operate and very expensive to construct, it is fading in significance. With its demise goes the hope of using OPOs as wide-band tunable coherent infrared spectrometers, at least for routine work. For specialised applications, however, the pulsed OPO is very attractive because of the high peak-powers which are available. Selective photochemistry where it is merely required that the output frequency be adjustable rather than smoothly tunable provides an obvious application. Thus Dai, Kung and Moore [843], using a Nd:YAG pumped $LiNbO_3$ OPO, have carried out a thorough study of the multiphoton dissociation of C_2H_5Cl in the 3·3 μm region. Other materials which have been used for pulsed OPOs are $LiIO_3$ [844], proustite [845] and CdSe [846]. The Nd:YAG pump is still very prominent, but other high-power infrared lasers permit operation much further into the infrared. A particularly attractive one is the HF chemical laser which when used with a CdSe non-linear crystal can provide tunable radiation in the interesting 14·1 − 16·4 μm region [847].

Another variant on the same sum/difference generation theme is provided by the modulator in which the simultaneous irradiation of a non-linear crystal by a powerful infrared laser beam and by microwave radiation can produce a coherent output tunable over several GHz each side of the laser line [848]. Modern versions of the modulator [849] can produce over 90% infrared to infrared conversion efficiency.

A still further variation on this useful theme is provided by up-conversion in which an infrared radiant field is mixed with a visible region laser beam to produce a resultant visible output at a higher frequency. There are two main applications here: infrared imaging and photon counting. The infrared imaging concept [850] owes its attractiveness to the possibility of directly converting an infrared scene into a visible one. It certainly works [851], but to get acceptable definition leads to a fall-off in conversion efficiency. This, together with the rapid advances which have occurred in scanning and staring imaging technology (see section 7.1), has led to a decline of interest in this particular way of rendering an infrared image visible. The infrared photon-counting application is, however, still very much alive [852]. The idea here is very simple. If one has a 100% efficient up-converter, then each low-energy infrared photon will be converted into one high-energy visible photon. When these converted photons are recorded by the very efficient detectors which are available in the optical region, photomultipliers, for example, a dramatic increase in the S/N will result. Up-conversion is mostly used by the near-infrared astronomers who can thus get noise equivalent powers of the order of 10^{-14} W at $3\cdot5$ µm. However, as is usual, a new technology tends to stimulate the older ones which it is attempting to replace and there have been some remarkable developments in conventional infrared detectors for the near infrared and these have made the position of the up-converter, which at the moment is far from being a 100% efficient device, rather problematical. To compensate, up-converters are now finding a new application as power monitors for infrared picosecond pulsed lasers [853].

6.3.3.3 Nonlinear applications of the Raman effect and of four-wave mixing

In the Raman effect, incident photons of frequency v_L are inelastically scattered to an emerging frequency $v_S = v_L \pm v_R$ where v_R is a characteristic resonance frequency of the medium. When the shift is to the red, one has Stokes scattering, i.e.

$$v_S = v_L - v_R, \tag{6.3.23a}$$

and when it is to the blue, one has anti-Stokes scattering, i.e.

$$v_{aS} = v_L + v_R. \tag{6.3.23b}$$

The anti-Stokes scattering, to conserve energy, has to involve the de-excitation of molecules already in an excited state of energy hv_R, so, under normal thermal equilibrium conditions, anti-Stokes lines are reduced in intensity by a Boltzmann factor $\exp(-hv_R/kT)$ compared with their Stokes counterparts.

This means, in practice, that anti-Stokes lines are only observed for small frequency shifts. The Raman effect was discovered in 1928 by Sir C. V. Raman who received a Nobel prize in 1930 for his work. It quickly established itself as a powerful weapon in the armoury of the physical chemist [854] despite its great weakness where the total scattered power is only of the order 10^{-6} of the incident. With normal spectroscopic observation methods and isotropic illumination, it follows that the power which can be collected and studied is going to be only of the order 10^{-9} of the incident power. The reasons why the chemists were so interested in so weak an effect were manyfold, but some of the more important were:

1. The symmetry of many molecules, and of many crystal unit cells, is such that not all of their vibrations are active in infrared absorption.
2. The Raman effect essentially shifts a resonance away from a region where the medium may be strongly overall absorbing say in the infrared, into a region, say the visible, where the medium may be transparent. Raman spectroscopy has thus made major contributions to the investigation of the spectra of aqueous solutions.
3. The scattered light shows polarisation characteristics because of the tensorial character of the polarisability—this reflecting essentially the tensor nature of χ. This phenomenon gives the possibility of determining the symmetry of resonant modes even for non-oriented media.

These factors combined to make Raman spectroscopy an essential complement to infrared spectroscopy for the investigation of molecular structure and the spectroscopists in the forties and fifties were prepared to make heroic efforts to ensure good Raman spectra. The exciting lines were produced by high-power, low-pressure discharge lamps such as the mercury "Toronto" arc which gave usable lines at 404·6, 435·8, 546·1 and 577·0 nm. High-luminous throughput Littrow type glass-prism spectrometers were developed to go with them and great ingenuity was devoted to the art of sensitising photographic plates. Later on the experimentalists lot was greatly improved by the development of photoelectric recording but it was the introduction of the laser, as an exciting source, which really revolutionised the art [855]. Nowadays laser Raman spectrometers are universal and their power is such that perfectly usable spectra can be obtained from quite microscopic specimens [856].

All of this was only of passing interest to the infrared technologist, who, despite the attractions of a frequency shifting technique, could not take seriously so weak an effect. However, in 1962 came the dramatic discovery that when Raman spectra are excited by intense Q-switched pulses, then the power transferred to the Raman line could be very high ($\sim 10\%$), provided the laser power was above a certain threshold value [857]. This was the celebrated phenomenon of *Stimulated Raman Emission*. After its discovery, ordinary Raman scattering came to be known, rather inaptly, as spontaneous Raman emission. Raman scattering can involve any form of elementary excitation but the usual ones are vibrational and/or rotational excitations. Stokes Raman shifts cover the full gamut of the corresponding frequencies, i.e. from fractions of a cm^{-1} to several thousand cm^{-1}. Very few of the spontaneous lines, however, can be obtained in stimulated emission. Basically, those that can are the totally symmetric, or A_1, varieties which tend to be sharp. Amongst these, the stronger the spontaneous line, the lower is the stimulated threshold.

Stimulated Raman emission, or the spontaneous kind for that matter, can be analysed within two different, but in the final analysis equivalent, frameworks. One can, at least formally, attribute it [858] to the third-order non-linear susceptibilities $\chi^{(3)}$ (ω_S, ω_L, $-\omega_L$, ω_S) and $\chi^{(3)}$ ($2\omega_L - \omega_S$, ω_L, ω_S) and one can also consider it to arise from the dependence of the molecular polarisability α, on the normal coordinate of vibration Q. This latter is Placzek's [859] approach and it is valid in the limit where the exciting line frequency is considerably less than that of any electronic resonance but considerably greater than that of any vibrational resonance. This condition is usually satisfied by transparent media excited by a visible region laser. The polarisability does, however, have resonance denominators in its formal dispersion theory expansion (see equation 6.3.14 for example) and the intensities observed in spontaneous scattering increase markedly compared with the Placzek theory predictions as the exciting line approaches a strong electronic absorption band. This is the *Resonance* Raman effect [860], and like the stimulated effect it varies in magnitude from line to line in the spontaneous spectrum. However it works just as well in the stimulated version and some of the lowest thresholds are observed for the vibrations of groups within a molecule which are chromophores for near ultraviolet bands. Thus the NO_2 symmetric stretching band in $C_6H_5NO_2$ ($\Delta\tilde{v} = 1345$ cm^{-1}) is very easy to get in stimulated emission. There are several other low-threshold lines in easily accessible liquids—e.g. C_6H_6, 992 cm^{-1}; C_6H_5Cl, 1002 cm^{-1}; CCl_4, 459 cm^{-1}; CS_2, 656 cm^{-1}; etc—and since one thus has a frequency-shifter in which the shift can be altered merely by changing the liquid in the cell, stimulated Raman emission has proven to be a very useful technique for the laser engineer. Stimulated Raman emission in gaseous hydrogen ($\Delta\tilde{v}$ = 4155 cm^{-1}), pumped by either the Q-switched ruby or Nd:YAG lasers, gives useful lines in the near to mid infrared.

The Raman scattering cross-section per molecule per unit solid angle ($d\sigma/d\Omega$) depends on the inherent molecular properties, i.e. ($d\alpha/dQ$) and also on the Stokes frequency v_S. The chemists naturally prefer to separate these effects and, using the Placzek theory, it is usual to express the total scattered power as a fraction of the incident power in the form [859]

$$(P_S/P_L) = \frac{16\pi^3 h N v_S^4 (\bar{\alpha}')^2 \, l}{9c^4 v_R [1 - \exp(-hv_R/kT)]} \left(\frac{6}{6 - 7\rho}\right), \qquad (6.3.24)$$

where N is the number of molecules per unit volume, l is the specimen length, $\bar{\alpha}'$

is the trace of the derived polarisability tensor and ρ is the degree of depolarisation. For most stimulated lines ρ is very small so the last factor can be ignored. To give a quantitative assessment of (6.3.24) one could consider the 459 cm^{-1} line of CCl_4. The polarisability derivative α' in terms of mass adjusted normal coordinates can be related to the derivative of the bond polarisabilities with respect to the bond length $\bar{\alpha}'_{C-Cl}$ by the relation [859]

$$\bar{\alpha}'_1 = 2\, M_{Cl}^{-1/2}\, \bar{\alpha}'_{C-Cl}, \tag{6.3.25}$$

from which with the experimental value, $\bar{\alpha}'_{C-Cl} = 2 \times 10^{-10}\,m^2$, one ultimately derives $(P_S/P_L) = 2\cdot4 \times 10^{-6}$ for an excitation wavelength of 500 nm, in good agreement with the figure quoted earlier. The fourth power dependence on ν_S, given in equation (6.3.24) derives from a third power dependence via the Einstein emission coefficient and a first power dependence which comes from the quantum-mechanical dispersion theory. A similar factor in Rayleigh scattering is responsible, as is well known, for the blue colour of the sky. In Raman scattering this factor severely militates against infrared excitation. Thus the scattering efficiency falls by a factor of sixteen in going from Argon ion to Nd:YAG excitation. In going to the CO_2 laser the factor is 10^{-5}! Raman scattering in the infrared can thus only be significant if the scattering cross-section increases to compensate. The best-known example occurs for semi-conductors where the resonance Raman effect permits not only spontaneous [861] but also stimulated Raman emission to be obtained. It should perhaps be remarked that there are very few known examples where there are no resonance effects observable. Thus for most transparent liquids [862], the total scattered Raman power varies more rapidly with ν_S than the fourth power.

The optical engineers are less concerned with these fine details and, as mentioned above, prefer to lump everything together into the differential scattering cross-section $(d\sigma/d\Omega)$ per unit volume which is defined very simply by

$$(P_S/P_L) = (d\sigma/d\Omega)\Delta\Omega z, \tag{6.3.26}$$

where $(d\sigma/d\Omega)$ is defined to be the scattering cross-section for 90° observation. To find the power at an azimuthal angle θ less than 90° it is usual to assume dipolar scattering and then P_S varies as $\sin^2 \theta$. As an example of the use of (6.3.26), the scattering cross-section $(d\sigma/d\Omega)$ for the CCl_4 459 cm^{-1} line has an experimental value of $5\cdot0 \times 10^{-6}\,m^{-1}\,ster^{-1}$ at the ruby laser wavelength of 694·3 nm from which, by multiplying by 2π, one finds a value of $(P_S/P_L) = 3\cdot1 \times 10^{-6}$ for a 10 cm long sample. This agrees remarkably well with the value given earlier which was derived from a completely different measurement scheme [863]. However, when the fourth-power dependence is borne in mind, the agreement is much less good. The discrepancy may be genuine, reflecting the severe difficulty of making these absolute measurements, or else it may be an artefact reflecting the effects of internal field corrections.

To pass from the simple spontaneous effect into the high power regime requires a different formalism. Several have been proposed [864], ranging from the relatively simple all the way to the formidable with full QED corrections, but an adequate account is possible within the not too difficult rate-equation approach developed by Yariv [817]. He writes the equation for photon conservation in the form

$$\frac{dN_S}{dt} = -\frac{dN_L}{dt} = S[C_0 N_L (N_S + 1) - C_1 N_S (N_L + 1)], \qquad (6.3.26)$$

where N_S and N_L are the photon occupancies of the Stokes and Laser modes respectively, S is a phenomenological scattering coefficient to be determined by experiment and C_0, C_1 are the respective probabilities of finding a scattering molecule in either its ground or first excited state. Yariv derives the factor in square brackets by treating the applied electromagnetic field as a combination of photon creation and annihilation operators operating on the modes which are represented as simple harmonic oscillators. By considering scattering into all the possible modes which lie within the line width Δv, Yariv shows that S may be related to the experimental quantity $(d\sigma/d\Omega)$ by the relation

$$S = \frac{v_L N c^4}{3v_S^3 n(v_S)^3 n(v_L) V \Delta v C_0} \left(\frac{d\sigma}{d\Omega}\right) \qquad (6.3.27)$$

where N is the number density of scatterers, V the volume and $n(v)$ is the refractive index of the medium. The next step is to consider various regimes for (6.3.26). The simplest is to assume that N_S is so small that it can be set equal to zero as far as N_L is concerned. One then has

$$(dN_L/dt) = -SC_0 N_L, \qquad (6.3.28)$$

which means that the laser beam suffers a Lambertian fall in intensity as it progresses through the medium with an absorption coefficient

$$\alpha = SC_0 n(v_L)/c. \qquad (6.3.29)$$

The "absorbed" power promptly re-emerges in the form of Raman emission at v_S but slightly reduced because each laser photon of frequency v_L is replaced by a lower frequency Stokes photon of frequency v_S. Assuming α to be very small, we can replace $(1 - \exp - \alpha l)$ by just αl and derive the emitted Raman power as

$$P_S = P_0 SC_0 v_S n(v_L)/c v_L. \qquad (6.3.30)$$

This equation, via the analogue of (6.3.27), eventually leads to equation of the type of (6.3.24) when a particular theory is used to evaluate S. In other words, and not surprisingly, we are discussing the spontaneous Raman scattering.

The next approximation is to consider N_S finite but not large enough to perturb C_0 and C_1 from their equilibrium values. One now has

$$(dN_S/dt) = S(C_0 - C_1)N_L N_S, \qquad (6.3.31)$$

so that the Stokes photon density grows exponentially with distance according to

$$I_S(z) = I_S(0)\exp(g_S z), \qquad (6.3.32)$$

where the gain factor g_S is given by

$$g_S = \left(\frac{d\sigma}{d\Omega}\right)\frac{Nc^2[1-\exp(-hv_R/kT)]}{3hv_S^3 n(v_S)^2 \Delta v}I_L. \qquad (6.3.33)$$

The gain in the single mode which phase matches with the laser beam thus depends on the laser power, the spontaneous scattering cross-section and inversely on the line width. These are all experimental facts of stimulated Raman emission as mentioned earlier. The beam will build up to macroscopic proportions provide the losses are less than the gain. The losses can be reduced by the use of a Fabry–Perot cavity but usually the gain in the Raman process is so high that large power conversions are observed from apparently super-radiant or "straight-through" systems. Actually there is usually slight feedback induced by the cell windows and this suffices. If this effective reflectivity is R and the cell length L then the threshold condition is

$$g_S = -(\ln R)/L. \qquad (6.3.34)$$

This combined with (6.3.33) gives the threshold intensity. Typical values with $R \sim 10^{-3}$ and $L \sim 5$ cm are of the order 50 MW cm^{-2}.

The final regime is when N_S becomes so large that C_0 and C_1 are perturbed. The Raman process then passes into a parametric instability mode in which there is an explosive build-up in the vibrational excitation of the medium. The Stokes power then saturates at a high value essentially because so much power is being converted out of the laser beam. New phenomena then occur such as the generation of the so-called "second", "third" etc. Stokes lines which in the simplest visualisation can be thought of as coherent stimulated Raman emission produced by the first Stokes Raman line acting as the exciting radiation—in other words Raman produced by Raman. These lines have frequencies $v_L - mv_R$ where m is an integer. It should be noted that these lines do *not* coincide with the overtones, which will experience slight departures from even spacing because of anharmonicity.

Anti-Stokes emission is not as important as is Stokes emission in an infrared context. It is distinguished from the Stokes emission by being confined to a very narrow conical shell around the exciting beam direction, whereas the Stokes emission is in the form of a much more diffuse cone. This is a consequence of the necessity of phase-matching for efficient anti-Stokes emission. Under conditions of normal dispersion, it is always possible to find angles which produce phase-matching at both the Stokes and anti-Stokes wavelengths. In the medium power or linear amplification regime of stimulated emission the anti-Stokes radiation can only be produced when *both* the laser radiation and the stimulated Stokes radiation are present. Thus, unlike the

Stokes case where the medium excited above threshold is *always* amplifying for the Stokes frequency, in the anti-Stokes case the medium is normally attenuating. In the simplest way of looking at it, power passes to the anti-Stokes line from the laser line via the Stokes line. This is reasonable since excited state molecules are required to generate the anti-Stokes radiation. Electromagnetically one can say that the driving of the medium at its resonance frequency $\nu_R = \nu_L - \nu_S$ by the applied laser field modulates the permittivity at this frequency and this modulation puts sidebands on the laser line. The upper sideband is, of course, the anti-Stokes line. However in the saturation or non-linear regime new phenomena arise and it is possible to get anti-Stokes radiation generated by three-body events of the type

$$\nu_L + \nu_L + \text{molecule} \rightarrow \nu_S + \nu_{aS} + \text{molecule}, \tag{6.3.35}$$

that is the simultaneous conversion of two laser photons into a Stokes–anti-Stokes pair.

Apart from simple frequency shifting, the stimulated Raman effect has been used to provide several forms of tunable coherant radiation in the infrared. The Raman spin-flip laser has been mentioned earlier but there are several others of which perhaps the most significant are polariton scattering in non-linear ionic crystals and resonance enhanced four-wave scattering from alkali metal vapours.

Polaritons are quantised entities which can propagate in ionic crystals and which can be regarded as mixtures of phonons and photons. The momentum of a polariton is thus partly mechanical and partly electromagnetic. The coupling of the mechanical vibrations to the electromagnetic field to produce polaritons has its origin in the very strong dispersion near the transverse optic frequency. One can thus combine the general result $\bar{k} = \omega/\varepsilon^{1/2}c$ with the dispersion relation for phonons (equation 5.3.17b) to give, in the loss-loss case, the dispersion relation for polaritons, namely

$$\bar{k} = \left(\frac{\omega}{c}\right)\left[\varepsilon_\infty + \frac{(\varepsilon_S - \varepsilon_\infty)\omega_T{}^2}{\omega_T{}^2 - \omega^2}\right]. \tag{6.3.36}$$

This function, plotted in Fig. 6.16, shows the expected gap between ω_T and ω_L but in addition there is an exclusion zone lying between the lines corresponding to the propagation of low-frequency and high-frequency photons. The result is the two branches shown in Fig. 6.16. In the lower branch the polariton is photon-like for low values of $\bar{k}$ whilst it is phonon-like for large values of $\bar{k}$. Polaritons in non-centrosymmetric crystals can be both infrared and Raman active so various possibilities arise in non-linear crystals, such as $LiNbO_3$, for the generation of infrared radiation as well as the shifting of the frequencies of visible and near-infrared lasers. Raman scattering by polaritons is particularly interesting because of the tunability possibilities. It was remarked earlier that Raman scattering, like infrared absorption, is a $\Delta\bar{k} = 0$ process. Now whilst this is nearly always so in practice it is not necessarily so in principle. All that is required in the scattering process is conservation of energy and momentum, so, taking the wave vector of the laser, the polariton and the Raman wave to be $\bar{k}_L$, $\bar{k}_p$ and $\bar{k}_s$, one must have

$$(\bar{k}_p)^2 = (\bar{k}_L)^2 + (\bar{k}_s)^2 - 2\bar{k}_L \bar{k}_s \cos\theta, \tag{6.3.37}$$

where θ is the angle between $\bar{k}_L$ and $\bar{k}_s$. In principle therefore by varying the angle of observation one can scan a small but nevertheless significant part of the dispersion diagram! In practice, of course, $\bar{k}_p$ will be of the same order as $\bar{k}_L$ and $\bar{k}_s$ unless θ is very small, and then, because of the asymptotic nature of the lower branch of Fig. 6.16 for $\bar{k}_p$ large, there will be no perceptible variation of frequency with θ. When θ is small, however, that is 4° or less, there is a rapid variation of frequency with angle so in the nearly forward scattering direction some very interesting phenomena might be expected. At these small angles one can replace $\cos\theta$ by $1 - \theta^2/2$ and then, by assuming only weak dispersion in the neighbourhood of the laser frequency, one can write (6.3.37) in the form

$$(\bar{k}_p)^2 = (\bar{k}_L - \bar{k}_s)^2 + \bar{k}_L \bar{k}_s \theta^2$$
$$= 4\pi^2 n^2 \bar{v}_p^2 + \bar{k}_L \bar{k}_s \theta^2. \tag{6.3.38}$$

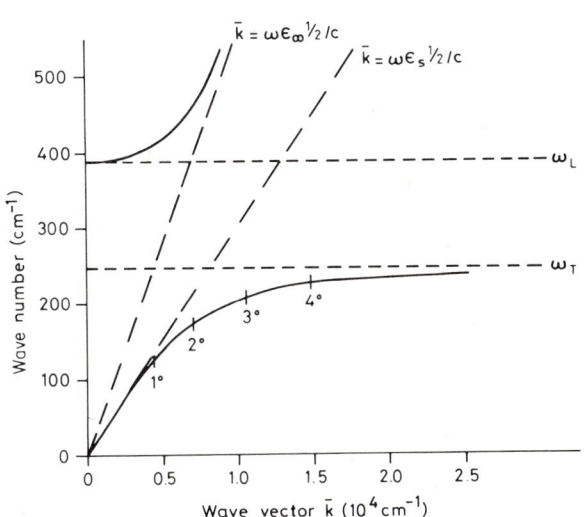

FIG. 6.16. Some calculated dispersion curves for the 245 cm^{-1} polariton in lithium niobate.

This represents a family of hyperbolae in the dispersion diagram and the intersections of these hyperbolae with the polariton dispersion curve give the frequencies of the Raman displacements for the appropriate angles. Some calculated intersections for LiNbO$_3$ are shown in Fig. 6.16.

The experimental results [865] confirm the above theory apart from the expected result that the polariton dispersion curve is considerably modified due to the finite width of the transverse optic resonance. The spontaneous Raman scattering shows the expected frequency variation in the near forward direction. The 248 cm^{-1} polariton in LiNbO$_3$ is readily obtained in stimulated emission when excited by a Q-switched ruby laser; the forward scattering displacement frequency is found [866] to vary from 248 down to 50 cm^{-1} as θ varies from 5 to 0·5°. The scattering at angles very close to the forward direction can formally be thought of as photon scattering by photons; it owes its intensity to the large non-linear susceptibility in LiNbO$_3$. It can also be considered to be a parametric process dominated by the nearby vibrational resonance. Because it is formally a parametric process, one can extract the idler radiation provided the pumping is intense enough. Yarborough and his colleagues [867] have succeeded in this way im producing intense pulses of tunable radiation in the far infrared.

Four-wave mixing processes, in general, are very useful for generating coherent infrared radiation. Thus the group at the Magnet Laboratory of MIT have used such processes in semiconductors to mix various combinations of CO$_2$ laser lines and in this way to produce many new coherent outputs in the 12–16 μm region [868]. A typical process would involve 2 photons from the 10·6 μm band mixing with one photon from the 9.6 μm band to give a long-wave resultant near 11·8 μm. These processes in germanium can be very efficient with conversions of the order of 1 %. Four-wave mixing in alkali metal vapours can give not only a very useful way of exploring the energy-level diagram under high resolution conditions, it can also provide a convenient tunable near-infrared source. The most celebrated experiment was that of Sorokin and his colleagues in potassium [869]. They focussed two orthogonally polarised beams, produced from two nitrogen-laser synchronously pumped dye lasers, into a heated cell containing potassium vapour. One beam was tuned to lie just above the 4s–5p resonance doublet at 24 701·4 and 24 720·2 cm^{-1} whilst the other was at a variable but lower frequency. The first beam, say of frequency v_L, induced stimulated Raman emission at a frequency v_S on the transition from the virtual state down to the 5s level. When the second beam of frequency v_p was also present, the four-wave process

$$v_{ir} = v_L - v_S - v_p \qquad (6.3.39)$$

led to powerful coherent emission in the 2 to 5·4 μm region. By varying v_p, v_{ir} could be scanned readily over the whole range. The necessary phase matching is produced, in the collinear geometry, by taking advantage of the strong dispersion near the 4s–5p resonance. Thus merely by adjusting v_L slightly, it is possible to get perfect phase matching and thus efficient conversion.

Stimulated Raman emission in hydrogen gas combined with four-wave mixing can produce some very useful lines. Thus, as an example, Byer and Trutna [870] in 1978 demonstrated 16 μm generation by scattering CO$_2$ laser radiation from the intense rotational excitation produced in the gas after stimulated Raman emission pumped by a Nd/YAG Q-switched laser. This radiation was not powerful enough to serve as the driver of a useful isotope separation machine but the promise is certainly there.

6.4 Optical materials for the infrared region

6.4.1 *Transmissive materials*

The various forms of glass, which are so useful in the visible and near ultraviolet regions of the spectrum, are of very limited use in the infrared because of the onset of intense absorption due to the amorphous analogue of the reststrahlen process. In fact ordinary glass is opaque beyond 2·5 μm and does not start to transmit again till well into the near-millimetre wave region [871]. Silica in its crystalline and fused forms is a little better, being useable out to 3·3 μm and again beyond 50 μm but since much infrared interest centres on the 8–14 μm region, even the crystalline variety has found only specialised applications. The main material which has been traditionally used for infrared work is crystalline rock salt, NaCl. It came originally from natural sources but is now grown artificially and is available commercially as specimens many inches in diameter and of considerable thickness. NaCl transmits to about 16 μm when heavy absorption due to the reststrahlen band centred at 61 μm sets in. The reststrahl frequency is determined mostly by the mass of the ions involved (see section 5.3) so the use of heavier alkali halides permits further extension into the infrared. Thus KBr which is transparent to 25 μm is widely used as is CsI which goes out as far as 70 μm. The main drawback to the use of alkali halides is that they are hygroscopic and their optical quality rapidly deteriorates if they are left in ordinary atmospheres at ordinary temperatures. Cell windows must therefore be stored in a dessicator whilst the windows and lenses of spectrometers etc. must be fitted with heaters to keep them 10 °C or so above ambient unless they are actually sited in an evacuated part of the instrument. Alkali halides are also soft and have relatively low refractive indices so other materials are required to complement them in applications where these aspects are important. Other ionic crystals are also used such as AgCl and BaF_2, but these are very expensive so they would only be used in applications where resistance to water was a first-order requirement. One can also use the reststrahl argument the other way round and use very light ions to yield materials with some degree of transparency in the very-far-infrared/millimetre-wave, region. The principal candidates here are LiF, BeO and SiO_2. Semiconductors provide some very important materials since in pure form they are quite transparent at all wavelengths up to that of the absorption edge. Germanium with $\lambda_g = 2$ μm is very widely used. It has a high index (~ 4) is non-hygroscopic and can readily be cut and polished. Unfortunately it is expensive and getting more so. Silicon has a very wide infrared window (1 μm $< \lambda <$ 300 μm) but impurity atom absorptions introduce some narrow regions where it is far from transparent so the whole window cannot be used continuously. Diamond is transparent throughout much of the infrared and is the hardest material of all. Unfortunately it is also the most expensive of all, but for some applications, like providing the windows for ultra-high-pressure infrared cells only diamond will do. Zinc selenide is very attractive in many respects and for some applications almost ideal but it is still rather costly. There are several further materials which are used in some other rather more specialised applications. A summary of the properties of some commercially available materials, taken from the sales literature of SPECAC† is given in Table 6.8. More details are available in a series of papers [872] produced by McCarthy. No one of these substances meets all the requirements put upon an ideal material but there are now sufficient of them available commercially to enable the infrared designer to reach a satisfactory compromise in most cases.

6.4.2 *Reflective materials*

For most infrared applications, glass blanks front-surface coated with aluminium provide very adequate mirrors. Reflectors can also be fashioned directly from solid metal. Copper is a favourite metal as also is aluminium and for particularly delicate applications a layer of gold can be deposited on the base metal substrate. The relectivity of a highly polished copper specimen in the infrared can approach unity so the gold plating would only be needed for those applications, the mirrors in long-path cells for example, where the highest possible reflectivity was required. Aluminium has the drawback that it instantly forms an oxide layer when the clean metal is exposed to oxygen. This layer of Al_2O_3 is quite heavily absorbing in the far to mid infrared.

The development of high-power lasers has put some new and rather stringent requirements on metallic mirrors. They must, for example, be highly reflecting yet have high damage thresholds and have high melting points. Molybdenum mirrors have proved very useful in CO_2 lasers. They have hard non-tarnishing polishable surfaces with reflectivities in excess of 98 % at 10 μm so a noble metal or dielectric coating is unnecessary. The thermal distortion is small and the laser damage threshold is the highest known typically 200 kW cm^{-2} for continuous-wave operation [873].

6.5 Commercial infrared detectors

Although the most sensitive and the most specialized detectors continue to be assembled by specialists primarily for their own use and for that of their colleagues, there is now available a wide range of commercial detectors and it is nearly always possible to buy one to do any particular job.

For wide spectral range detection, thermal detectors have to be used. The

† The Spectroscopic Accessory Company Ltd., Unit 3, Lagoon Road, St Mary Cray, Kent.

four kinds available are bolometers, thermopiles, pyroelectric detectors and Golay cells. Bolometers operate at cryogenic temperatures. They are sensitive and moderately fast but they are also expensive and temperamental. For these reasons they tend to be reserved for frontier work in specialised research laboratories. The thermopile represents almost the opposite pole. It works at room temperature, is not very sensitive but is robust and cheap. Despite the development of newer detectors, the thermopile continues to be used in commercial infrared spectrometers for the middle to bottom range of the market. The main problems with thermopiles are to get effective absorption of the radiation and to get a medium-level/high-impedance signal out, suitable for driving the preamplifier of the detection electronics. The radiation receiver is often made from blackened gold foil which also serves to complete the electrical circuit. The very low output of the pile (which may in fact be a single junction) is converted to a more acceptable value by means of a high-ratio transformer. Pick-up is then a very real hazard but this can be obviated to a large extent by the use of effective mu-metal screening. The working parts of the thermopile are usually mounted in vacuum to increase the sensitivity and reduce the noise. Infrared radiation is then incident via a suitable window. The initial construction of a commercial thermopile is a delicate matter but once it is made its lifetime can be very long indeed provided it is handled with reasonable care. Some of the thermopiles introduced in the immediate postwar years are still working away perfectly happily. Thermopiles, and most other detectors for that matter, are available commercially from the Oriel Company.

The Golay cell, which was described earlier in section 4.4.4.10, is the most sensitive room-temperature detector—in fact it is very close to being thermal noise limited. The original design due to Marcel Golay [319] which did such sterling service in the early days of far-infrared spectroscopy has naturally evolved as new technologies have been introduced. The crucial device is the absorbing pneumatic membrane and this has been greatly improved in recent years as new polymeric films became available. This, combined with the use of modern clean-room facilities, has led to near optimum performance. The light source, which used to be an incandescent lamp, is now an LED and the light-beam detector is a photo-FET. Naturally all the subsequent electronics is solid state, mostly based on planar transistors. The heat generated within the detector envelope is greatly reduced by these innovations and this in turn produces better stability and a much longer life. The detector is also much more rugged and, needing much less power, can be operated in a much more stable mode. Because of the very broad-band response, some care is needed in specifying the sensitivity but with a black-body source at 500 K, a chopping frequency of 11 Hz and a bandwidth of 1 Hz, the NEP has a typical value of 10^{-10} W. Commercial versions (supplied by Cathodeon under the code name IR50) are supplied with entrance apertures of either 3 or 5 mm and a wide range of window materials is available. This includes KBr, CsI, KRS-5, Diamond, Quartz and Silicon. Quartz-windowed Golays are cheap and there

TABLE 6.8

Material	Useful range (μm)[a]	Refractive index at 5 μm	General information
Fused SiO_2: ultraviolet grade infrared grade	0·17–2·7 0·25–3·3	~1·45 at the sodium D line wavelength	Unaffected by most solvents, transmission at 0.2 micron. 98% for ultraviolet grade, 40% for infrared grade
MgF_2	0·11–8	Birefringent: n_o 1·378 n_e 1·389 at 0·59 μm	Slightly more soluble in water than CaF_2. Useful material from vacuum ultra-violet to infrared region: infrared and ultraviolet polarizers
LiF	0·12–7 120–∞	1·33	Slightly soluble in water, suitable for vacuum ultraviolet and near infrared
CaF_2	0·13–12 250–∞	1·40	Insoluble in water, resists most acids and alkalis. Its high mechanical strength makes it particularly useful for high-pressure work. Does not fog.
BaF_2	0·15–15	1·45	Insoluble in water, soluble in acids and NH_4Cl. Very sensitive to mechanical and thermal shock. Good resistance to fluorine and fluorides. Does not fog
Irtran 2 (ZnS)	2–13	2·22	Insoluble in water, normal acids and bases, and virtually all organic solvents. Reacts to strong oxidising agents. Good resistance to thermal and mechanical shock. Suitable for work in temperature range $-200\,°C$ to $+800\,°C$.
NaCl	0·25–16	1·52	Soluble in water and glycerine, slightly soluble in alcohol. Most common infra-red material due to low cost and good transmission. Fair resistance to mechanical and thermal shock and can easily be polished.
ZnSe	0·5–20	2·4	Insoluble in water. High resistance to chemical attack. Organic solvents, dilute acids and bases have no effect. Because of its low absorption in infrared, ideal for ATR work. Low absorption at 10·6 μm–hence popular window material for CO_2 lasers

Material	Transmission range	Refractive index[a]	Properties
AgCl	1–25	2·00	Insoluble in water, soluble in acids and NH_4Cl. Very sensitive to mechanical shock is malleable and will cold form. Corrosive to metals and alloys. Good resistance to thermal shock. Sensitive to strong ultraviolet radiation, will darken with prolonged exposure
Ge	3–23	4·01	Insoluble in water, soluble in hot sulphuric acid and aqua regia. Suitable for ATR work where high-pressure contact is required.
KCl	0·3–20	1·47	Similar to NaCl but with extended transmission, lower solubility and lower reflection losses
KBr	0·23–25	1·54	Soluble in water, alcohol and glycerine, slightly soluble in ether. Hygroscopic. Fairly good resistance to mechanical and thermal shock
AgBr	0·45–30	2·30	Similar to AgCl but with extended transmission
KRS-5	1–38	2·38	Slightly soluble in water, soluble in bases, not soluble in acids. Not hygroscopic. Good transmission range, ideal for ATR work
CsBr	0·24–40	1·66	Soluble in water and acids, soft and hygroscopic
CsI	0·24–70	1·74	Soluble in water and alcohol, soft and extremely hygroscopic. Very useful because of wide transmission range
Si	1·2–300 not continuous	3·4	Very hard and inert crystal with high refractive index. Useful for far-infrared spectroscopy in the region $30\text{–}400\ cm^{-1}$
Polyethylene	16–∞	1·52	Inexpensive far infrared window material. Insoluble in water, tends to swell and become contaminated with some organic solvents. Cannot be used for high-temperature work (mp $\sim 110\,°C$)
TPX	50–∞	1·43	Similar to polyethylene for far infrared but has the advantage of being transparent in the visible region

[a] Dependent on sample thickness.

are no problems with moisture take-up but on the other hand they can only be used beyond 50 μm. The alkali-halide windows are sensitive to water vapour but are cheap and very transparent. KRS-5 has all the well-known advantages and disadvantages of this material. In particular it is soft and tends to flow slowly under gravity. Beyond any doubt the diamond-windowed Golays are the most desirable but their high, and rapidly increasing, price makes them beyond the reach of many laboratories. Golay detectors are still microphonic and still much bulkier than all the competing thermal detectors but for those applications where the last ounce of S/N is needed over a wide spectral range from a room temperature device they cannot be beaten.

Pyroelectric detectors, despite considerable development work, are still somewhat inferior to the Golay cell in sensitivity but for many applications they more than make up for this by their shorter response times, much smaller size and much greater ruggedness. As an illustration virtually every commercial infrared spectrometer in the medium to high price bracket now features a pyroelectric detector. The commercially available types (the Plessey ones are typical) are high impedance devices based on either doped TGS or else on a lead zirconate ceramic. This element is followed by an FET operational amplifier and the high-level output from this combination can then be directly processed in conventional electronics and digital equipment. The lead zirconate element has a high Curie temperature (230 °C) so it can be used for high radiant levels; lithium tantalate is even better with a Curie temperature of 620 °C. The absorptivity of all these materials varies considerably with frequency so to ensure a uniform response in radiometric connections it is advisable to blacken the element. TGS is hygroscopic, has a very low Curie temperature, 49°C, and has a more structured response than the ceramics. It is also rather fragile but where the ultimate in sensitivity is required it is essential. Typically, ceramic pyroelectric detectors have NEPs of the order 2 $\times 10^{-9}$ WHz$^{-1/2}$ at 10 Hz modulation frequency whilst the TGS types better this by a factor of 2. The NEP increases by an order of magnitude for lead zirconate in going to 1 kHz but LiTaO$_3$ and TGS hold up rather better. This response to change of the modulation frequency does depend on the heat capacity and hence the size of the detecting element. So, if sensitivity can be sacrificed, it is possible to get very short response times. Detectors having an active area of 1 mm^2 or less and driving a current amplifier can be used for laser pulse analysis.

The narrow-band detectors used in the infrared are naturally much more sensitive than the broad-band ones and, in addition, they are usually much faster. In the very near infrared ($\lambda < 1$ μm), photomultipliers with S-20 or S-1 photocathodes can be used. These give best results when cooled, but cooling too much is detrimental. A good compromise temperature seems to be about -30 °C and coolers based on the Peltier effect are available to reach this sort of temperature. The S-20 photocathode is really a near-infrared-extended visible-region composition and when available (possible suppliers are RCA and

Mullard) it is better to use the S-1 type. This is based on a caesium coating deposited on an oxidised silver substrate (i.e. CsOAg). It shows a peak in sensitivity at 0.8 μm with a variation of as much as ± 0.1 μm. Like all photomultipliers there is considerable variation in performance from tube to tube and where possible the infrared practitioner will usually be prepared to pay the extra cost for a selected one. The NEP values at room temperature range between 10^{-10} and 10^{-11} W at 0.8 μm but a cooled selected tube can approach 10^{-12} W. With good post-detector electronics, photon-counting is then quite feasible. The S-1 photomultipliers are very fast with a time resolution characteristic of the order 5 ns. Similar, or better, time resolution is available now from a range of fairly sensitive photodiodes. The vast majority of these are based on silicon technology. They peak near 1 μm with a slow fall-off into the visible but a very rapid fall-off into the infrared. By about 1.2 μm their response is essentially zero. Vast numbers of silicon photodiodes are made and sold. The cheapest ones are low-sensitivity/low-response speed types used in such mundane things as burglar alarms and movement sensors in general. The most expensive are the high-sensitivity high-response speed types used in long-distance fibre-optic telecommunications. For this latter application, new types of sensitive photodiodes, peaking in response at the desired wavelengths of 1.3 and 1.55 μm, are now under intensive commercial development.

For applications extending further into the near infrared, the outstanding detectors are the PbS and PbSe photoconductors and the InSb photovoltaic device. Lead sulphide and lead selenide detectors are nearly always used cooled. A Peltier cooler is normally employed built in to the detector encapsulation. The useful wavelength range is 1–5 μm and the available sensitivity is high ($D^* > 10^{10}$). Detectors of this type are supplied by the Wentworth Laboratories in Sandy, Bedfordshire, England, and by Optoelectronics in the USA. The InSb photovoltaics can be used either at ambient temperature when the wavelength range is 1–7.5 μm with $D^* = 2 \times 10^8$ or cooled when, at liquid-nitrogen temperature, the wavelength range shrinks to 1–5.3 μm and the D^* becomes 5×10^{10}.

In the mid infrared, and particularly in the important 8–14 μm atmospheric window, the two outstanding narrow-band detectors are the mercury cadmium telluride ($Hg_x Cd_{1-x} Te$) photoconductor and the lead tin telluride ($Pb_x Sn_{1-x} Te$) photovoltaic. Both usually have the junction formed in epitaxial material and operate at liquid nitrogen temperature. The wavelength of peak response depends on the stoichiometry parameter x and can fairly easily be varied between 9 and 11 m. The sensitive area of these detectors is small (300 μm $\times$ 300 μm) so the D^* values are not particularly high, but the other side of the coin is that they are very useful for thermal imaging work and their rapid response times (usually RC limited to a few nanoseconds) makes them near ideal for work with mid-infrared pulsed lasers. The need merely to cool to liquid nitrogen temperatures makes it not at all difficult to use these detectors in the field. Cryogenic liquid cooling is the norm, but when this is impossible it is feasible to use Joule Thompson coolers to reach the optimum temperature.

At longer wavelengths still, liquid helium cooling becomes essential. The most widely used far infrared photodetectors are the Putley [874], the Rollin [875] and the hot-electron bolometer [876]. The latter two are available from QMC Industrial Instruments and the former from Mullard. All of these detectors depend on absorption by free carriers in slightly impure indium antimonide. In this material the effective mass is very low ($m^* = 0{\cdot}014\,m_e$) so even for nearly intrinsic material (10^{-19} m^{-3}) there will be significant overlap because the Bohr orbits are so large (10^{-7} m). The carriers then form an incomplete conduction band lying just below the conduction band. At very low temperatures ($T < 4$ K), thermal excitation into this band is suppressed and in addition the coupling of the carriers to the host lattice becomes very weak. The incoming radiation then "sees", in essence, a dilute electron gas and the electron mobility will rapidly increase as the inflowing energy is absorbed. The increased mobility can be observed, in the usual way, by means of the consequent increase in the current flowing in the external circuit. These detectors are very sensitive, NEPs as good as 3×10^{-13} WHz$^{-1/2}$, but since the impurity concentrations are very low the plasma frequencies lie in the very long wave region and detectivity is essentially confined to the region below 50 cm^{-1}. They are thus all far-infrared "hot-electron" photo-conductive detectors—the differences lie in the methods chosen to ensure an impedance match with the external electronics. The Rollin form [875] features a superconducting step-up transformer immediately following the detecting crystal. It is the most sensitive of the three but its response is limited to about 20 cm^{-1} and below. The Putley [874] version usually features a superconducting solenoid wound round the detecting crystal. The magnetic field "freezes-out" the continuum of allowed levels in the conduction band into the allowed Landau levels plus forbidden gaps—in this way the impedance is greatly increased. The operation of the Putley detector is very complicated: three different types of detection mechanism tending to operate simultaneously. Firstly, the freeze-out introduces conventional extrinsic photoconduction, secondly the overlap remaining between the levels of finite width results in hot electron photoconductivity, and thirdly the presence of the magnetic field produces electron cyclotron resonance photoconductivity. The relative importance of these three is determined by the size of the magnetic field and so by varying the field one can to a certain extent achieve trade-off between sensitivity, spectral detection range and speed of response. At its widest the Putley can respond from microwave frequencies all the way to 50 cm^{-1}. In the past it has proved to be a very versatile and useful detector—it certainly played a crucial role in revolutionizing far-infrared spectroscopy—but nowadays it is tending to be replaced by the hot-electron bolometer which achieves the spectral and impedance matching by carefully controlled doping of the InSb crystal.

The modern infrared detectors, which are available commercially, are very good indeed and there are nowadays few occasions when the infrared researcher needs to manufacture his own. Undoubtably new kinds will come on the market as time goes in, perhaps arrays based on a wide range of semiconducting materials, but the very fact that there is so healthy a market for infrared detectors reflects the generally healthy state of infrared science and technology in general.

6.6 Commercial infrared spectrometers

6.6.1 Commercial dispersive infrared spectrometers

Laboratory infrared spectrometers were developed in primitive form both in Germany and in the USA before the Second World War but the really rapid development came after that conflict when much of the infrared technology developed for military and defence purposes became available for civilian applications. Automatically operating instruments were then offered commercially by Perkin Elmer and Beckman in the USA and by Grubb–Parsons, Hilger and Pye Unicam in England amongst several others. The original instruments featured alkali halide (in fact usually sodium chloride) prisms but after a while these were gradually replaced by the higher-performance (and much cheaper!) replica gratings which were then becoming readily available. Infrared spectrometers with their ease of operation became very popular in chemical analytical laboratories and such instruments as the Perkin Elmer Model 21 and later the various forms of infracord became ubiquitous in their applications.

The basic layout of these instruments was that of a double-beam ratiometer with the two beams produced from a common source (usually a Nernst glower). One of these traverses the specimen, whilst the other provides a reference usually either by passing through an equivalent but empty space or else by passing through an identical cell containing pure solvent. The reference channel also contains an attenuating comb which can be automatically moved in and out of the beam to ensure that at each spectral frequency which is being investigated the two beams are of equal intensity. The two beams emerging from the sample chamber are then recombined by off-axis mirror systems at a reflective chopper, one beam, say the sample, being transmitted by the chopper and the other being reflected. The two beams will thus be modulated but 180° out of phase with one another. The combined beams then pass through the entrance slit of the monochromator which is usually laid out in the Fastie–Ebert configuration to minimise aberrations. The grating, usually plane, is mounted on a table which can be rotated and as it rotates the dispersed radiation sweeps over the exit slit. The radiation emerging from this slit is condensed into a suitable detector, usually, at least in the early days, a thermopile. Thermopiles are slow in their operation so low chopping speeds are obligatory. At these low speeds, the output from the electronic system, that is the PSD, is of a very low frequency so the best mode of operation is a null one in which the output signal is coupled via a servo to the motor controlling the attenuating comb and the servo operates until the output from the PSD (i.e. the voltage at the modulation frequency) is zero. This will be so when both beams are of equal intensity. The out-of-balance servo voltage can then be displayed on a pen-recorder whose rate of progress is coupled to the grating drive. In this way one gets a completely automatic presentation of a transmittance spectrum. In more modern instruments and especially in the more expensive versions, fast detectors such as the pyroelectric type are used and with these it is possible to use purely electronic ratio-recording and thus do away with the attenuating comb and its servo. Ratio-recording provides much better radiometric precision, particularly at low transmittance values. However, even ratio-recording cannot deal with the problem that over the range for which the instrument is designed (usually $4000-180 \, cm^{-1}$), the variation in spectral power output from a black-body source (see Fig. 2.7) is enormous. This problem is usually handled by having the entrance and exit slit widths automatically controlled by a second servo system so that the average spectral power flowing to the detector is constant. The drawback to this solution is, of course, that the spectral resolution varies with wave number but this need not be a major nuisance in analytical chemistry provided the spectra being compared are produced either by the same or else by closely similar instruments.

Another problem stemming from the wide spectral range which the instrument has to cover is that blazed gratings, which can be extremely efficient at the chosen wavelength, fall off in performance as one goes to longer or shorter wavelengths. This is usually coped with by having more than one grating on the table and at a series of fixed wavelengths, automatic change to a more appropriate grating takes place. If only two gratings are being used, as might be the case in a cheaper instrument, they can be mounted back-to-back and interchange assured by a quick "flipover" mechanism. "Stray light" is a major hazard with all these instruments so filters mounted on an automatically operating wheel are used to restrict the operation to the appropriate range for the grating currently being used. Stray light of a quite different kind arises when one is trying to study specimens at either a higher or a lower temperature than that of the detector since radiation flow, which will be modulated by the chopper, will then occur in one direction or the other and the reported transmittance values will be incorrect. To overcome this the more expensive commercial instruments feature double-chopping in which an extra chopper operating at a different frequency and mounted near the source is used.

During the 1970s two major technological advances provided serious challenges to dispersive infrared spectrophotometers and to their traditional mode of operation. The first of these was the development of commercial Fourier transform instruments and the second was the development of cheap high-performance microelectronics. The response to these challenges has brought about revolutionary changes in the mode of operation of the "top-of-the-market" dispersive instruments. Thus if one considers a modern instrument such as the Perkin–Elmer model 983 one sees at once that *all* of the operation of the instrument is controlled by a microprocessor and that the operator communicates with the instrument entirely via a keyboard and a VDU whose display is controlled by a fully interactive software routine. The grating rotation, instead of being controlled mechanically by means of a set of profiled cams driven by a constant speed electric motor, is produced by a stepper-motor which goes at a varying rate determined by the MP which for this purpose has access to a "look-up" table stored on a ROM. Merely by software controls it is possible to change the operation of the instrument to several different modes, and one has constant wave number, constant wavelength etc. Grating changes likewise can be selected to suit the operator's convenience rather than always occurring at the same values of the wave number. In this way an interesting region will not be marred by a grating change. The slit widths are again controlled by small stepper-motors driven by the MP consulting a "look-up" table. Ratio-recording is used, of course, but the output from the PSD is digitised and stored in the computer's memory. The ordinate presentation thus has all the sophistication made possible by digital rather than analogue operation. Thus Savitsky–Golay [877] digital smoothing can be employed and the output can be plotted in any of several different modes, $\%T$, absorption coefficient, ln (τ), etc. The computer both controlling the instrument and processing its data flow can react intelligently to different spectroscopic problems. Thus there is a buffer store so that the spectrum can be plotted more slowly than it is being observed and, should the buffer be in danger of overflowing, the scan speed will be automatically slowed down. Spectra stored in the computer's memory can be averaged to give improved S/N. Spectral subtraction and scale expansion can be invoked to reveal the presence of very weak bands and thus enormously enhance the analytical possibilities of the instrument. The MP can also "talk-to" an external main-frame or minicomputer and in this way consult vast "libraries" of stored spectra with a view to the positive identification of an unknown. All of this was, of course, an obvious response to the analytical challenge thrown down by the FT instruments whose unavoidable digital mode of operation gave them an enormous advantage when it came to analysis at the state-of-the-art level. Part of this advantage still exists since there is no possibility of abscissa back lash with an FT instrument, whereas this is still a possibility even with a computer-controlled dispersive instrument. However by using materials, e.g. polymeric films, with richly structured spectra as occasional calibrants, it should be possible for the computer to keep its look-up tables constantly under revision and thus achieve a virtual "lock-in" of the abscissa scale. In this way subtraction and scale expansion would become quantitatively reliable. At the moment the 983 has a wave number repeatability of

0.005 cm^{-1} and an ordinate accuracy of 0.1% which make it very competitive. Its spectral resolution is 0.4 cm^{-1} which is adequate for most analytical work. Unlike most of the older generation of infrared spectrophotometers, the 983 is readily adapted to emission work and the emission can be ratioed against a theoretical or experimental spectrum of a black body at the same temperature stored in the computer memory. This ability to compare a spectrum with others recorded on other occasions or else even obtained from a library adds a whole new dimension to the double-beam concept. It has been used for example to determine dichroic ratios in stretched polymers but so far this cannot be done dynamically because the scan speed is still too slow. FTIR instruments (see next section) have, however, adequately fast scan speeds and these have been coupled to dynamic "stretchers" to show the time variation of the dichroism.

6.6.2 *Commercial Fourier transform infrared spectrometers*

Far-infrared aperiodic Fourier transform infrared spectrometers have been available since the early 1960s from companies such as Grubb–Parsons, RIIC and Specac, in the United Kingdom [878]. Later instruments were produced by Coderg in France [879] and Polytec in Germany [880]. These were all based essentially on the concepts developed by Gebbie and his group at the National Physical Laboratory in England. The layout of one version of the modular cube interferometer used by the NPL group [881] is shown in Fig. 6.17. Aperiodic instruments for the mid and near infrared [882] have been developed by several groups and have been used to provide very good spectra but so far aperiodic commercial instruments have not been developed, mostly because the periodic form (see later) has proved so successful. The mechanical problems which arise at shorter wavelengths are solved by using "cats-eye" retroreflectors as the mirrors and by using laser referencing to ensure accurate sampling. At these shorter wavelengths, it is necessary to use the conventional compensated form of Michelson beam-divider with the two plates made from a suitable transmissive material, usually an alkali halide, but at the longer wavelengths it is more usual to use stretched polymeric films usually made from polyethylene terephthalate ("melinex" in the UK, "mylar" in the USA). All the commercial far-infrared FT instruments feature this form of beam-divider usually in the form of a metal former over which the film is stretched. The drawback to film beam-dividers is that they have zeroes of performance at those wave numbers where the multiple-beam interference [883] leads to a zero reflectance. Compensatingly, of course, this same phenomenon leads to a performance much higher than one might expect from the relatively low (~ 1.69) refractive index of the polymer, at the corresponding multiple-beam enhanced reflection maxima, which lie in between the minima. This alternation of performance requires that more than one thickness of film be used if a wide spectral range is to be covered. This can be a nuisance because water vapour

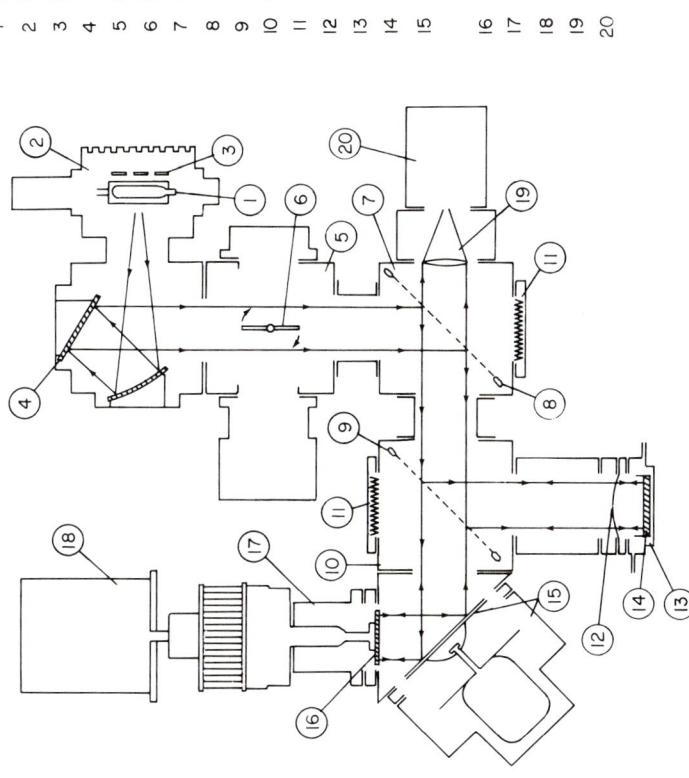

1   Mercury lamp
2   Water cooled lamp house
3   Low-speed chopper ($16\frac{2}{3}$ Hz)
4   Two-mirror collimator
5   Cube 3
6   High-speed chopper (800 Hz)
7   Cube 2
8   Wire grid polariser/analyser
9   Wire grid beam splitter
10 Cube 1
11 Radiation dump
12 Polystyrene window
13 Liquid cell
14 Stainless steel mirror
15 45° mirror and phase modulator assembly
16 Scanning mirror
17 Micrometer
18 Stepping motor
19 Focussing lens
20 Detector

FIG. 6.17. One possible optical layout of the modular far-infrared Michelson.

absorption compels the experimentalist to evacuate his far-infrared interferometers and breaking the vacuum to change beam-dividers is not a trivial
operation. Recent commercial interferometers, for example the BRUKER,
have introduced automatic beam-divider changing, under vacuum, to get
round this difficulty.

Commercial near-infrared interferometers are universally of the rapid-scan
type. The main problems encountered are two-fold. Firstly it is necessary to
have the continuous oscillatory motion of the mirror smooth enough to
preserve interferogram quality over the entire span of the motion. Secondly
one must choose a design which not only ensures that the mirror moves such
that its reflecting surface stays rigidly parallel to its initial position, but also
ensures that the interferometer stays in alignment for extended periods. The
most popular solution is the use of an air-bearing. Other possibilities, however,
such as the tilt-insensitive rotating system developed by James [884] will
probably also come into use as these instruments become more common and
their price drops. These mechanisms have proven performance for routine
low to medium resolution work, but for high-resolution work, even the air-
bearing becomes less than satisfactory. The solution for these applications,
introduced by BOMEM, is not to seek still more sophisticated mirror drives,
but instead to correct for the tilt errors of the moving mirror by altering the tilt
of the fixed one. The referencing helium/neon fringes can be used to provide
the necessary information to the correcting servo. The beam-dividers are
normally of the two-plate compensated type. The base material is usually KBr
with a suitable evaporated coating, for example germanium. Water vapour
attack on the hygroscopic beam-dividers and other optical components can be
prevented to a large extent by the use of pin-hole free polymeric coatings such
as parylene [885]. The detector used in nearly all commercial interferometers

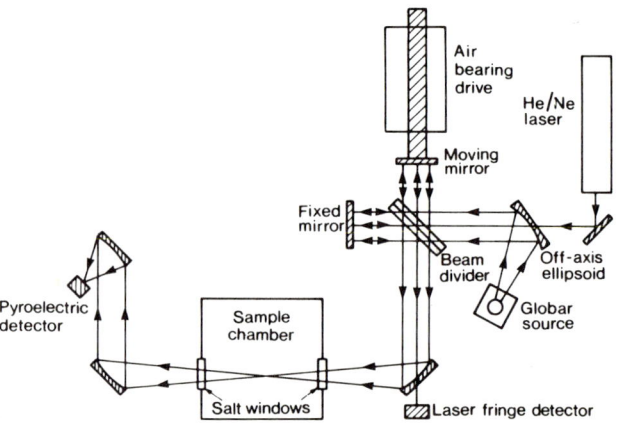

FIG. 6.18. Schematic of the optical layout of a commercial mid-infrared interferometric spectrometer.

is the TGS pyroelectric type. This has ample response speed and sensitivity to give high-grade spectra in single runs. For those difficult cases where the signal is very low, averaging of either interferograms, spectra or both will eventually give S/N values of any desired magnitude. The computer system controls the operation of the instrument, transforms the interferograms and processes all the data. In fact with most of the modern commercial instruments it is very easy to forget that one is doing anything so complicated as Fourier transform spectroscopy! The computer is fully interactive and guides the user, via prompts and questions on the VDU, to the best approach for any particular problem. Instruments are offered by several manufacturers: some of the best known are Analect, Bomen, Bruker, Digilab, Nicolet and Perkin–Elmer. A schematic layout of the optics of a modern commercial interferometer is given in Fig. 6.18.

Chapter 7
Practical Uses of Infrared Radiation

7.1 Infrared imaging and thermography

The use of infrared radiation to form images, that is for "seeing", is restricted for practical reasons to the atmospheric "windows". One has therefore very near infrared systems operating at wavelengths less than 2 μm, near-infrared systems working in the 3–5 μm window, mid-infrared systems working in the 8–14 μm "window" and far-infrared systems working at wavelengths greater than 200 μm. Infrared systems are always going to be less sensitive and less easy to use than visible region systems, so one would only use an infrared imager for those applications where the visible system would fail. These would include

a. Convert surveillance of a scene, especially at night and particularly for military and security purposes.

b. Monitoring slight differences of temperature, from a distance, by observing the variations of mid-infrared emission. This has widespread uses in areas ranging from medical diagnosis to the rapidly growing modern field of energy conservation.

c. Seeing through fog and mist—this relies on the rapid fall off of Mie scattering when the wavelength gets larger than the mean diameter of the scattering particles.

d. The early detection of plant disorders by means of monitoring the near-infrared reflection. The observations are usually carried out from aircraft and rely on the strong near-infrared absorption bands of chlorophyll. Slight changes in the chlorophyll, due to disease or mineral deficiency, produce little observable effect in the visible region but quite dramatic effects in the near infrared [886].

e. Searching for missing persons at night and in water or snow. This is very effective for finding people who are still alive since the warm human stands out well from the cold background.

f. Finding buried objects, particularly bodies, hidden in shallow graves or under bracken. These may be quite invisible visually but stand out clearly in the infrared where ordinary forms of camouflage fail.

g. Finding unconscious people trapped in smoke-filled rooms. Such people, lying on the floor, may be quite invisible to the firefighters from even a few feet away yet be perfectly obvious from several yards away with the help of an infrared imager.

h. Detection of wood rot. The emissivity of wood changes drastically when it is wet or rotting and the extent of damage to wooden structures is readily assayed with an imager.

i. Monitoring of high-voltage electricity supply. Incipient failures in overhead high-voltage cables, especially at the joints, are easily seen from the ground whereas *in situ* inspection would be difficult, time consuming and costly. Modern systems feature aluminium cables and these are difficult to join so failures are becoming more common.

j. Fault-finding in printed circuit boards. This is one of the most effective and cost saving applications of thermal imagers since a component which is overheating is immediately obvious.

k. Monitoring the heat balance of buildings. This is a rapidly growing area. Modern buildings often feature large span roofs which are insulated with fibre glass. For one reason or another, the fibre glass is often left out of a section and the detection of the omission by conventional methods would be too costly to contemplate. Imagers can detect the faulty section very quickly and with certainty.

l. Detecting incipient fires in mines. These are not only a great danger to the miners, they can also threaten the loss of a great deal of expensive equipment. Early detection can save literally millions of pounds.

m. Military operations. These are mostly classified but seeing through conventional camouflage and through smoke-screens is one obvious application. The problem with present-day smoke screens is that the particles phosphorus pentoxide, for example, are quite large (~ 10 μm) and even at 10 μm wavelengths the depth of penetration is not great. At longer wavelengths the atmosphere becomes opaque again so a satisfactory solution to the problem of "seeing" through smoke-screens will probably rest on the development of imaging, that is millimetre-wave, radars. A major difficulty with *forward looking infrared* (or FLIR) systems is that the primary optics, nearly always made of germanium, are very susceptible to damage; in fact germanium is a rather soft material. In battle situations the lenses would have only a short lifetime. It has even been calculated that for a fighter plane flying at Mach 2 and entering a rain cloud the lens would only last for a matter of seconds! Various ways of protecting the surface have been devised, one of these is via the deposition of carbon from a hydrocarbon (butane) plasma. This gives a surface which is very hard—it has been likened to a two-dimensional diamond!

Infrared imaging systems differ in several ways from their visible region counterparts but one of the more important differences stems from the nature of the infrared "scene". In the visible, for the reasons spelled out in section 2.2, it is rare to find self-luminous objects. Nearly all visual scenes have to be illuminated, either artificially or else naturally by some source such as the sun or moon. The contrast in the scene can therefore easily approach 100% (from pure white to pure black). In the infrared on the other hand and especially in the 10 μm region where lies the peak of the black-body radiation corresponding to ambient temperatures, the intrinsic illumination of the scene can easily exceed the extrinsic, if this is present at all, and the contrast can be rather low. Features in an infrared scene are distinguished from one another then not so much by colour or texture but by temperature and emissivity. This puts rather stringent sensitivity requirements on an acceptable infrared imaging system.

The available means for realising an infrared imager fall into various categories. The oldest type was the evaporagraph in which absorption of the radiant heat caused a volatile liquid spread out on a black surface to evaporate and the drying pattern which was a crude representation of the energy distribution could readily be seen by eye. Nowadays there are several lineal descendents based on phenomena such as the absorption edge of thin films, variations in the surface tension of liquids and most recently the scattering and polarisation of liquid crystal films. This latter stems from a close relative of the evaporagraph, the use of certain crystals which undergo a phase transition near room temperature involving a change of colour. Kneubuhl and his colleagues in Zurich [887] have used liquid crystals whose helicity varies with temperature to produce thermograms of the field patterns of submillimetre lasers. These detectors which are the infrared analogues of photographic plates and which are wavelength independent could have several applications for those situations where the infrared intensity is reasonably high. Systems such as this do have the merit of simplicity but they have the demerits of being slow, insensitive, very subject to random fluctuations of local temperature and liable to blurring by lateral conduction. For these reasons, they have only seen limited uses in some rather specialised applications.

At the very shortest infrared wavelengths, one can take advantage of imaging based on the external photoelectric effect. Thus one can use infrared sensitive film [888] and straightforward photography—straightforward that is except that one must be careful not to use optics bloomed for the visible region! It is possible to use vidicon tubes with S-20 photocathodes to perform infrared television but this has so far only been done for producing special effects on rare occasions. The most widely used near-infrared imaging system is the wavelength-changing image intensifier in which the original infrared image is transformed into a corresponding spatial variation of the photocurrent and this current when suitably multiplied is finally incident on a phosphor screen to give a visible image corresponding exactly to the infrared one. These devices are widely used for night vision purposes by the military, often in the form of infrared binoculars and they are also used for gunsights. The wavelength to which they can be used is limited by the wavelength sensitivity of the initial photocathode. Presently this means about 1.2 μm but there are several promising photocathodes in the research stage which may give image-intensifiers working out to nearly 2 μm. The natural level of near-infrared radiation at night is rather low but this can be supplemented by artificial illumination from a near-infrared laser, for example GaAs at 0.85 μm, without necessarily giving the game away.

At longer wavelengths where efficient photocathodes are not available, imagers have to rely on conventional infrared detectors either of the thermal or else of the internal photoelectric kind. Presently available imagers divide naturally into two distinct strains which are sometimes referred to as the "staring" and the "scanning" versions. In a staring imager, each element of the picture in the image plane (that is each "pixel") is associated in a one-to-one fashion, either with its own detector or with its own place on a vidicon receiving screen. In a scanning imager, on the other hand, the number of detectors is much less than the number of pixels so the picture has to be scanned opto-mechanically over the detectors and their outputs have to be electronically processed to give an acceptable picture. At the moment it is not possible to make two-dimensional arrays of detecting elements plus processing electronics with enough elements to give a high resolution picture and, even if one could, one would have severe problems in ensuring detector element uniformity so the main type of staring imager is the vidicon and in particular the pyroelectric vidicon. The early forms of scanning imager all featured a single detecting element and this is still common practice. The opto-mechanical scanning moved the picture back and forth and up and down over the detector but one can think just as well that it is the detector which is moving regularly, line by line, over the picture image. It will be seen at once that this is a very inefficient process since each pixel is observed (or "interrogated") only once per complete frame. The use of detector arrays either linear or rectangular can improve this situation dramatically. One immediate gain is that if the array contains N elements in a horizontal (or serial) format then there is available a signal-to-noise ratio improvement by the factor $N^{1/2}$ without any need to get particularly good detector element uniformity. If, on the other hand, the array contains N elements in a vertical (or parallel) format, then one can get a very welcome reduction in the required scan speed, in fact by the factor N^{-1}, but detector element uniformity becomes more important. This scan rate reduction can be a very important consideration when one is trying to match the high frame rates needed for a TV compatible system. Clearly both improvements are available if one has a rectangular array but of course to ensure either does demand individual processing of each signal from each detector in the array. This is achieved by having separate preamplifiers and time-delay-and-integrate (TDI) electronics all assembled on the same solid-state package. A recent advance in this area is to use bias current control of the hole/electron drift velocity to ensure a match with the rate of motion of the pixel down a line of the array. In this way it is unnecessary to have separate discrete circuits for each detector and the cost of manufacture is greatly reduced. This work is mostly associated with RSRE from where have come the TED (or *Tom Elliott Device*) and its newer descendants such as SPRITE [889]. Undoubtedly, for all these reasons, arrays will become more and more prominent as time goes by but at the moment most commercial equipment still uses single element detectors—usually indium antimonide for the 3–5 μm band and mercury cadmium telluride (or MCT) for the 8–14 μm band. The single detector has the merit that it can be extremely sensitive since particularly good crystals can be selected and it is also much easier to screen the element from unwanted thermal radiation from outside the scene area. On the other hand, the scan requirements are much more severe and especially so if TV compatibility is desired. The usual arrangement is to have two rotating mirrors or prisms, one to do the line scan and the other to do the line shift, but some imagers feature a single rotating polyhedron with each face slightly inclined to its predecessor to do both tasks. At the moment no single-element detector camera can offer a full 625 line TV compatibility. Some scan about 70 lines in a frame and then do this a further three times displaced progressively downwards. This gives a total of 280 lines and requires therefore a special TV display but it can be used with conventional video cassette recorders. Others scan alternate lines of the 625 raster so that they have full CCIR compatibility and then use the unscanned lines for other purposes—for example displaying temperature profiles across the picture. The newer cameras, however, will scan all the 625 lines but not in real time. In this case, an intermediate memory system or buffer store based on charge-coupled devices (CCDs) may be used and this memory system is interrogated by the fully CCIR compatible display. All of these still require the opto-mechanical scanning and this adds to the weight and to the cost of the imager. The lack of real time operation is, however, quite acceptable since thermal imagers are seldom used to look at rapidly changing scenes. Apart from TV compatible systems which are mostly used as fixed institutions in a laboratory or workshop, the other main requirement is for lightweight hand-held imagers. In these the usual practice is to use the scanning of the prism or mirror system to reconstruct the picture automatically via an LED array, operating in the red region of the visible, whose individual diodes are controlled by the outputs of the corresponding elements in the detector array. The MEL imager is a good example and its design nicely illustrates the compromises that have to be struck in designing an imager. Thus to achieve low cost and light weight calls for sacrifices of picture quality and a 12 element CMT vertical array is used, scanned in parallel eight times to give just 96 lines.

The pyroelectric vidicon is a very exciting development for two reasons.

Firstly, since it is based on a thermal detector, there is little wavelength dependence of the sensitivity and the camera can be used right out into the millimetre waveband [890]; secondly, since the response is based on the rate of change of radiant power, the camera does not respond to the large (if unvarying) background so one gets good contrast "pictures". In this tube, a thin slice of pyroelectric material (usually triglycine sulphate, TGS) acts as the target both for the illumination on one side and for the probing electron beam on the other. Because of the structural similarity to the ordinary photoconductive vidicon, this infrared tube is also loosely called by the same name but it is very important to note the operational differences which arise because the detecting element is an insulator. The probing electron beam, whose function is to return the target to cathode potential, will clearly carry out this task quickly, unless something is done about it, and the camera sensitivity will fall to zero. Two ways round this difficulty have been proposed. In the first [891] the vacuum in the tube is not very hard so ion bombardment will deposit a uniform positive charge on the target. In the second [892] the target is made to go positive again during the normal scan fly-back period by bombarding it with an electron beam sufficiently energetic to produce secondary electron emission. Further details of the vidicon operation have been given by Stillwell [238] and Hadni [893]. The position now is that the pyroelectric vidicon has been developed to the point where television-like thermal pictures of scenes can be produced which show minimum resolvable temperature differences of the order of 0·2 °C and resolutions of the order of 300 TV lines. Pyroelectric vidicons have low power consumption and are in principle lightweight devices.

The operating wavelength band of an infrared imager is dictated by the need to strike a compromise between reasonable atmospheric transmission, good contrast, that is temperature resolution in the image and acceptably good signal-to-noise ratio. The thermal radiation from the ambient background peaks near 10 μm, so from a signal-to-noise consideration the wide mid-infrared window is ideal. However, the contrast between objects of slightly different temperature is poor in this region whereas it is good at 5 μm where one is in the Wien regime. The ambient background is so low in the very near infrared (~ 1 μm) that imagers, as mentioned above, can only be used in conjunction with some artificial illumination, for example that from the sun, the moon or an infrared laser. In all the accessible windows, however, the wavelength is so much smaller than the sizes of the features in the image plane that optical and mechanical shortcomings provide the limitation to picture quality so considerable further improvements could be achieved if thought desirable and cost effective. The definition of the image in any optical system with a given aperture falls as the wavelength increases (see section 3.2). One should therefore think in terms of increasing the aperture as one goes further into the infrared but present-day systems are nowhere near the diffraction limit (quite unlike their visible region counterparts) and large apertures are not necessary on this account alone. However the incident infrared intensity will usually be low and the detectors which are available compare adversely with those available at optical frequencies, so one needs fast optics in order that the overall system will have an acceptable signal-to-noise ratio. The chief noise source with semiconductor detectors is the thermal promotion of carriers into the conduction band. Cooling of the detector is therefore desirable and the longer the wavelength the lower should be the operating temperature. Cryogenic liquids and in particular liquid nitrogen are widely used for laboratory work but they are usually undesirable for field work especially under the rather rough conditions which frequently prevail, so various forms of refrigerator are used instead. Joule Thompson cooling has been used for a long time but is neither cheap nor light weight. The Stirling cycle seems a better proposition but nevertheless the engine weighs 2 kg and the cool down time is as long as 10 minutes. Thermoelectric coolers are light weight and quick but unfortunately are limited to temperatures above 190 K. This is adequate, however, for the 3–5 μm band. Pyroelectric vidicons have to operate at ambient temperature but the multiplex (or $N^{1/2}$) advantage makes up for the lack of cooling.

Infrared optics are nearly always made from single crystal germanium and since germanium is a relatively rare element this is becoming rather worrying to the military who make extensive use of it. In fact germanium is now regarded as a strategic element! The art of fashioning lenses from germanium and giving them suitable antireflection coatings has progressed very far, under the stimulus of military contracts, and nowadays there is little difficulty in getting hold of quite complex aspheric elements for infrared optical systems.

Medical uses of infrared imaging (usually in this context called "thermography") had a considerable boost some years ago when it was mooted as a possibly cheap and effective means for mass screening of women for early breast cancer. This hope has faded since it has been shown to be reliable only when used by a skilled clinician and that rather defeats the object of the exercise. The basic point of thermography is that heat is generated deep within the body and flows to the surface both by conduction and by transfer in the blood flow. Any abnormal condition deep in the tissues may therefore be revealed by local regions on the skin of enhanced or depressed temperature. The important matter, as far as the thermographer is concerned, is the minimum resolvable temperature difference (or MRTD) of the instrument. Currently this is about 0.1°C. Since the final display will be on a television screen, it might just as well be in colour. One can therefore use contour plateaus each representing a defined temperature range and each assigned a particular colour. There doesn't at the moment seem to be any agreed colour-code for international usage and merely seeing a colour coded picture doesn't give much extra information unless one has the key. The colour code adopted emphasises *differences* of temperature so the colours are chosen on the basis of their contrast with one another rather than on the basis of a spectral criterion. Thermography works best for the appendages, the legs, arms, neck and head because these have the simpler vasculation and are further from the central thermal "core". With present sensitivity, thermography can readily pinpoint circulatory and vasospastic disorders in the limbs and it is very effective for showing up rheumatoid and arthritic conditions. It has been used for medical research purposes, for example in investigating the theory that those people who readily put on weight even when on quite spartan diets are deficient in energy disippating adipose brown fat and has shown promise for monitoring drug treatment, but it is mostly still a diagnostic tool. It is proving very good at burns prognosis work where it can be relied on to show areas where skin grafting is necessary. It is especially good for the diagnosis of varicocele a testicular condition that can make men infertile and it is proving very valuable in the investigation of thrombosis. Patients for thermography have to abstain from alcohol and nicotine for the previous twenty four hours and must not, of course, use any talc or body lotions which might affect their emissivity. The examination is usually carried out in a warm room between 18 and 20 °C. Routine work is confined to the usual atmospheric windows but research work is under way to investigate the use of microwaves. These can penetrate much more deeply than the infrared which is essentially a surface only tool. One possible application of microwave thermography might be in the location of deep brain tumours.

The commercial uses of thermography rest on its unique ability to give an immediate picture of heat flow. As an example one can scan a house, in winter, and immediately locate all the points where the insulation is inadequate. Increasing energy shortfall will be the normal future situation and energy economy will be vital. Thermography which can diagnose heat loss and monitor the effectiveness of the steps taken to correct it will be an essential tool for heating engineers. Aerial thermography can be used to study the flow of warm water from power stations. Oddly enough, warm water discharged into an otherwise cool lake can be a form of pollution, thermal pollution, and it can adversely influence the natural ecology of the lake. Thermography should have many industrial uses, for example in heat treatment plants, and it also has potential as a low-temperature form of pyrometry. An increasingly important application is the non-destructive testing of fabricated metal pieces to detect stress concentration build-ups [894]. These are potential sources of failure when the piece is in use and although it is, in principle, possible to analyse the design and fabrication mathematically to see whether it is a good one, this is not always possible in practice. The engineer looks for a reasonably quick and inexpensive test and thermography seems very attractive. The piece is painted black to give it a good emissivity and it is then cyclically stretched in a suitable machine. At the peaks of the applied force the temperature everywhere will rise but the temperature rises will be maximal at the regions of high stress. A thermographic camera can readily detect these hot spots. An interesting side line is that if a hot spot does not appear where one is expecting it, one can conclude fairly safely that the piece has cracked and has thus relieved the stress!

The use of infrared imaging systems for seeing through fog is potentially very valuable indeed. The basic point is that the competing, very much longer wavelength, radar systems cannot give, at least at the moment and at least at reasonable cost, recognisable images. A 10 μm infrared scanning telescope, on the other hand, can readily "see" several kilometres across the dense mist, haze or fog which is commonly encountered in major cities like London or New York. There are obvious applications in city management, traffic control for example. Airports are particularly prone to disruption by fog. Aeroplanes, in the air, can be controlled perfectly well with conventional S- or X-band radar but an aeroplane on the ground is a quite different matter. The risk of collisions between buses, cars and slowly taxiing aeroplanes is very real and because of this all the vehicles have to move very circumspectly and that means slowly. Infrared scanners by enormously enhancing the visibility could get things moving more quickly and thus help the airport to function normally despite the fog. A difficulty is that the skin of the aeroplane and the surrounding fog may be at the same temperature and then the contrast in the picture will depend solely on the ratio of the emissivities. This situation could be improved dramatically by the use of auxilliary illumination and here the CO_2 laser could be ideal. The combined system would be fairly bulky and certainly expensive and would probably only be justified for emergency vehicles such as ambulances and fire engines.

The prospect of infrared television throughout the whole spectral region has brightened considerably with the recent development, as mentioned earlier, of the pyroelectric vidicon tube. The principal drawback, at the moment, is that these cameras are not very sensitive so they can only be used for very bright infrared illumination. They have been used, however, very effectively to study the far-field patterns of infrared lasers. Another possibility for imaging, so far only in the laboratory curiosity class (see section 6.3), is to use a very intense laser, ruby for example, to up-convert the infrared signal into the visible region by the process of parametric amplification [895]. This though is a very costly operation and it will probably stay in the curio class until some application arises where the conventional infrared imaging techniques fail.

7.2 The use of infrared radiation in chemical analysis

From the account, given in section 5.2.3, it will be realised that the infrared spectrum of a given molecule is unique to that molecule and if one therefore records the spectrum of an unknown pure substance one should be able to identify the substance immediately by reference to a library of previously determined spectra. Chemical analysis, of this nature, is enormously important in industry, in medicine, in research and in many other fields and since, prior to the development of infrared spectroscopy, there was no way of doing it other than the use of wet chemistry which is an inconvenient and slow technique, the arrival of automatic infrared spectrometers in the 1950s made dramatic advances possible. It is almost certainly true that chemistry in its various guises provides the majority of uses for infrared spectroscopy and by the 1960s the industry, set up to provide the chemists with their quick-acting automated analytical spectrometers, was a large one in its own right. Nowadays there are several rival techniques, mass spectrometry, gas and liquid chromatography, nuclear magnetic resonance etc., but nevertheless a chemical analysis laboratory of any size will still have one or more hard-worked infrared spectrometers in its armoury. The development of computers has had a major impact on the problem of identification since the computer can store vast numbers of standard spectra in its memory bank and scan through these very rapidly to find a match with a spectrum just recorded. Computers are also very useful for analysing the spectrum of a complex mixture. Infrared spectra, in the form of a plot of absorption coefficient versus wave number are, by Beer's law, linearly additive so one can in principle break down the spectrum of a mixture into its constituents and hence deduce the amounts of each pure substance present in the original mixture. The computer is at its best when the constituents of the mixture are always the same and only their relative amounts vary. The development of Fourier transform spectrometers (see section 6.6.2) has had a considerable impact here since the spectrometer has an in-built computer which, although used primarily to transform the observed interferogram into the desired spectrum, can also be used to sort out the spectrum into its constituents. The interferometers have another advantage over dispersive instruments in this connection, since the abscissa (frequency or wave number) axis is rigidly defined, there being no possibility of "back-lash", drift or other causes of wave number imprecision. The radiometric precision of the interferometers is also of paramount importance when "scale-expand and subtract" techniques are being used to reveal the presence of minor components.

All the other variant techniques of infrared spectroscopy can be used in analytical applications. A particularly important example is the use of ATR (see section 3.7.3) to analyse surface coatings. This task is also being tackled by photoacoustic FTIR techniques [397]. Straightforward reflection spectroscopy is also important, especially for moisture meters which work by comparing the reflectivity at a wavelength (e.g. 1·94 µm) where water absorbs strongly with that at one (e.g. 1·70 µm) where it does not.

These various analytical techniques work well down to about the 1 % level but if one is looking for a constituent present at still lower levels, additional tricks become necessary. Basically one must either increase the observing path length, the concentration of the desired constituent, or both. With gaseous samples, only an increase of path length is usually required since the absorption bands of the interfering majority constituents will be very sharp, but for liquid samples this may not work because the interfering bands, being broad, will eventually "black-out" the desired spectral band. In this case some form of pre-concentration must be used. Gases can be studied using the well-known White long path cell [896] and liquids in variable path cells such as the model 7000 made by Specac. Solids may either be dissolved in a suitable solvent or else dispersed homogeneously in a suitably transparent solid host matrix. Potassium bromide is widely used for this purpose and presses for making K Br pellets are commercially available from several manufacturers. Far-infrared applications of the same technique usually invoke pellets made from compressed polyethylene powder.

The fact that the infrared spectrum is so molecule specific makes possible the use of detectors in which there is no attempt made to separate the spectrum either physically, as in a dispersive spectrometer, or mathematically, as in a Fourier spectrometer. The best known of this class of detector is the Luft, described in Chapter 4, but the rapid development of opto-acoustic detectors is providing a rival. Luft detectors make ideal passive monitors which can be used either to protect personnel from contact with dangerous gases or else as part of automatic control equipment in factories. They are limited, however, in their sensitivity to about the part per million level. To do better than this requires some of the techniques described in the next section.

Infrared spectroscopy is thus almost irreplaceable for quantitative chemical analysis of mixtures but its usefulness does not end there since throughout its history it has constantly proved to be a most powerful tool for working out the possible structure of a completely uncharacterised compound. This is because of the strong correlations which exist between the chemical structure of a molecule and its infrared spectrum. Thus all molecules containing methyl groups show absorption bands in the 3·45 μm region, all ketones show an absorption band at $\sim 5\cdot9$ μm, all nitriles an absorption band at $\sim 4\cdot45$ μm etc. These correlations are so good that a skilled spectroscopist can often hazard a very good guess as to the chemical structure of a new compound knowing only its infrared spectrum. An exhaustive account of this branch of the art is available in the well-known textbook by Bellamy [5]. The basic physical reasons for these correlations lie in the nature of the molecular force field (see Appendix 6). The diagonal elements of the force constant matrix are determined entirely by the chemical nature of the bond, interbond angle etc. in question and will therefore be closely similar for similar groups in similar environments. The interaction (off-diagonal) elements are also correlated with the chemical nature of the vibrating group and, furthermore, rapidly fall to zero as the distance between the two displacements in question increases. These two facts lead to the situation that one can think, to first order, of the spectrum of a complex molecule as made up of the isolated vibrations of its constituent parts and then to second order of perturbations brought about by the non-zero off-diagonal elements in G or F. This treatment is the more exact the higher the vibration frequency: C-H stretching modes, for example, can usually be regarded as entirely localised. However for larger molecules at low frequencies it is not an acceptable approximation and one then has modes which involve completely delocalised motion. The far-infrared spectrum of a chain molecule therefore gives much information about the chain, but the other side of the coin is that the far-infrared spectra of nominally similar chain molecules may in fact be very different as illustrated, for example, in Fig. 5.14. For this reason structural analyses mostly rest on features in the 2–15 μm band which has come to be called the "fingerprint" region. It is also fortunately the region where cheap spectrometers based on the sodium chloride prism or else on mass-produced gratings can be used.

7.2.1 *The use of infrared radiation in pollution monitoring*

7.2.1.1 *Introduction*
One of the less desirable aspects of the highly industrialised societies of the developed world is the increasing pollution of the atmosphere which is produced as a by-product of their consumer-oriented economies. Thus both coal and natural gas contain sulphur compounds and when these important fossil fuels are burnt to provide power the sulphur is converted to sulphur dioxide and this is liberated into the atmosphere. SO_2 itself is unpleasant; it causes severe irritation of the respiratory tract, for example, but in the atmosphere it is also slowly converted into sulphur trioxide and this, combining with atmospheric water vapour, forms sulphuric acid! When this falls to the ground as a dilute solution in rain, machinery, buildings and wildlife can be extensively damaged. Acid-rain of this type is currently causing devastation to many of the forests of northern and central Europe. Another example of pollution is provided by the discharge of lead compounds from the exhausts of motor vehicles. The lead is introduced into the gasoline fuel as lead tetraethyl, an "anti-knocking" compound, and the concentrations of lead near motorways and motorway intersections can become high enough to pose a serious threat to children living nearby, since, if it is ingested via their lungs and stomachs, it can cause retardation of brain development. In city centres where the traffic is often moving only at a snails pace, one has not only lead pollution but also extensive carbon monoxide pollution. Carbon monoxide is particularly dangerous since it complexes with the haemoglobin of the blood and thus seriously impairs oxygen transport to and carbon dioxide depletion from the tissues of the body. The primary pollutants produced by automobile engines, CO, unsaturated hydrocarbons, partly oxidised hydrocarbons etc. can react with oxygen and water vapour under the influence of ultraviolet light from the sun to produce a very poisonous and corrosive substance, photochemical smog, which contains, for example, the peroxides of some hydrocarbons. In certain places, Los Angeles is a notorious example, particular atmospheric conditions can cause these products to be trapped in the lowermost layers of the atmosphere and it can become dangerous even to venture forth from one's home!

Pollution of the lower atmosphere, that is the troposphere, can have immediately obvious results but pollution of the stratosphere, whilst not noticeable at once, can lead in the long run to much more serious consequences. This is because the stratosphere contains a layer of ozone, mostly distributed between 24 and 30 km in altitude, and this layer absorbs all ultraviolet radiation of wavelength shorter than about 350 nm incident on the earth from the sun. If this radiation were not absorbed it could have very deleterious effects on people and animals exposed to it—skin cancers for example. The situation is shown (very schematically!!) in Fig. 7.1. The absorption of incoming radiation by the stratospheric ozone also causes a warming of the stratosphere and produces therein a positive temperature gradient. The stratosphere is therefore basically stable unlike the troposphere whose negative temperature gradient makes it unstable and produces all the familiar phenomena of "the weather". Clearly, quite apart from increased health hazards, there could be very detrimental climatic changes if there were to occur any significant reduction in the amount of ozone in the stratosphere.

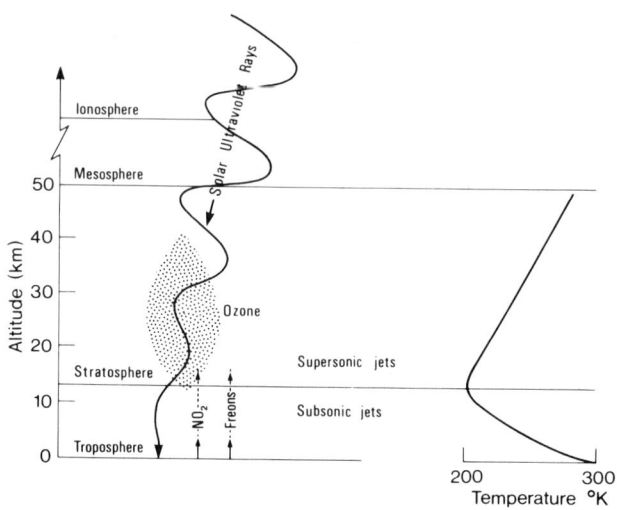

FIG. 7.1. Schematic of the structure of the atmosphere and of the temperature variations within it.

The two pollutants which are currently giving rise to the most concern are NO_x, that is a mixture of unspecified nitrogen oxides, in fact mostly NO_2 and the halogenated methanes and ethanes. The former are produced by the engines of supersonic aircraft which, because of skin-heating effects, have to fly in the low-density stratosphere. The latter are widely used in spray-cans as propellants and are for all intents and purposes indefinitely stable in the troposphere. Vast quantities have been liberated over the last twenty years, mostly CCl_2F_2 and CF_3Cl, and these gases can readily be detected in our present-day atmosphere by infrared spectroscopy over quite moderate path lengths. From the lower atmosphere these gases—their commercial names are freons, fluons, arctons etc.—are inevitably diffusing into the stratosphere. Both NO_x and the freons are potentially dangerous because they can catalyse the decomposition of ozone to oxygen which is transparent to the hazardous middle UV radiation. Oxygen in fact does not start absorbing till almost 180 nm. The catalytic destruction of O_3 by NO_x takes place via an exceedingly complex series of competing reactions but rather simplistically, one can write

$$O_3 + NO \rightarrow O_2 + NO_2$$
$$NO_2 + O_3 \rightarrow NO + 2O_2 \qquad \text{etc.}$$

The halogenated paraffins are not harmful in themselves, their presence in the lower atmosphere causes no apparent hazard, but in the stratosphere they can be decomposed by short-wave radiation liberating chlorine atoms and these atoms are potentially very dangerous indeed. Again there is a series of catalytic reactions of the general type

$$O_3 + Cl \rightarrow O_2 + ClO$$
$$ClO + O_3 \rightarrow Cl + 2O_2 \qquad \text{etc.}$$

Pollution of all kinds is obviously a bad thing and one can understand why it causes so much concern and even occasionally legal intervention; inevitably there is a conflict of environmental and economic interests. Thus if power stations were compelled to ensure that *no* SO_2 was liberated into the atmosphere, the price of electricity would inevitably rise. Likewise many people would be reluctant to give up the advantages of supersonic travel and even more the luxury of instant action spray cans. In this conflict situation, a balance has to be struck and it is here that infrared technology can make a contribution. Governments and international agencies decide where the balance is to lie and set upper limits for the concentrations of specified pollutants in the atmosphere. That done, some means is now required for monitoring the concentrations and seeing that the limits are not exceeded. Point sampling methods, usually automated, which rely on wet-chemical analysis of samples of air can make a contribution but of their nature they are inflexible and slow in their operation. Also to cover a large area effectively by these methods would be very costly. The infrared techniques on the other hand are flexible and quick and are just as capable of doing point sampling as they are of giving integrated concentrations over long paths. They are therefore, at least in principle, cheap, adaptable and easy to use but in addition they can monitor places, the tops of tall chimney stacks, for example, which would be quite inaccessible, to the point sampling technique.

7.2.1.2 Infrared monitoring of lower atmosphere pollution
All the various ways of using infrared radiation for pollution monitoring rely on the highly structured and molecule specific infrared spectra which are

encountered. The situation is basically the same as in analysis except that instead of working at the 1 % level one is now working at the 10^{-6} level and below. Considerable enhancement of sensitivity is therefore required. One can achieve this in three ways. Firstly, as mentioned earlier, one increases the path length either by observing over a long direct atmospheric path or else by using a folded long-path("White") cell. Secondly one can use some form of preconcentration. Thirdly, however, one can note that the line widths of the absorbing lines in question will usually be only of the order 1 GHz, so a considerable increase in sensitivity will be available if one has a tunable quasi-monochromatic probing beam at one's disposal. These considerations point to one of the most popular tools for atmospheric pollution monitoring the infrared diode laser. These lasers (see section 6.2.9 for more details) can be roughly tuned to a given spectral region by suitable choice of stoichiometry and then can be fine-tuned over a range of about 1 cm^{-1} by varying the temperature of the diode or else by subjecting it to hydrostatic pressure. Temperature variations are usually achieved by varying the driving current. A scan of a narrow region of the infrared using a diode laser [897] is shown in Fig. 7.2. The oscilloscope presentation in real time illustrated in this Figure was not possible before the introduction of injection diode lasers. It is achieved by superimposing a current produced by the oscilloscope time base ramp voltage, suitably attenuated, onto the driving (i.e. the tuning) current of the laser. The output from a diode laser is much more divergent than that from other lasers, essentially because the cavity is very short with a low Q, but with suitable optics

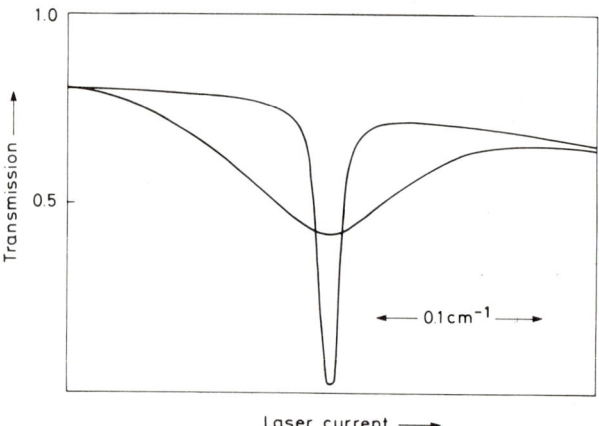

FIG. 7.2. Observation of the P(5) line of CO present to the extent of $\frac{1}{2}$ % in N_2. The spectrum was recorded in real time and the diagram is taken from a polaroid photo of the screen of a storage oscilloscope. The broad line corresponds to atmospheric pressure and the narrow line to a much lower pressure produced by pumping out the cell. The transition from a pressure-broadened to a Doppler-broadened regime is nicely illustrated. (Diagram kindly supplied by J. R. Gott [897].)

the beam spread can be reduced to the point where the beam can be sent over atmospheric paths of the order of hundreds of metres. A suitable arrangement is shown in Fig. 7.3. Apart from outdoor pollution monitoring, this set-up has important industrial applications in factories. Thus it can be used to detect leakages of the dangerous carcinogenic vinyl chloride (CH_2:$CHCl$) in PVC plant and by detecting sudden fluxes of carbon monoxide it can be used as an automatic fire monitor. An effective way of using the laser in automatic monitors is to drive it by square-wave pulses so that its operating wavelength jumps back and forth between the peak absorption wavelength of the chosen pollutant line and a nearby wavelength where the pollutant does not absorb. If the returning beam from the retroreflector is received by a suitable detector and the output from this is phase-synchronously rectified (see section 4.1.3) one will get a resultant signal only when there is a measurable quantity of pollutant present. Over path lengths of the order of hundreds of metres the system is potentially very sensitive ($\sim 10^{-9}$ volume mixing ratio in favourable cases) but there is always the drawback that what one measures is the integrated absorption along the full path travelled by the beam. In other words one is measuring only the column density and whilst one may get a safe reading from, say, a vinyl chloride monitor this may be because the distribution of the gas is very patchy with pockets of dangerously high concentration separated by long stretches of clean air. So a danger reading should always be believed, but a safe reading should be treated with caution. The only certain way of avoiding this difficulty is to supplement the long path monitoring with point monitors, preferably mobile. Suitable optics can readily match the divergent beam from the laser to a White cell and this cell can be carried, round the area to be monitored, on a truck. In this way, admittedly rather laboriously, a point-by-point map can be built up of the pollutant distribution. In a factory, however, one would probably have a good idea of where the high risk areas were and place one's point monitors accordingly. The long-path-cell point monitor is not unfortunately free from problems of its own. At the concentrations envisaged, sorption and desorption from the walls can be very

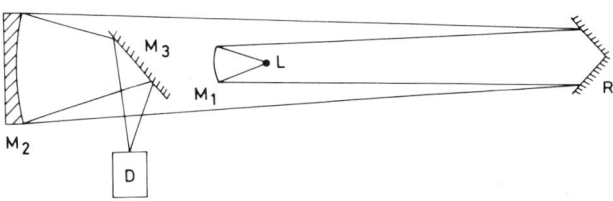

FIG. 7.3. Apparatus used for long-path atmospheric monitoring. The radiation from the laser L is made quasi-parallel by the convex mirror M_1, it is returned along its path by the retroreflector R, converged by the focussing mirror M_2 and then deflected by the 45° mirror M_3 on to the detector D.

significant causes of systematic error. Another point common to both techniques is that to turn the instrument readings into parts per million concentrations one needs accurate laboratory calibrations of the pollutant line strengths. This points up one of the most useful applications of diode lasers, the establishment of precise laboratory standards both material and "on paper". In the same connection one also needs a much more broad-band yet reasonably high-resolution spectrometer to give an overall view of the region of interest. Spin-flip lasers (see section 6.2.10) were at one time considered ideal candidates for this job since they cover at high resolution the heart of the finger-print region. However the difficulties which have been encountered with their tuning characteristics make them nowadays less attractive and the current favourites are interferometers (see section 6.6.2) either rapid-scan or aperiodic. Modern interferometers have such high radiometric precision that it is possible to use scale-expansion and subtractive techniques to achieve a sensitivity much better than 0.1% in percentage transmission units. When coupled to a long path cell, one has a very sensitive spectrometer. In the USA, the Environmental Protection Agency (EPA) has used mobile interferometric systems to carry out field measurements but these instruments also have a major role to play in setting up laboratory standards.

Another branch of atmospheric pollution monitoring using lasers involves the use of backscattering techniques. An intense pulse of laser radiation is sent out along a chosen atmospheric path and sensitive detectors are used to record the time variation of the power which returns. The analogies with RADAR are strong and this branch of the art is often known as LIDAR for this reason. The technique is difficult and costly but it does have the enormous advantage of providing range-resolution, that is one can determine the concentration of pollutant all the way along the path out to a distance from whence the returning signal is too weak to measure. If one can tune the laser to be resonant with an absorption line of the pollutant, then the technique of Differential Absorption Lidar (or DIAL) can be used. In this, two pulses are sent out, one resonant with the line and the other not. The backscattering is different for the two pulses and if one has available an accurate laboratory determination of the absorption strength of the line in question, it becomes possible to determine pollutant concentrations absolutely. This method is very useful for checking on the emission levels from tall chimney stacks. So far measurements have been restricted to the near infrared and the 8–14 μm atmospheric windows. The near-infrared approach has the advantage that powerful tunable lasers are available but the disadvantage that few pollutants have strong absorption lines there. The mid-infrared approach is attractive because virtually every poll-utant has intense lines in that window but on the other hand the best laser, the CO_2, is not easily tuned. When there is a good coincidence, however, the system can be extraordinarily sensitive because one can use another CO_2 laser to provide a local oscillator for a superheterodyne receiver. The transmitting

CO_2 laser is usually a pulsed TEA type (see section 6.2.2) and one can put the very stable continuous-wave local oscillator tube into the same cavity. A promising approach to the tuning problem is to use high (10–15 atm) pressure CO_2 lasers which because of the pressure broadening give out a quasi-continuous spectrum so one can choose the output frequency by means of cavity length adjustment, the use of etalons or both. Exact coincidence can therefore be achieved but one then has a local oscillator problem. Diode lasers represent a possible solution here since they too can be tuned into exact coincidence with the resonant frequency. Another laser which is starting to be used for this sort of work is the DF laser which radiates a series of lines near 3·7 μm in the 3·5 − 5 μm atmospheric window. With any of these systems it is in principle possible to determine the concentrations of more than one pollutant. In fact if one has N pollutants one needs only N probing frequencies to determine them all. The mathematical process is matrix inversion but provided one has chosen the frequencies well and provided the experimental imprecision is not great, the matrix should be well conditioned and the inversion proceed smoothly.

The power of infrared techniques to monitor pollution of the lower atmosphere is thus demonstrated and as more powerful tunable lasers become available the sensitivity is going to get better and better. This presents a dilemma since no sooner has the sensitivity been demonstrably increased but the politicians, driven on by the ecological lobby and anxious not to make a mistake, set this new level as the maximum tolerable level. As an example, infrared techniques can now readily detect carbon monoxide at the 10^{-9} volume ratio level and there is pressure to have this level legally enforced as the maximum permissible. This despite the fact that no medical authority would assert such levels to be harmful—in all probability the human race has evolved against concentrations of this order as the natural background! The resolution of this dilemma lies however with politicians, not infrared technologists.

7.2.1.3 *Infrared monitoring of stratospheric pollution*
Infrared monitoring of stratospheric pollution consists of two operations— the relatively easy one of seeing whether the concentrations of the majority and minority constituents are varying with time, and the much more difficult one of trying to detect the build-up of man-made pollutants and the presence of their reaction products. One has the basic choice of attempting to use either the pure-rotation spectrum in the far infrared or else the vibration/rotation spectrum in the mid infrared. At first sight it might be thought that the pure rotation region would be unattractive for monitoring atmospheric composition because the spectra lie on top of one another, whereas for the vibration/rotation regions each spectrum will be well separated from the others because of the widely varying values for the vibration frequencies. However the pressure is so low in the stratosphere that pressure broadening

becomes negligible and the combination of low temperature and low frequency makes Doppler broadening likewise negligible and the observed lines are very sharp indeed. Therefore, provided one has sufficient resolution, one can sort out the great forest of lines quite adequately and do a good job of quantitative analysis. There are strong similarities here to the well-established use of microwave spectroscopy to do quantitative analysis, but fortunately for the stratospheric observer the ultrahigh resolving power of microwave spectroscopy is unnecessary and adequate monitoring is possible using the resolutions of between 0·06 and 0·01 cm^{-1} provided by good quality Fourier transform instruments. In the mid infrared still lower resolution is adequate because of the separation and grating instruments are frequently used. At the moment both regions are being used extensively, though perhaps the far infrared more so for the majority components and the mid infrared more so for the trace components, and both are yielding some very valuable observed data (see for example Fig. 4.25). For both mid- and far-infrared observation one can choose to work either in absorption or in emission. Absorption spectroscopy requires the availability of a source—usually the sun—so one is restricted to daytime observation. Emission spectroscopy can be carried out at any time but it requires the use of at least a cooled detector and possibly even the use of an entire spectrometer cooled to a cryogenic temperature. The principal experimental problems arise from the opacity of the troposphere which means that it is not possible to monitor the stratosphere from the surface of the Earth. The intervening tropospheric levels in fact radiate for most of the infrared as a black body at about 290 K (see Fig. 2.21) and even if one searches in the so-called atmospheric "windows", for example that at 8–16 μm, the only effect is that the black-body temperature falls to about 220 to 230 K. No detail of the very sharp stratospheric lines can therefore be observed from ground-based stations. The higher one can go, the clearer becomes the view and some classic work in this direction was that of Gebbie [898] who reported far-infrared atmospheric spectra recorded at the Jungfraujoch high altitude station (3450 m) in 1957. The spectra were certainly provocative but even Jungfraujoch was not high enough for effective stratospheric spectroscopy. All modern work therefore has used high-flying aircraft or else balloons. In the near future, these observations "from underneath looking up" will be supplemented by the "from on-top looking down" observations made from orbiting space stations and satellites.

As was mentioned earlier, the normal stratospheric constituent of most interest is ozone O_3 whose stratospheric mixing ratio is typically 7×10^{-6}. Extensive programmes for monitoring this gas have been carried out since the mid 1960s using the prototype Concordes and, more latterly, balloons. The prominent Q-branches in the pure-rotation spectrum (see Fig. 5.8) are readily observed. The ambient temperature of the gas is about 220 K so it is possible to work in "emission" using a warm (~ 300 K) detector or in conventional emission with a cryogenically cooled detector. Most modern work has been along this latter line using a Rollin-type detector operating at pumped liquid helium temperature. An example of a spectrum obtained in this way is shown in Fig. 7.4.

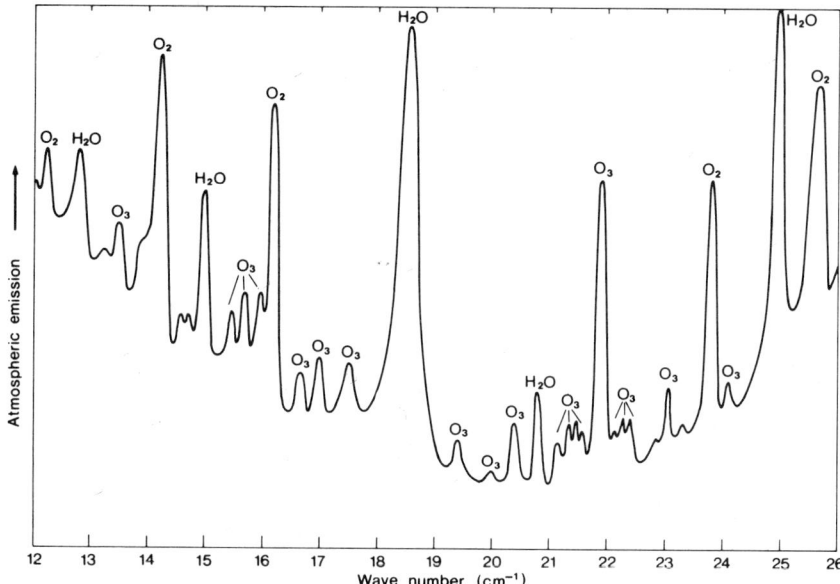

FIG. 7.4. Stratospheric emission spectrum obtained by Harries and his colleagues [54] in 1971 from on board a Comet aircraft at an altitude of 12 km. The detector was a Rollin type and the resolution achieved was 0.06 cm^{-1}.

The other stratospheric majority components which absorb strongly in the far infrared are oxygen and water vapour. Oxygen absorbs via a magnetic dipole allowed pure-rotation spectrum (see section 5.2.2) which is fundamentally a weak process but there is so much oxygen in the atmosphere and the path lengths involved are so long that the observed triplets are quite intense. Water vapour is an inherently strong absorber with a large dipole moment and even though its mass mixing ratio is much less than that of oxygen (typically 2×10^{-6} in the stratosphere) its spectral lines dominate the atmospheric absorption in the far infrared and it is the somewhat chaotically distributed lines of the asymmetric top H_2O which are responsible for the "blacking-out" of the far infrared at the surface of the earth. The so-called atmospheric "windows" are essentially the regions between the H_2O lines but the line-shapes observed are somewhat anomalous with higher than expected intensity in the wings, so these "windows" are rather murky. Fortunately the scale height for water vapour is much less than that for the other gases and the bulk of the water vapour is confined to the troposphere. The water vapour lines arising in the stratopshere are much narrower than those arising in the troposphere so the net effect is to produce the clarification of observation with increasing altitude that was referred to earlier. Some simulated atmospheric transmissions as a function of altitude are shown in Fig. 7.5.

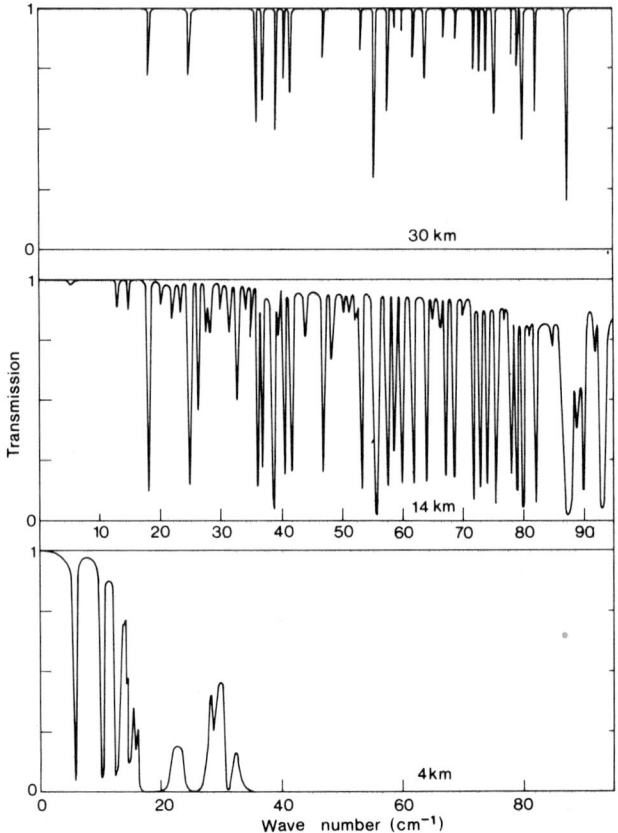

FIG. 7.5. Estimates of the atmospheric submillimetre transmission as a function of altitude [899].

The sharpening and weakening of the water vapour lines with altitude makes it possible to observe the very much weaker features of the minority components. The most important of these, from the pollution viewpoint, are the oxides of nitrogen and their reaction products which typically might be present in mixing ratios of the order 10^{-9}. Nitric oxide, NO, and nitric acid, HNO_3, are relatively easy to observe because their spectra consist of regularly spaced lines (HNO_3 is an accidental symmetric top) but the spectrum of NO_2 is made up of an immense array of randomly placed lines. Occasionally the lines of NO_2 do blend to make a stronger and therefore potentially observable feature. The best known example is the pseudo Q-branch at 37.8 cm^{-1}. A far-infrared emission spectrum, due to Harries and his colleagues [900], which possibly shows this feature is given in Fig. 7.6. One says "possibly" because unfortunately weak lines of the majority components or strong lines of their isotopic variants might explain many of the so far unassigned lines in the stratospheric spectrum. P and R lines of ozone with positions and intensities that are not known very well are particularly troublesome. The isotopic forms of H_2O have been investigated by Fleming [901] and by Partridge [412] so their lines are now well characterised and should not cause any difficulty. Bearing in mind these caveats and also bearing in mind that the pollutant lines lie virtually at the limits of detectability, it is nevertheless possible to hazard guesses as to how the pollutants vary with height. Nitric acid, for example, seems to peak at 25 km, whilst NO_2 seems to have its peak above 30 km. These studies have shown that the pollution of the stratosphere by high-flying aircraft is a much less serious risk than was at first thought. The stratosphere seems to be basically stable to the injection of oxides of nitrogen—a perhaps not too surprising conclusion since nitrogen and oxygen are abundant there and in the presence of hard ultraviolet their compounds might be produced quite naturally. It seems that the formation of nitric acid is a natural

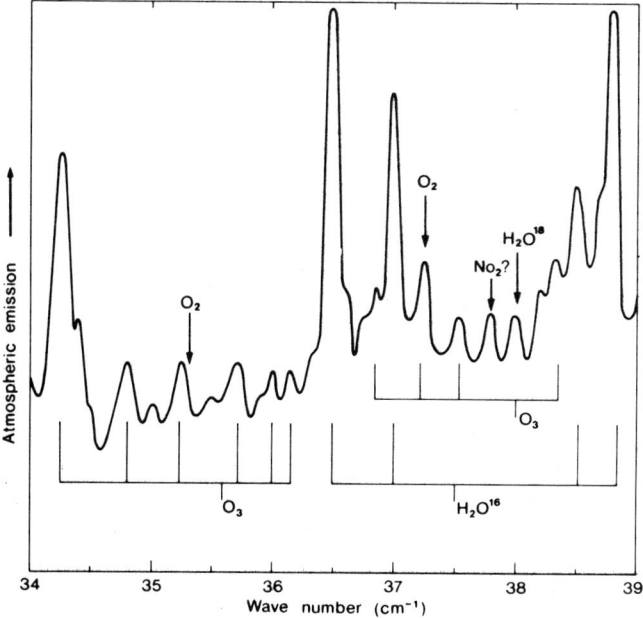

FIG. 7.6. Portion of a far-infrared emission spectrum due to Harries et al. [900] observed at an altitude of 35 km and with a spectral resolution of 0.07 cm^{-1}.

mechanism for removing NO_x from the stratosphere—and HNO_3 is some-times referred to as a "sink" for this reason. The HNO_3 is fairly stable to ultraviolet light and eventually falls to the surface in rainwater to the benefit of the nitrate hungry plants on the surface.

The monitoring of freon pollution of the stratosphere has turned out to be a much more difficult task. The far-infrared spectra of the freons [902] are very complex and their presence amongst the normal forest of stratospheric lines is exceedingly difficult to detect. Claims have been made for the presence of HCl, pure rotation lines in the stratosphere but the evidence is so far equivocal. Attention has therefore switched to the mid infrared and to the vi-bration/rotation spectrum. The freons are readily observable by mid-infrared spectroscopy over quite moderate tropospheric paths but so far the evidence for their presence in the stratosphere is tantalisingly inconclusive. The mid-infrared spectra have, however, shown conclusively the presence of HCl (along with NO_2, HNO_3, HF etc.) in the stratosphere. Despite the present unclear situation, speculation has been rife as to whether there is a "sink" for stratospheric chlorine. One possibility is chlorine nitrate $ClO\,NO_2$ and the far-infrared spectrum of this substance has been carefully studied by Fleming [903].

Although most stratospheric monitoring has been done using broad-band spectrometers, a significant amount has been carrried out using narrow band, essentially single-frequency instruments. Prominent amongst these have been selective radiometers which respond to the emission signal from a single chosen gas. Houghton at Oxford [904] and Smith at Heriot–Watt [905] universities have been very active in this field. The selective chopper and pressure-modulator radiometers developed by these workers operate either by chopping between an empty and a full cell or else by periodically modulating the pressure of a small amount of gas in a single cell. The output of the PSD will in either case contain information only about a narrow band of spectral frequencies close to the line centres of the gas in the radiometer and one has therefore a high degree of selectivity. Harries and Chamberlain have developed a third type [906] in which a Michelson interferometer, set accurately to physical zero path difference but containing two identical cells in the partial beam arms, is used. If a gas is put into one of the cells there will be a variation of refractive index near the absorption lines of that gas (see Fig. 3.8) and the optical path difference in the interferometer will vary. The returning signal towards the source, which would be zero in the absence of the gas, becomes finite but *only* for frequencies corresponding to the absorption frequencies of the gas and again one has a very selective device. These selective radiometers have proved very valuable indeed for monitoring majority components, especially H_2O, but their sensitivity does not seem, at the present, high enough to detect minority components and there are severe problems with chemically unstable or very corrosive reference gases.

The sensitivity of the monitoring system can be enhanced still more if one can use a swept essentially monochromatic source in what amounts to a microwave spectrometric system. Indeed microwave spectrometers have been flown in balloons [907] but the most promising line of investigation here seems to be the possible use of tunable infrared lasers. Patel has, for example, flown a spin-flip Raman laser aboard a balloon and detected stratospheric NO [908]. There are severe logistic problems here but if one could imagine two satellites, widely separated but communicating via a tunable laser beam passing through the upper reaches of the stratosphere, the sensitivity would be much better than the 1 in 10^9 level and all possible monitoring tasks could be carried out. What is clear is that infrared spectroscopy will continue to play a major role in the global task to prevent irretrievable damage to the protective ozone layer in the stratosphere.

7.3 Infrared astronomy

Astronomical observations in the infrared can give two distinct types of information [51]. One can firstly detect phenomena which are only manifest in the infrared, for example "infrared stars", and secondly one can study, more clearly, objects, ordinary stars for example, whose radiation is severely attenuated on its way to us by absorption and scattering by the interstellar dust. The principle difficulty which prevents extensive use of the infrared is that the earth's atmosphere is opaque (see Fig. 2.21) throughout much of the region and ground based observations are restricted to the atmospheric "windows" at 1·0, 1·2, 1·6, 2·2, 4·4, 8–16 μm and beyond 300 μm. The principle cause of the absorption is water vapour and since this has a smaller scale height (i.e. the height at which the partial pressure has fallen to 1/e of its value at sea level), 2 km, than the principal gases, i.e. 8 km, considerable advantages accrue to working in dry climes and from as high as altitude as possible. Observatories are located therefore in such places as Tenerife and Hawaii and from these places very much valuable data has been gained. However, still more information is available from higher up and the ground-based observations have recently been supplemented by observations from balloons and from high-flying aircraft and soon these will be joined by observations from stable orbiting platforms. When one bears in mind how much has been discovered with the limited visibility we have had so far it is quite sanguine to expect dramatic developments when the whole infrared is available and over the whole sky with a telescope of high aperture and good spatial resolution. However, these techniques will always be expensive and there will be a continuing rôle for ground-based observations, especially when the terrestial astronomers confine themselves to investigations particularly suited to their limitations. This is illustrated by the recent direct detection of the rings of Uranus. The difficulty which had previously prevented direct observation is that the rings have a low albedo, presumably due to their being composed of rocklike fragments, and shine rather dimly. They therefore cannot be seen against the glare of the very bright primary. However, when observations are made at 2·2 μm where strong absorption in the methane clouds gives the planet likewise a low albedo, the rings can then be detected.

7.3.1 *Studies of and through the interstellar dust*

The galaxy, whose diameter is of the order $1·2 \times 10^{21}$ m and whose mass is about $2·2 \times 10^{41}$ kg, is made up of a more or less spherical and very ancient structure composed of very old Type II stars plus globular clusters and a very much flatter (thickness $\sim 1·6 \times 10^{19}$ m) spiral structure made up of gas clouds and very much younger Type I stars. Stars are condensing from the gas clouds all the time and we can in fact detect several good candidates for protostars in nearby clouds. The sun is located in a spiral arm (the Orion material arm) at a distance of $3·1 \times 10^{20}$ m from the centre and one of the best studied of the gas clouds, that associated with the trapezium nebulosity, lies in the constellation of Orion. Stars once formed from the interstellar material go through their evolutionary cycle but the lifetimes (and initial brightness) depend strongly on the mass, the variation goingly roughly as $M^{-4.5}$. Thus massive stars are of spectral types 0 and B (blue supergiants $T_{\text{surface}} \sim 30\,000$ K) and have very short lives. In their final phases they undergo mass loss and hence return material into the interstellar medium. The material so returned has, however, by nucleosynthesis been transformed into helium and heavier elements such as carbon, nitrogen, oxygen, silicon etc. and these elements can condense into solid matter, graphite, silicates etc. This solid matter in the form of particles about 1 μm in diameter forms extensive lanes of dust in the spiral arms. The dust causes severe obscuration of visual observations in the plane of the spiral structure and prevents optical telescopes from seeing further than about 2×10^{19} m in the galactic plane. This is the reason why it took so long for the spiral structure of the galaxy to be inferred and it was only with the development of 21 cm radioastronomy, which depends on the magnetic dipole allowed transition between the hyperfine states of the ground level of the hydrogen atom, that this structure was proved. The Mie theory of scattering by dielectric particles shows that the attenuation falls quite quickly when the observing wavelength gets large in comparison with the size of the scattering particles, so infrared observation can penetrate much further into the dust. The 21 cm observations are essentially unaffected by the dust and detailed maps of hydrogen atom concentration have been drawn up with their aid. The galactic centre lies in the constellation of Sagittarius some $3·1 \times 10^{20}$ m away and is for all intents and purposes unobservable at optical wavelengths. However, in the far-infrared and millimetre wave regions of the spectrum it is extremely bright and furthermore shows detailed structure which is currently the object of intense investigation. The gas clouds in the spiral arms have number densities of molecules (mostly H_2) much higher than average. Thus the average number density for the galaxy as a whole is about 1×10^6 m^{-3} but in the gas clouds it can be as high as 1×10^{15} m^{-3}. The gas clouds always contain dust and there seems to be some, as yet imperfectly understood, symbiotic relationship between the dust and the molecules present in the cloud [25]. These molecules can be quite complex and examples as large as C_2H_5OH, $C_2H_5NH_2$ and HC_9N have been detected. The largest molecular cloud is near the galactic centre—it is called Sgr B—and millimetre-wave radioastronomy [909] has, by picking up the characteristic micro-wave rotational lines, proved the existence there of more than forty complex molecules, including those quoted above. What is particularly fascinating is that some lines, particularly of H_2O, OH and SiO, are seen in stimulated emission and one has point-like maser sources in the sky! Other clouds, including the famous one associated with the Orion nebula (OMC-1), which is very much bigger than the visible nebulosity, show similar features. However, since OMC-1 is so much nearer ($\sim 1.2 \times 10^{19}$ m) than Sgr-B, it is possible to study the features in it with much better precision. Thus OMC-1 is known to contain a compact infrared nebula, called the Kleinmann–Low nebula (KL), which lies behind the trapezium group which itself is on the near side of the cloud to us. KL appears to contain all the maser sources of the cloud. Current opinion is that KL is a protostar in the early stages of gravitational contraction.

Some of the molecular clouds are cold and dark, but the majority contain bright young stars (the trapezium group in the Orion nebula for example). In addition the clouds may contain either visible stars which show a large infrared excess or else, and even more spectacularly, objects which can only be detected in the infrared. These are thought to be either very young stars completely surrounded by dust, or else protostars which are still in the gravitational contraction phase. The best known example is the BN object, named after its discoverers Becklin and Neugebauer, which lies in the Kleinmann–Low nebula. BN is apparently less than 200 astronomical units (1 au is the mean earth–sun distance 1.5×10^{11} m) in diameter but its infrared emission is some thousands of times the total power output from the sun. It therefore looks very much like a massive object condensing down towards being a supergiant star but the problem, as always, is the dust: one cannot be sure that BN is not already an O-type supergiant but surrounded by a dusty mantle which is thermalising its radiation down into the infrared. Even more intriguing is the recent work on the star MWC 349 which belongs in the Cygnus OBII association. This again shows strong infrared excess which is apparently associated with there being, around the star, a large dusty ring, which we see face-on. This disk has a mass of about 1/50 that of the sun and a radius of 10^{11} m so it could be a prime candidate for a protoplanetary system.

The actual process of mass loss can be studied in isolated late type stars of spectral types M and cooler, especially those which are Mira-type variables. A particularly interesting special case is provided by the planetary nebulae in which the material forms a visibly observable disk. In the extreme case of mass loss, the star's photosphere may not be visible at all and the star will only radiate perceptibly in the infrared. Two of the brightest infrared stars in a survey carried out by Caltech were found to be evolved stars—these were called NML Cygni and NML Tauri, the initials standing for the authors of the survey, Neugebauer, Martz and Leighton [910]. Considerable interest attaches to the nature of the dust clouds round these stars. The ones which show a featureless black-body emission spectrum (for example C-stars) are probably surrounded by graphite and/or metallic particles. This is the case for NML Cygni, but for most stars the infrared emission profile shows features which can be identified with specific compounds. These features in order of increasing wavelength are 3·9 μm SiC_2, 9·7 μm silicates, 11·2 μm SiC and 18 μm silicates. It has been shown that a "dirty" mixture of carbon, silicates and metallic particles could provide enough optical depth to thermalise the star's emission. From a detailed model one can derive the size of the disk. The simplest model (not applicable in any real case) would be to assume a photosphere temperature of T_s, a disc temperature of T_D, star radius of R_s and disk radius of R_D and *strict* black-body behaviour. Then

$$R_D/R_s = (T_s/T_D)^2. \qquad (7.3.1)$$

The dust in interstellar clouds, produced by mass-loss processes, is of obvious interest since planets such as the earth are made from it but it is also of interest because it almost certainly acts as a heterogeneous catalyst for the formation of the complex molecules in the clouds. How this happens is only slowly becoming clear but it will be very surprising if future far-infrared observations have little to add to the flow of information from millimetre wave observation.

7.3.2 *Non-thermal sources of infrared radiation*

All cosmic objects radiate in the infrared, by necessity, but they will only be of particular interest if they show an infrared excess, that is if the infrared output is much more than would be expected by extrapolating the visible luminosity. Stars surrounded by dusty envelopes will, as discussed above, show infrared excesses by essentially taking high-temperature black-body radiation and transforming it to lower-temperature radiation from a larger area. This is a thermal process. There are, however, two non-thermal processes which can lead to infrared excess: these are "free-free" or plasma emission and synchrotron emission. The "free-free" emission occurs in highly ionised regions of the interstellar medium where the hydrogen atoms have dissociated into free electrons and protons. These are usually called H II regions to distinguish them from the neutral unionised HI regions. The ionisation in H II regions is produced by the radiation from hot young O and B stars embedded in them and these regions are some of the most beautiful objects which can be seen through a telescope. They show brilliant colours principally due to the recombination radiation in the Balmer series of hydrogen but with other species making significant contributions. From a quantum-mechanical point of view, the solution of the electron/proton problem includes not only the bound states [equation 5.1.5], but also non-bound states which lie above the ionisation limit and form a continuum. Transitions within the non-discrete states in the continuum can take place virtually at any frequency. One can also analyse the situation classically since a continuum is involved and one then thinks of an electron being decelerated (that is "braked") in the field of a proton and hence emitting Bremsstrahlung radiation. The emitted spectrum is, however, not strictly white because one has to take into account how the absorption coefficient (and hence the blackness) varies with frequency. The solution for an idealised plasma is

$$\alpha(v) = \frac{v_c}{c}\left(\frac{v_p}{v}\right)^2\left[1 - \left(\frac{v_p}{v}\right)^2\right]^{-1/2}, \qquad (7.3.2)$$

where v_c is a collision frequency and v_p is the plasma frequency defined by equation (2.2.3). Equation (7.3.2) indicates that waves do not propagate in the plasma for $v < v_p$ and this is reinforced by considering the refractive index

$$n(v) = \left[1 - \left(\frac{v_p}{v}\right)^2\right]^{1/2}, \qquad (7.3.3)$$

which becomes imaginary for $v < v_p$. This means that the plasma is essentially black below v_p but above v_p its blackness depends on v and also on v_c. The collision frequency can be calculated for simple plasmas, for example hydrogen, but here we are only interested in the functional form and it will be taken to be a constant even though this is not quite true and it does show a weak dependence on v. The plasma frequencies in all astronomical situations are very low, in fact they lie in the radio frequency region and therefore in infrared observations we will always have $v \gg v_p$ and the radical in (7.3.2) can be neglected. The absorption coefficient then depends inversely on the square of v. The observed emission spectrum will thus start off with the v^2 dependence of a Rayleigh–Jeans black body, will then enter a plateau region where the v^2 dependence of the energy density cancels the v^2 dependence of α but then, when the Wien region is encountered, will fall off even more rapidly than does that of a black body. This is illustrated in Fig. 7.7. The kinetic temperatures of most H II regions can be several thousand degrees Kelvin so the Rayleigh–Jeans region extends well into the infrared and the flat plateau region should be easily observable. From the form of the curve the electron concentration can be inferred and also the electron kinetic temperature. These are very valuable contributions, especially since at these long wavelengths dust absorption can be ignored and the values deduced can be relied on.

Within the galaxy there are usually magnetic fields present and electrons circling in the magnetic field will also give off Bremsstrahlung radiation (see section 2.5) and this does in fact contribute quite markedly to the radio-frequency "sky-noise". However if the electrons are relativistic and in intense magnetic fields, synchrotron radiation will also be emitted. This has the characteristics that the intensity falls according to a power law

$$I \sim v^{-\alpha} \tag{7.3.4}$$

and that the radiation is strongly angularly confined. The synchrotron radiation shows self-absorption effects above a certain cut-off frequency given by

$$v_R = 1 \cdot 6\, N_e / B \varepsilon_0, \tag{7.3.5}$$

where N_e is the electron density and B is the magnetic field, and this cut-off effect can lead to an observed spectrum which looks like a black-body curve. If there is a lot of dust present to help with thermalisation it can look still more like a black-body curve. However, unlike the synchrotron sources which are used on earth, those in the depths of space have no inherent stabilising mechanisms and therefore can vary rapidly. This is the usual touchstone for identifying synchrotron emission. Strong infrared emission produced by the synchrotron mechanism is observed for the Seyfert and other very active galaxies where up to 80 % of the total radiation may be in the infrared. There is speculation that the activities of these galaxies, which may in fact be closely related to the quasars, is due to their containing a massive black hole which is swallowing material. Certainly synchrotron radiation has been observed from the few black-hole candidates. Cygnus X1 for example, known so far in our galaxy.

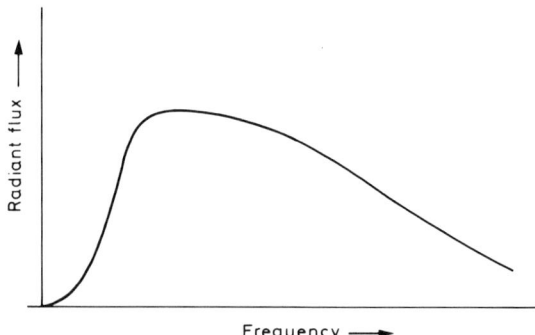

FIG. 7.7. Schematic diagram of observed emission from a "free-free" plasma.

7.3.3 *The cosmic background radiation*

The existence of the isotropic cosmic background radiation, which is closely similar to that of a black-body source at $\sim 2 \cdot 9$ K, was mentioned in Chapter 2. Considerable attention has been devoted to measurements of this radiation, despite the formidable experimental difficulties, because of its fundamental importance. It is after all the only direct evidence we have as to the state of the Universe before the galaxies condensed. Richards and his colleagues have had considerable success in the measurements and their liquid helium cooled interferometer flown abroad a balloon is a wonderfully sophisticated piece of equipment [46]. The principal difficulty which has to be overcome is the liability of the receiving system to see a warm object, the earth for example; even the thin plastic sheet used to prevent all the liquid helium boiling away can give a measurable signal when compared with a source at 3 K! Nevertheless Richards and his colleagues have managed to confirm beyond doubt the earlier work of Robson *et al.* [46] which showed the maximum (at 6 cm^{-1}) which is the hallmark of a thermal curve and furthermore have followed the descending line into the far infrared. In this connection it is interesting to consider the various sources of isotropic signals coming from the sky. This is illustrated schematically in Fig. 7.8.

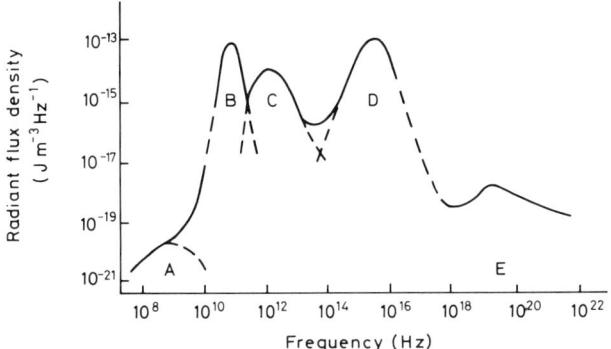

FIG. 7.8. Contributions to the overall "sky-noise". A is the radioemission due to galactic plasma, B is the 2.9 K background radiation, C the infrared emissions from all sources including stars and heated dust, D is the ultraviolet radiation from very hot stars and E is an X-ray continuum whose nature is something of a mystery. (After Winnewisser *et al.* [25].)

What this figure brings out is how lucky we are in the vicinity of the sun to have the background radiation resolved from the other components so that it can readily be studied almost in isolation. A particularly fascinating observation which has been made from a high-flying aircraft is that the background radiation seems less intense (and therefore cooler) in the direction of the constellation Virgo. The cooling amounts to some 3 mK and if interpreted as a Doppler shift due to the motion of the earth with respect to the primordial plasma, would indicate an "absolute" velocity of 3×10^5 m s^{-1}.

7.3.4 Discrete line-spectra in the infrared

Sharp lines due to transitions in atoms and molecules occur in the infrared just as much as in the visible region so the infrared observations can complement the visible ones, or do rather better than this if dust obscuration is a problem. One characteristic, however, that distinguishes astrophysical spectra from laboratory ones is the very different relationship between transition moment and observability. In the laboratory, where path lengths are necessarily small ($<$ 1 km) fairly high gas pressures must be used ($>$ 10 mtorr) and it follows that unless a transition is fairly strongly allowed it will not be seen because the competing collisional de-excitation process will win. This is why only electric dipole selection rules are usually considered. Even for the next strongest member of the hierarchy the magnetic dipole transition, very special techniques (resonant cavities for example) are needed if observation is to succeed. In molecular clouds, the collisional rate can readily be of the order of one per century and it follows that "forbidden" lines may be seen with the same order of intensity as allowed ones. Put in simple terms if an atom or molecule is taken up to an excited state it must eventually come down again and, whether it does this at once or after a time delay, there will be produced in both cases just one photon per excitation. Therefore if the excitation rate is slow compared to the de-excitation rates, "allowed" and "forbidden" lines can appear with comparable intensities. An extreme case is provided by atomic hydrogen. Transitions from $2P_{\frac{1}{2}}$ to the ground state $1S_{\frac{1}{2}}$ are strongly allowed—Lyman α— but the $2S_{\frac{1}{2}} \rightarrow 1S_{\frac{1}{2}}$ transition is absolutely forbidden for any one photon process. The absorption process $1S_{\frac{1}{2}} \rightarrow 2S_{\frac{1}{2}}$ can be induced in the laboratory by double-photon absorption from intense laser radiation—a very non-linear process—but the converse double-photon emission process must also take place in interstellar space since a high concentration of $2S_{\frac{1}{2}}$ atoms does not build up. Of course there are several two-stage processes which could take $2S_{\frac{1}{2}}$ down to $1S_{\frac{1}{2}}$ but nevertheless it is thought that the double-photon transition does occur. Forbidden lines are therefore very good probes for the conditions in the emitting regions and if enough lines of enough different species are available a very good picture can be built up of the electron density, the temperature and the collisional rates. Forbidden lines are known in the visible region—for example the lines in the green once attributed to "Nebulium", now known to be due to O III, but the infrared is a particularly favoured region because the "spin-flip" magnetic dipole allowed transitions within fine-structure multiplets occur here. There is also the additional advantage that infrared observations can probe deeper into the dust. The full advantage of infrared observations have not so far been realised because of atmospheric absorption but the lines of Ar III at 8·99 μm, S IV at 10·54 μm, Ni II at 12·81 μm, S III at 18·71 μm and O III at 88·35 μm have been observed from a wide range of H II regions.

Atomic hydrogen has been known for a long time and as mentioned above we now have a very good understanding of its distribution in the galaxy and of its interaction with the stars. However it has become abundantly clear in the last few years that the cold molecular clouds contain mostly H_2 molecules and since it has also become clear that these clouds are not only the birth places of stars but that they are also intimately bound up with the generation of the spiral structure, the study of H_2 molecules in the clouds is of considerable interest. The difficulty is that the H_2 molecule being a closed shell molecule with a centre of symmetry has no dipole allowed transitions in the infrared. The hydrogen can be studied indirectly by the effects it has on the spectra of molecules, CO for example, which do have allowed rotational and vibrational transitions. This is because the H_2 molecules, being enormously the most abundant, will be entirely responsible for any collisional broadening. The first few lines of CO (see section 5.2.2) occur at

$$\left. \begin{array}{llll} J = 0 \rightarrow J = 1 & 115\cdot271\,\text{GHz}, & 3\cdot845\,\text{cm}^{-1}, & 2\cdot6\,\text{mm} \\ J = 1 \rightarrow J = 2 & 230\cdot538\,\text{GHz}, & 7\cdot690\,\text{cm}^{-1}, & 1\cdot3\,\text{mm} \\ J = 2 \rightarrow J = 3 & 345\cdot796\,\text{GHz}, & 11\cdot535\,\text{cm}^{-1}, & 0\cdot867\,\text{mm} \end{array} \right\} \quad (7.3.6)$$

and are amongst the best characterised of the millimetre-wave/far-infrared lines observed from the depths of space. They have even been observed from other galaxies. There is thus ample observational material for studying H_2 indirectly but, of course, where possible direct observations would be preferred. In the presence of hot stars, the H_2 molecules can become vibrationally excited (by prior excitation to an electronically excited state via an allowed transition, followed by radiative decay down to the ground state) and then there is the possibility of seeing the H_2 vibration/rotation spectrum made active in emission by the quadrupole allowed transitions. The energy levels (in cm^{-1}) of H_2 are defined by the relations

$$T_v = G_v + B_v J (J+1) - D_v J^2 (J+1)^2, \qquad (7.3.7a)$$

where

$$\left. \begin{array}{l} G_v = 4400\cdot390(v+\tfrac{1}{2}) - 120\cdot8148(v+\tfrac{1}{2})^2 + 0\cdot724\,19(v+\tfrac{1}{2})^3 \\ B_v = 60\cdot8409 - 3\cdot017\,74(v+\tfrac{1}{2}) + 0\cdot028\,55(v+\tfrac{1}{2})^2 \\ D_v = 0\cdot046\,84 - 0\cdot001\,706\,(v+\tfrac{1}{2}) \end{array} \right\} \quad (7.3.7)$$

so, bearing in mind the quadrupole selection rule $\Delta J = \pm 2$ which gives $O(\Delta J = -2)$ and $S(\Delta J = +2)$ branches, one would expect to see the transitions

$$v = 1 \rightarrow v = 0 \left\{ \begin{array}{llll} S(0), & J' = 2 \rightarrow J'' = 0, & 4497\cdot804\,\text{cm}^{-1} = 2\cdot22\,\mu\text{m} \\ S(1), & J' = 3 \rightarrow J'' = 1, & 4712\cdot830\,\text{cm}^{-1} = 2\cdot12\,\mu\text{m} \quad (7.3.8) \\ S(2), & J' = 4 \rightarrow J'' = 2, & 4916\cdot700\,\text{cm}^{-1} = 2\cdot03\,\mu\text{m} \end{array} \right.$$

and these have, in fact, been observed from several cosmic objects including the planetary nebulae such as NGC 7027. The quadrupole selection rule can be derived from the correspondence principle by noting that the quadrupole goes into itself after a rotation through $180°$ so the quadrupole rotates twice as fast as does the molecule. Undoubtedly, when observations are made above the atmosphere still more lines of this band will be seen and more information about the excitation/de-excitation balance will become available.

One of the most interesting problems to which millimetre-wave/infrared observations can be turned is that of isotope abundances. The various routes of nucleosynthesis can give rise to varying ratios of for example D/H and $^{13}C/^{12}C$ and one can hope in principle to find out how these ratios do in fact vary across the galaxy. Deuterium is very labile in stellar interiors and one would expect that near the galactic centre, where the interstellar medium will have been processed several times over, it would be rarer than it is in the vicinity of the sun. On the other hand ^{13}C is produced in the distended atmospheres of late-type giants and Mira variables and would therefore be expected to be *more* abundant near the galactic nucleus. The solar system was essentially decoupled from the interstellar medium some 4.6×10^9 years ago, so measurements on active regions of the medium can tell us what has been going on in the realm of galactic nucleo-chemistry since then. The difficulties of getting this information are two-fold. Firstly, although the rotational lines of the rare species will be optically thin and the column density can be directly evaluated (section 2.2.2) in nearly all cases, the lines due to the abundant species will be optically thick and the column density information is not forthcoming. The second difficulty is that in the highly exotic < 100 K chemistry of the molecular clouds it does not follow at all that the concentrations of isotopes in molecules will be the same as their abundances—in other words there may be preferential concentration. Given enough information it is possible despite these difficulties to arrive at some sort of answer (see Winnewisser *et al* [25] for details). Thus deuterium is far more abundant on earth than it is even in the "local" interstellar medium around the sun and ^{13}C certainly falls in concentration as one moves out from the galactic centre. Again, when it is possible to study the important pairs $^{12}CO/^{13}CO$, HCN/DCN at shorter wavelengths much more precision will be given to these statements.

7.3.5 *Miscellaneous astronomical applications of infrared measurements*

Apart from the thermal analysis of the planets given in section 2.2.6, not much more has been done on infrared studies of the planets. The atmospheres of the giant planets can be probed by infrared spectroscopy quite effectively and it was this which gave us our understanding of the composition of, say, Jupiter's atmosphere but the prime components H_2 and He cannot be detected in this way. Interplanetary probes are more likely to provide the missing information than will ground-based infrared spectroscopy. The variation of the apparent black-body temperature of the sun with wavelength can be well understood in terms of the "free-free" process described above [911]. The number density of the electrons near the photosphere will be much higher than anything we have discussed previously and the plasma frequency will lie in the far infrared. Thus as the wavelength lengthens one sees less and less into the sun's atmosphere—in other words only to higher and higher levels. This is why the temperature falls but at sufficiently long wavelengths where one is looking only at the corona ($T \sim 10^6$K) the temperature begins to climb rapidly again.

The infrared study of extragalactic objects is only in its infancy. The normal galaxies have received scant attention, chiefly because they are very faint objects and secondly because it is felt that their spectra will be quite similar to that of the milky way. Quasars, however, are quite another matter. These extraordinary objects, which are found at the nucleus of many remote galaxies, are extremely compact, say of the size of the solar system or less, yet give out more than 100 times the radiant power of an entire galaxy. It is hypothesised that a quasar is associated with the formation of an extremely massive ($>$ one million solar masses) black hole and that quasar formation may well be a normal phase in the evolution of galaxies. Some twenty quasars have, so far, been investigated but no general principles have emerged. Each appears to be an individual and there does not seem to be any obvious correlation between infrared and radio emission. The infrared is valuable, however, in determining the red-shifts and hence the presumed distance. All objects beyond the local group of galaxies show a red-shift of their spectral lines which is usually interpreted as a Doppler shift due to the expansion of the Universe. This expansion is given by Hubbles law

$$v_R = Hd \tag{7.3.9}$$

where v_R is the velocity of recession, H is Hubbles constant and d is the distance. Astronomers usually prefer to use their familiar unit of distance, the parsec (3.1×10^{16} m) and then Hubble's constant is found to be 65 (km s^{-1}) per Mps. In SI units its value is 2.11×10^{-18}s^{-1}. The universal expansion is usually interpreted as the result of the birth of the Universe from an initial "Big Bang" and within this scenario, one would say that a particle, initially ejected with a velocity v_R would, after a time T, have reached a distance $d = v_R T$. Hubble's constant is thus simply the reciprocal of the lifetime of the Universe. The value deduced in this way, namely 15×10^9 years, is in good agreement with that derived from other physical evidence such as the abundance ratios of the longest lived radioactive isotopes. At the distances of the majority of the quasars, i.e. 1000 Mps or more, the velocities of recession are becoming comparable to that of light and it is then necessary to use the relativistic version of the Doppler formula, viz.

$$\frac{\Delta\lambda}{\lambda} = \sqrt{\frac{c+v}{c-v}} - 1 = z. \tag{7.3.10}$$

As an example, for the famous quasar 3C273 whose distance is thought to be 460 Mps, the velocity of recession would be 25 300 km s^{-1} and z would be 8.46×10^{-2}. Paschen α (1.8746 μm) would therefore be seen shifted to 2.033 μm and this has been confirmed experimentally. Still more dramatic is the case of PKS 0237–23 where Balmer α (0.6561 μm) has been seen shifted to 2.2 μm. This gives a z value of 2.35, a velocity of recession of 0.84 c and a distance of 4560 Mps. The record at the moment is set by the very remote object PKS 2000–13 whose z value of 5.2 corresponds to a speed of recession equal to 0.95 c! Clearly, with this sort of recessional velocity, all the optical lines will be shifted well into the infrared. Here again, as in so much earlier, one can expect dramatic developments when large satellite-borne telescopes coupled to sensitive ^{3}He cooled detectors become available. A major step in this direction came with the successful launch of the first infrared satellite-borne telescope in January 1983. This instrument—IRAS—is equipped with four filters which enable it to make whole-sky surveys in four narrow bands: 8·5–15 μm, 19–30 μm, 40–80 μm and 83–119 μm. The lifetime of its helium cooling system is estimated to be only seven months but the mission has been so successful that within this time slot it has proved possible to complete all the scientific tasks. Amongst the triumphs of the flight was the first clear observation of the birth of stars from collapsing cool dust clouds not only in our own galaxy but also in the larger Magellanic Cloud.

Ground based infrared astronomers have not, however, entirely been put out of business by the flight of IRAS. They still have much better spatial resolution and observations in the atmospheric "windows" are still leading to major discoveries. Thus observations from the Anglo–Australian telescope in New South Wales, at 1·2, 1·6, 2·2, 3·8, and 4·8 μm have, by penetrating galactic dust clouds, revealed detail in the intriguing object IRS-16 which lies near the centre of the galaxy and permitted for the first time the direct observation of the nucleus of the giant radio-galaxy Centaurus A. Nearer home these observations have proved that the giant planets Jupiter and Saturn do indeed have internal heat sources. Near-infrared radiation measurements using the infrared telescope on Mauna-Kea have enabled the diameters of the four satellites of Uranus to be determined for the first time to high precision. The results are Ariel (1330 km), Umbriel (1110 km) Titania (1600 km) and Oberon (1630 km). These values are considerably larger than the previous estimates obtained from visual observations. At longer wavelengths it is possible to make some observations through the not very transparent atmospheric window from 8 to 12 μm. However, the attenuation is so high and the signal strength accordingly so low that the ultimate in detector sensitivity is required. Fortunately the availability of high-power continuous-wave lasers in this region makes heterodyning a practical possibility and heterodyne ground-based infrared astronomy is already giving some useful results. Thus ammonia has been identified, via some lines of its ν_2 band, in the extended atmosphere of the planetary nebula IRC − + 10216 and the successful further investigation of the variation of concentration of this gas with distance from the central star augurs well for future astronomical spectroscopy in the mid infrared.

7.4 Technological uses of infrared lasers

7.4.1 *Infrared laser photochemistry*

The average dissociation energy of a chemical bond, 3–4 eV, lies, from a spectroscopic point of view, in the ultraviolet so it might be thought that irradiating molecules with intense mid-infrared radiation would only be tantamount to heating the gas with a bunsen burner! Dissociation might eventually take place but only because the gas would have got very hot through absorbing a lot of energy. It turns out, however, that this simplistic view is quite wrong. When gases such as SF_6 or CF_3I are exposed to an intense pulse of CO_2 laser radiation, the molecules dissociate and the kinetics prove that the dissociation has occurred through individual molecules absorbing enough photons (30–40!!) to reach the dissociation limit. Collisions between molecules play only a small role. This is strikingly confirmed by the observation that if the laser is tuned to coincidence with a $^{34}SF_6$ line then that species is strongly preferentially decomposed compared to the much more abundant $^{32}SF_6$. It is this surprising ability to carry out isotope selective photochemistry—with all its promise for cheap isotope separation—that has provided so much emphasis to this new art. However it may well turn out to be a major tool of future preparative chemists [912]. Already reactions such as

$$
\begin{array}{ccc}
\begin{array}{c} Cl \\ \diagdown \\ \end{array} C = C \begin{array}{c} H \\ \diagup \\ \end{array} & \longrightarrow & \begin{array}{c} Cl \\ \diagdown \\ \end{array} C = C \begin{array}{c} Cl \\ \diagup \\ \end{array} \\
H \diagup \diagdown Cl & & H \diagup \diagdown H
\end{array}
\qquad (I)
$$

and

$$
\begin{array}{c}
O \text{———} O \\
| | \\
CH_3 - C \text{————} C - CH_3 \\
| | \\
CH_3 CH_3
\end{array}
\qquad \longrightarrow \quad 2(CH_3)_2\,CO \qquad (II)
$$

have been carried out under laser irradiation. These chemical reactions prove that it is not peak power which is important but rather the energy fluence (that is the number of Joules absorbed in a cross-section of 1 m^2). Thus the ion $[(C_2H_5)_2O]H^+$ has been dissociated by laser irradiation of only 1 W cm^{-2} but lasting for 1 second. This underlines the mechanism of progressive photon absorption mentioned above.

The major difficulty facing anyone who attempts to interpret this phenomenon of "multiple-photon" dissociation is that molecules are anharmonic (see section 5.2.3). Thus even if one has tuned the laser to a given vibrational/rotational transition of the ground state, the next step ($v = 1 \rightarrow v = 2$) will not coincide and all apparently that one can do is to pump molecules from the ground state to the $v = 1$ state. This phenomenon is of course used in the passive Q-switching of the CO_2 laser using an intracavity cell containing SF_6. Now if one were considering a diatomic molecule (Fig. 7.9(a)) this argument would be unavoidable and significantly no one has yet reported the photodissociation of a diatomic molecule using an infrared laser. With a polyatomic molecule, however, which has several modes of vibration (see section 5.2.3.1), the number of modes increases rapidly with the excitation.

Thus if we denote the number of quanta of excitation by n then for a triatomic molecule we would have for $n = 1, 2, 3$ etc. the number of levels 3, 6, 10 etc. Clearly for large n the energy levels will be densely packed, especially when anharmonicity is taken into account. When further one considers the rich rotational substructure of a level and the natural width of these levels, it is clear that for large $\mathscr{E}$ there will be a quasi-continuum of states and this provides a natural mechanism for the progressive absorption of quanta leading finally to a transition into the true continuum and hence to dissociation. In the quasi-continuum, the amplitude of the atomic oscillations within the molecule will be very large and this provides the mechanism for surmounting the high potential barrier separating *cis/trans* isomers and thus allows isomerisation reactions such as (I) to take place. The isomer when formed can rapidly de-excite since its lower discrete levels will not be in resonance with the pumping laser. This selectivity is, however, only relative for many molecules since a reasonably large molecule will have so many combination and overtone tones that it will absorb weakly virtually everywhere. So, if the laser power is high enough, a resonance with an allowed line is not absolutely necessary to get started onto the ladder of photon absorption leading to reaction. It still remains true that to get effective action and especially to get highly selective action— isotope separation for example—one looks for a resonance and moderate laser power. The SF_6 selective dissociation is a particularly favourable case because the relatively large mass change in going from ^{32}S to ^{34}S causes a big shift (17 cm^{-1}) of the v_3 infrared active mode. The isotope enrichment that is attracting most interest however—that of UF_6 is a far less favourable case because (a) the vibrational isotope shift is very small ($< 1 \text{ cm}^{-1}$) and (b) there are no powerful lasers yet in the 16 μm region where lies the v_3 fundamental of UF_6. Apart from the atomic energy aspects of the separation of ^{235}U from ^{238}U, there is a not inconsiderable industry involved in producing and using non-radioactive isotopes as tracers. Indeed there is a considerable lobby which

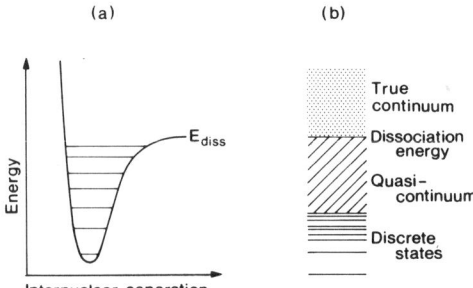

FIG. 7.9. The vibrational energy levels for (a) a diatomic molecule and (b) a polyatomic molecule. For the latter the number of levels in a given energy increment $\Delta\mathscr{E}$ increases rapidly with $\mathscr{E}$ leading to a quasi-continuum.

presses for the use, wherever possible, of non-radioactive isotopes in medicine. The laser method promises to produce a wide variety of separated isotopes at a relatively cheap price. Laser photochemistry can also be used as a means for ultra purification—by destroying the impurities present—and as a means for displacing equilibria in a desired direction.

Chemists, however, would like to take the method still further and to invoke bond-selective photochemistry in which one would selectively rupture one chosen bond in a molecule. If this is to be possible one has to know the make-up of the normal modes of vibration of the molecule (that is the harmonic analysis of the atomic motion) and how the particular modes in question couple anharmonically to all the remaining modes. The anharmonic coupling essentially connects the levels of the pumped mode to a "heat sink" made up of all the other modes. There will then be damping of the pure mode oscillation in the usual way with a T_1 (lifetime) broadening and a T_2 (dephasing) broadening. The relative magnitudes of T_1 and T_2 can only be estimated at the moment and their absolute magnitudes are unknown but it does seem that they will both be less than 30 ps. Selective bond rupture is therefore going to involve extremely short extremely intense laser pulses and is probably beyond the reach of present technology. Also laser photochemistry is confined at the moment to gas-phase reactions since the strong coupling to the environment which occurs in condensed phases and which is manifest as considerable line broadening leads to very short (< 3 ps) relaxation times.

7.4.2 Cutting and machining with infrared lasers [913]

As a tool for cutting, welding, polishing and heat treating materials the CO_2 laser has many advantages, thus:

a. Its radiation at 10.6 and 9.6 µm is not seriously absorbed over moderate distances through the atmosphere.

b. The laser can be operated continuous wave or pulsed or even Q-switched.

c. The power available is high—several kilowatts continuous wave, several megawatts pulsed and the energy conversion efficiency is high.

d. The beam being coherent can be focussed down to a very small spot ~ 10 µm in diameter.

e. The beam, unlike any conventional cutting tool, has no mass and no inertia and can be moved virtually instantly by the simple means of moving a mirror.

f. The laser because of its mode of operation and the ease of adjustment of its beam is easily incorporated into an automated system (for example a car production line) and is very suitable for computer control.

g. The laser is quieter than conventional machines, does not generate vibration or dust and one does not need to keep up-to-date a large stock of expensive cutting tools.

Cutting of materials is the easiest application for the laser; thus in cutting through sheet metal, it is only necessary to melt the metal and to blow away the molten metal with the conventional rare gas or dry nitrogen jet. If oxidation of the exposed surfaces causes no difficulties, then oxygen gas can be used as the jet and the heat produced by the exothermic reaction between the molten metal and the oxygen can help the laser with its cutting. Some typical results for cutting steel plate is that with plate thickness of 1/8 inch and laser power of 3 kW it is possible to cut at a rate of 100 inches per minute. The cut faces are usually very straight and clean and the thermal effects due to the cutting do not penetrate very far into the bulk of the metal. In cutting metal the laser has the huge advantage that it cuts hard and soft materials with equal success—there is not the problem of tool wear and slow cutting when rough materials are to be handled. This advantage is even more marked when fibrous or composite materials are to be cut and laser cutting is making great inroads in this area despite the fact that the initial outlay to buy a laser system is much higher ($\sim \times 5$) than for a conventional system. The ultimate in this direction is the cutting of cloth where the speed of the laser, its ready automation and its ability to start working without a clearly defined edge as reference, give it an enormous advantage. In fact the cutting of men's suits out of whole cloth by an automated laser is so efficient that the cost differential can be recovered within a year. The tailor merely needs to measure his customer and feed the results into the microprocessor controlling the laser cutter and the various pieces of cloth ready to be stitched can be available in a matter of minutes.

In welding metal, the laser is more efficient than the conventional methods and can perform not only the conventional butt-weld but can also do the much deeper (and stronger) "key-hole" weld. The absence of a flame is a great advantage in the welding of special materials such as aluminium and titanium alloys. The ease of control of the laser beam makes it ideally suited for welding together pipe sections from the inside. However the laser sytem really comes into its own when small welds are required in rather inaccessible places. The previous technique, much used in the aerospace industry, was electron beam welding but this was very costly and time consuming because the piece to be welded had to be first mounted in a vacuum chamber. Even in the mass production industries, such as the car industry, laser welding of precision parts, for example synchro gear rings, is being widely adopted.

Heat treating to make the surfaces of moving parts very hard is a most important industrial process. Essentially a narrow strip on the outside of the piece is heated to a temperature where it undergoes a phase transition and it is then rapidly cooled so the high temperature—and harder—phase is "frozen-in". An alternative way of getting the same result is to clad the surface with a harder alloy. Lasers have proved very suitable for both these processes and the high speed with which the laser delivers its heat leads to improved results compared with conventional processes. The absorption of heat from the laser beam can be aided by prior painting of the piece with a carbon black based coating. Laser hardening and surface treating goes very well indeed with automated production lines, because time-consuming "by-pass" lines are not needed, and it goes particularly well with flexible manufacturing systems (FMS) where by making full use of robots and computer control it is possible to have a line which can be switched quickly and efficiently from one type of product to another. FMS is already enabling medium-sized enterprises to compete in the market with the giants of mass production and it is also enabling companies to respond rapidly to changes in consumer demand because small-batch production can now be an economic proposition. Lasers are more, however, than highly convenient substitutes for otherwise awkward production processes. They are increasingly being used for operations which would be almost impossible by conventional methods. An example is *in situ* alloying where one can either coat with melted alloy—similar to cladding—or else make the alloy exactly where it is required by adding the requisite elements in powder form. The laser heating can be used to produce a controlled diffusion of the alloying elements into the bulk and highly non-thermodynamic equilibrium situations can be produced leading to materials with very novel properties.

The one common industrial practice where the laser encounters difficulties is drilling. This is because the only way the metal can be got out of the hole is to vaporise it. One therefore needs much more energy and in addition the vaporised metal forms a plasma which is virtually opaque to the laser radiation. As a result the laser cannot penetrate further into the hole. To get round this difficulty one can go over to pulsed operation with the plasma hopefully decaying between pulses. It turns out fortunately that with high enough energy there is an explosive development of metal vapour which blows the excess out of the hole. Even so it has been found advantageous to use shorter wavelength lasers (Nd/YAG; ruby) for the drilling of small holes since these can penetrate the plasma better. CO_2 lasers can be used, when much larger holes have been made by conventional drilling, to ream them out, that is to remove the burrs. Thus the place of infrared lasers in machining and processing plant is assured and as the complexity of the jobs to be done increases they will become still more commonplace [914].

7.4.3 *Infrared lasers in surgery* [915]

Continuous-wave CO_2 laser beams focussed to a point with a germanium lens have several advantages over the conventional scalpel for surgery. Thus the cut being contactless minimises tissue dislocation and reduces the risk of spreading infections or tumour metastases. The laser beam essentially melts the flesh away and the heat cauterises all the opened blood vessels and the minor air filled spaces. The surgeon therefore has a bloodless field in which to work. CO_2 laser surgery has been used for many types of operation but those on the lungs (where there are paranchyma to be sealed) and on the kidney where there is a plentiful blood supply have proved particularly apposite. Healing of tissue after surgery has been investigated by Dittrich [916] who finds that the laser gives a cosmetically better scar than does scalpel surgery but is otherwise similar. Both however are much better than conventional thermocautery. A practical problem with laser surgery is that the usual tactile "feedback" which the surgeon gets from the scalpel is completely missing and surgical technique may well have to be relearned. A further problem is that the CO_2 laser beam is invisible, but this particular difficulty can be overcome completely by incorporating a low-power helium/neon laser beam concentrically, to act as a marker for the higher power CO_2 laser radiation. With this advance, laser surgery can become a powerful technique especially since the optical depth of human tissue at 10 μm is less than 10^{-2}mm and highly sensitive discrimination is therefore possible.

In the hospital environment, however, CO_2 lasers have the considerable drawback that they are bulky and hence difficult to fit into the crowded space of a normal operating theatre. Mirror optic systems are therefore usually employed to bring an overhead beam into the theatre from a laser housed conveniently in an adjacent ante-room. The beam can then be directed to where it is needed by means of a focussing arrangement mounted above the operating table. There is little doubt that the surgeon's job would be made considerably easier if fibre-optic techniques were available to guide the radiation to the point where it was needed. Unfortunately silica-based glasses are completely opaque in the 10 μm region so all the extensive technology developed in the context of near infrared telecommunications is unavailable. Research has therefore been directed towards making fibres from glass-forming systems which do transmit in the mid infrared. The classic example is provided by arsenic trisulphide, As_2S_3, but there is available an almost perfectly continuous stoichiometry based on IV/V/VI mixtures. These chalcogenide glasses have considerable promise—the most promising being perhaps the three exotics, $As_{13}Ge_{30}Se_{27}Te_{30}$, $As_{15}Ge_{30}Se_{55}$ and $As_{38}Ge_5Se_{57}$—but the lowest losses achieved so far, 1–5 db m^{-1} for the quaternary glass, restrict the lengths which can usefully be used. Even so real applications in surgery do arise. There should also be applications in remote sensing, image relaying and infrared radar.

A possible problem of unknown seriousness is that, just as high-intensity millimetre-wave radiation is thought to be carcinogenic, there is a similar risk of cancer induction by high-power CO_2 radiation. At the moment this unquantified risk is of merely academic interest since laser surgery is only used for very serious conditions but should it be extended—into the cosmetic field for example—some assessment of the possible dangers would be highly desirable.

7.4.4 Infrared lasers in fusion research

A very considerable effort is currently under way in all the technically advanced nations to find a way of harnessing thermonuclear fusion for peaceful purposes. The search has become even more urgent with the gradual realisation that a major energy shortfall is drawing near due to the rapid consumption of the readily available fossil fuels. Already most western nations produce more than 10% of their energy requirements from nuclear power stations but the basic material itself, uranium, can also be regarded as a fossil fuel (a fossil from a supernova explosion) and it too will eventually be exhausted. By contrast, if a way could be found to fuse hydrogen and to extract the heat so produced, an energy source would be available which would be virtually inexhaustible. There are two main routes being explored. The older is to use the containment and compression of a very hot hydrogen isotope plasma in a magnetic "bottle". This has led to a fairly satisfactory line with the development of the Tokamak, a toroidal device, and at the moment several very large machines of this type are under construction—for example the Joint European Torus, or JET, at Culham in the United Kingdom. The more recent approach is to use very high power pulsed infrared lasers to rapidly heat the hydrogen up to the temperature where the nuclear reactions can take place. For both kinds of machine it is unlikely that they will operate initially on a pure hydrogen fuel since the temperatures for the proton fusion reaction (60 $\times 10^6$ K) are inaccessibly high. Instead, mixtures containing the two heavy isotopes, deuterium and tritium, will be used. Deuterium, as mentioned in section 7.3, is very labile and temperatures of the order 20×10^6 K will suffice to achieve the reaction

$$D + D = He^4 \qquad (7.4.1)$$

There are two immediate difficulties with the laser method. Firstly, at the power levels required, many terawatts, the radiation pressure is so strong that the pellet containing the hydrogen would be simply blown away if it were hit by a single beam. This can be overcome by arranging to have the pellet struck simultaneously by many beams symmetrically disposed in space. The second problem is that as soon as a certain amount of power has been absorbed the pellet will be vapourised and turned into a plasma. The plasma will act as a mirror preventing penetration of the laser beam into the gas. The solution is to use as short a wavelength as possible and to get the power in as fast as possible, that is to use a Q-switched mid- or near-infrared laser. So far only three types of laser have been used: the neodymium in glass, the CO_2 and the iodine lasers [764]. In order to reach the required power levels, it is necessary to have the beam from a primary laser pass through several amplifiers and some very tricky questions of timing and pulse shaping arise not to mention the problems of damage to the optical elements. Nevertheless the preliminary set-ups have worked satisfactorily and the builders see no insuperable difficulties in getting the full-scale installations working and delivering the specified power. As an example, the ZETA Nd/glass machine of the University of Rochester has worked satisfactorily with its initial six beams and is now working well with its full twenty-four beam capacity. The final installation, to be called OMEGA, will deliver 30 TW of power into 100 μm diameter glass spheres containing mixtures of deuterium and tritium. The force of the intense pulse will compress the gas mixture to densities some 10^2–10^4 times that of solid hydrogen and the temperature of the gas will correspondingly shoot up to somewhere between 50 and 100 million K. At these temperatures, the very easy nuclear reaction

$$D + T = He^4 + n \qquad (7.4.2)$$

will take place freely and it can be monitored by detecting the neutrons produced. The experiments so-far are very promising, but it must be stressed that there is a long way to go before we have a practicable reactor. In fact it will probably be several years before the energy obtained from laser-fusion machines even equals that required to extract the deuterium, prepare the tritium and fire the laser.

A quite different application of lasers in fusion research is their use to monitor the various parameters, temperature, electron-density etc. of the plasma in a tokamak. This relies on the phenomenon of Thomson scattering, which can be regarded as a virtual inverse of the process of photon emission by an accelerated charge. The incoming photon interacts with the charged particle and is then scattered from the interaction region. The thermal distribution of energies which the particles display is transferred to the scattered photons and if one has a virtually monochromatic input beam, then the scattered beam will emerge with a Gaussian spectral profile whose width gives immediately the plasma temperature. The main experimental difficulty is that the cross-section for Thomson scattering is very small so very intense lasers are required in order that the scattered wave-field be detectable. Such lasers have low repetition rates so it is not possible to follow the full-time time evolution of the temperature during the brief lifetime of the plasma. The ideal wavelength for the present generation of fusion machines is about 200 μm but far-infrared lasers are, at the moment, limited in their power output [712], so work tends to go on in parallel using the far from ideal but very much more powerful ruby and neodymium lasers. Thomson scattering measurements have yielded many valuable results and in particular their ability to give absolute numbers is very commendable, but it is fair to say that they are always carried out at near to "state-of-the-art" operation and because of this several other techniques are also being used for plasma diagnostics.

One of the most promising of these is the rapid-scanning far-infrared interferometric spectroscopic version developed by Costley and his colleagues [917] which relies on measuring the time and space variation of the electron-cyclotron emission from the plasma.

7.4.5 *Infrared lasers and mass-spectroscopic analysis*

Mass spectrometry is a very convenient and very accurate method for analysing the composition of metals—the problem though is that a small amount of the metal has to be first vapourised from the surface to provide a sample for the mass spectrometer. High-power Q-switched pulses from a Nd/YAG laser provide a very neat way of providing these samples. The laser radiation is focussed down to spots of diameter 8 μm and each pulse delivers about 20 mJ over a pulse time of the order 20 ns. The laser pulse vapourises the material from a pit about 1 μm and thus delivers about $10 \times 10^{-18}\,\mathrm{m}^3$ of material into the entrance port of the spectrometer. This is a convenient amount for high-grade high-resolution analysis. The method thus permits microanalysis of surfaces and it has proved therefore very suitable for studying hydrogen embrittlement of the high-strength steels used for pressure vessels and for investigating the possibly deleterious effects of intense high-energy radiation on the metals used in nuclear power stations. It also has applications in the microelectronics industry since it permits the investigation of the diffusion of dopants in the semiconducting substrates.

7.5 Metrological uses of infrared radiation

7.5.1 Infrared distance measurement

Non-contact distance measurement divides naturally into two quite different classes, namely the measurement of distances greater than and less than a metre or so. In the long-distance category one either relies on goniometry, that is the measurement of the slightly different angles subtended by a remote object at two separated, but accessible, points, or else one relies on measuring the time taken for a pulse of radiation to travel to and return from the remote object. For short distance measurement one nearly always relies on interferometry, usually of the Michelson variety. Black-body sources are valuable for short-distance measurement because the unique maximum in the "white-light" fringe pattern produces a very easily recognised fiducial mark. However for long-distance measurement a coherent source is almost indispensable since there is much less angular spread and it is much easier to detect a returning weak signal if it is narrow band coherent. Near-infrared lasers are particularly valuable in this context because the radiation suffers less Mie scattering in traversing murky atmospheres so longer ranges can be reliably covered and secondly these lasers can be very intense.

Visible region broad-band or quasi-monochromatic radiation is often used for short-distance measurements of static objects manufactured to a very high finish, but for distance measurements in manufacturing environments where the objects are usually far from static and where the finish can be relatively rough infrared radiation can be very useful. The object to be measured is used as one "mirror" of a Michelson interferometer and its distance is determined by finding the distance the other mirror has to be moved to produce a zero-path difference fringe. The absolute precision of this technique is of the order of $\lambda/10$

so, to ensure ordinary workshop tolerances, that is $\pm 25 \mu m$, a peak wavelength of 250 μm is perfectly acceptable. An instrument working in the far infrared at this sort of wavelength, the "Teramet", has been described by Gebbie and his colleagues [918].

The choice of wavelength for a long-distance measurement instrument has to be a compromise between various factors. Thus it should be as long as is possible whilst still having good transmission but on the other hand one also wants the radiation to be as intense as possible and to be readily modulated. Nearly all the commercial instruments use the gallium-arsenide injection laser whose operating wavelength is about 0.85 μm. Instruments of this type are naturally of considerable interest to national standards laboratories and a particularly accurate form, the Mekometer developed at NPL [919], has been used for such fascinating experiments as measuring the annual increase in the size of Iceland. Military rangefinders tend to work at longer wavelengths because maintaining good range despite bad atmospheric conditions tends to be the overriding consideration. Holmium lasers operating at 2·064 μm are finding useful applications in rangefinders. At this wavelength, the clear air absorption coefficient is only of the order of 0·1 neper km^{-1}.

7.5.2 Infrared lasers and the fundamental standards

The original "fundamental" standards, used in commerce and in international trade, were physical entities such as bars of metal to act as standards of length, lumps of metal to act as standards of weight (or mass) and standard pendulums or even the Earth itself to act as standards of time. Such standards served trade well for thousands of years as one is readily reminded when visiting ancient market places where one frequently finds a metallic bar equal in length to one ell or other unit, fixed in a conspicuous place on the wall. The merchants could then test their rules (or "secondary" standards) against this primary standard and the customers would know they were getting a fair deal.

As measurement science advanced and as new technologies emerged, more accurate and reproducible standards were needed for length mass and time and new material standards were needed for such quantities as voltage and resistance. Invar rods were introduced for length since these were relatively insensitive to changes of temperature, platinum/iridium ingots were introduced for mass since these were dense and relatively insensitive to corrosion and Weston cells and standard resistors were introduced to cope with the needs of the electrical engineer who wanted to calibrate his day to day measuring instruments. However, at the turn of the century the practical problems of using material standards—for example the need for each country to obtain and maintain a copy of the defined international standard—began to become more and more formidable and it was also becoming obvious that the accuracy of measurement used by scientists was getting greater than the reproducibility (~ 1 in 10^4) of the appropriate standard. The need to develop absolute standards which would be independent of any arbitrary material definition was becoming very clear indeed.

Michelson was among the first to suggest that the properties of atoms should be used to define standards—in particular he suggested that the wavelength of a suitable emission line should be adopted as the basis for a definition of the metre. Such a standard would be very attractive since presumably the properties of atoms do not vary with place on the globe and each country could therefore realise its fundamental measurement system independently but concordantly. Michelson's suggestion was particularly apt since his famous two-beam interferometer provided the ideal link instrument to connect the fundamental standard atomic wavelength with the material sub-standards—measuring rods—which would be used for the everyday measurement. The emission lines from low-pressure atomic vapours have the ratio $v_0/\Delta v$ of about 10^5, so an order of magnitude improvement in accuracy was possible. This accuracy can be improved still further if structure due to isotopic composition and structure due to hyperfine interaction are not present in the chosen line. Michelson suggested the mercury green line at 546·1 nm because it is intense, sharp and lies in the region where the eye is most sensitive but when it eventually became possible to obtain separated pure isotopes, the $2p_0 \rightarrow 5d_5$ line from ^{86}Kr at 605·780 211 nm was adopted as the fundamental standard since it is essentially structureless, arising, as it does, from a pure isotope with zero nuclear spin. The wavelength of this line is *defined* (to a precision of nine significant figures) but it is now known that it is only realisable to ± 4 parts in 10^9. The major residual cause of broadening is the Doppler effect and here, of course, a heavy atom like Krypton is an advantageous choice since in the red region of the visible, the broadening Δv_D is, from (2.6.13a), only 0·662 GHz and one has $\Delta v_D/v_0 = 1·338 \times 10^{-6}$. Good though this is, one clearly cannot hope to define the line centre to better than two orders of magnitude higher precision unless one is prepared to reduce the Doppler broadening still further by going over to an atomic beam source. Such sources give very weak radiant output and even if one were prepared to live with this, one would not get over a new difficulty which has emerged, namely that the Krypton line, for some not understood reason, is apparently asymmetric [920] so the fundamental definition becomes ambiguous. For all these reasons, conventional atomic line standards of length cannot ever be expected to be reproducible to much better than 1 part in 10^8.

The development of lasers, from 1960 onwards, aroused great interest amongst the metrologists because the radiation from these devices was powerful, highly monochromatic, almost non-divergent and could be mixed in a non-linear detector to yield frequency information directly. Lasers were used almost at once for long-range distance measurements, as has been already mentioned, perhaps the most spectacular example being that of the accurate determination of the earth–moon separation, but as standards of length they were not acceptable. The reason for this is that although their instantaneous line width is very small, for a continuous-wave laser perhaps only some kHz, and one can therefore define the centre frequencies to extraordinary precision, e.g. 1 part in 10^{11}, this centre frequency is tunable anywhere within the spontaneous profile where there is gain. Also for a He/Ne laser longer than about 30 cm, even with the cavity length stabilized, the output can be on any of a number of longitudinal modes of separation $c/2L$ spanning a range of several GHz. A laser alone is therefore best regarded as a slightly tunable oscillator rather than as a useful frequency standard. When, however, ways were found to stabilise lasers by locking them to the Lamb-dips of coincident molecular or atomic absorption lines (section 2.4.2), interest revived since the combination now provided a potential fundamental standard [921]. There were two main thrusts here:

1. To provide *de facto* length standards which were at least an order of magnitude more reproducible than the ^{86}Kr standard.
2. To provide frequency or wavelength standards for the infrared and visible region.

The length standards application is said to be *de facto* because although these stabilised lasers (with present accuracies between 1 part in 10^{10} and 1 part in 10^{11}) are in fact now used preferentially over ^{86}Kr for realising the metre and are also being strongly urged as bases for a *redefinition* of the metre, the metrologists are showing considerable caution about such a redefinition. One reason for this is that all radiant definitions of length depend on the idea of a co-phasal surface and in practice that means plane waves. Present experimental interferometric systems for realising the metre are limited, by the lack of planarity of the waves traversing them and by associated problems, to a precision of about 1 part in 10^{11}. Even if much higher grade optical systems were developed one could only expect perhaps an order of magnitude improvement in the practical realisation and this would be still far below the precision to which time can be realised (3 parts in 10^{14}) using the caesium clock. Metrologists are uncertain therefore which way they should move. They could adopt a new laser standard and redefine the metre (though this process would always be subject to the threat of a still better laser being found!) or more radically they could abandon a primary length standard altogether and just use the caesium clock frequency standard [94] plus a *defined* (that is "fixed") value for the speed of light *in vacuo*. Probably this latter approach will eventually be adopted especially as studies on pulsars have shown that the velocity of electromagnetic radiation is independent of frequency, at least to an uncertainty much less than that of present frequency measurements. For the moment though, stabilised lasers are being used to realise the metre in practice even if they are not in theory. There are several infrared candidates but the metrologists show an overwhelming preference for a visible region laser. It is not entirely a matter of preferring a beam with lower diffraction effects, it is also a matter of preferring a beam, and even more importantly the interference patterns derived from it, which they can see, because this enormously simplifies the setting up of the apparatus and the detection of any malfunction. Locking visible region lasers turns out to be an easy task because the iodine molecule has an exceedingly rich spectrum in the visible region and the chance of a coincidence with a laser line is very high. One can therefore use either an intracavity or else an external cell, perhaps heated to increase the vapour pressure or the population of vibrationally excited states, and servo the laser frequency to the Lamb-dip of a chosen line. The I_2 absorption maximum is in the green—which is why iodine vapour is purple—but there are lines densely sprinkled virtually everywhere. The absorption is dominated by the $B^3\Pi o +_u \rightarrow X'\sum o +_g$ electronic transition which is made up of a set of sub-bands corresponding to vibrational excitation each of which has the usual PQR structure associated with rotational excitation. Natural iodine has only the single isotope I^{127} which has a spin of 5/2 so each ro-vibrational line has a complex hyperfine structure. To the resolution required by laser-locking, it is necessary to analyse the hyperfine interaction not only to the quadrupole level but also to the nuclear magnetism/molecular rotation level—the so-called $I \times J$ interaction. The result is a set of a 21 closely spaced hyperfine lines. These are usually designated simply by the letters of the alphabet, a, b, c . . . t, u, since any more informative notation would of necessity be rather cumbersome. The 632·9 nm He/Ne line lies within the profile of the blend line produced by, $v'' = 3 \rightarrow v' = 6$, P (33), and, $v'' = 5 \rightarrow v' = 11$, R (127) and within this blend it is virtually coincident with the hyperfine component i of the R (127) line. The helium/neon laser locked to the Lamb-dip of this component has a vacuum wavelength of 632·99139 nm. The laser prefers to run on this particular component of the 3s → 2p band but if an intracavity prism is introduced to frustate 633 nm emission, several other components can be persuaded to lase. The known emission lines are at 594, 605, 612, 629, 633, 635, 640 and 731 nm. The second easiest of these to obtain is the so-called "orange" line at 612 nm. This too can be locked to an iodine line, $v'' = 2 \rightarrow v' = 9$ P (47), hyperfine component o. In the region of the orange line, several forbidden lines occur and there are also found some level crossing lines. In addition the hyperfine components are readily power broadened ($\Delta v \approx 7$ MHz). For these reasons it is preferable to lock the laser with an external cell. The cavity length is then controlled with a piezo-electric element fed from a third harmonic feedback loop. Transit time broadening can be reduced by the use of a beam expanding telescope and the need to use low powers (~ 200 μW) can be compensated for by the use of energy storage in an external cavity which serves also to reduce the laser noise. With all these refinements it is possible to lock the laser frequency so that its variations are no more than a few tens of kHz (that is 1 part in 10^{11}) and it then has a vacuum wavelength of 611·970 771 ($\pm 4 \times 10^{-6}$)nm [922]. This line is emerging as an excellent candidate for a redefined standard of length. Another good candidate is the argon ion laser line at 515 nm. This can be locked to the iodine line $v'' = 0 \rightarrow v' = 43$ P (13), hyperfine component a_3, and has then a vacuum wavelength of 514.673 467 ($\pm 4 \times 10^{-6}$) nm [923]. Iodine has only one other reasonably stable isotope, namely I^{129} (half-life 1.7×10^{6} years) and this has a nuclear spin of 7/2. The pure isotope is available reasonably cheaply since it is produced in nuclear fission and pure I_2^{129} can also be used to lock visible region lasers [924]. The hyperfine structure is even more complex and much of it has not so far been assigned.

Infrared lasers may have only a small part to play in the field of wavelength standards but when it comes to frequency standards a very different situation prevails. The basic reason for this is that time and frequency are defined in terms of the caesium clock, that is in terms of a hyperfine ground state transition in the neutral caesium atom at 9·192 631 770 GHz. This is a microwave frequency and since there is, at the moment, no way of either directly multiplying it up to visible region frequencies or else of down-dividing visible region frequencies into the microwave region, there is no option but to use transfer oscillators, spanning the infrared, if one wishes to connect the time standard to the length standard. The first determination of the frequency of an infrared laser came in 1967 from Hocker and Javan [66]. They measured the well-known 337 μm line of HCN and found a value of 890·7607 GHz. Their experiment was subsequently repeated by amongst others Knight and Bradley [925] who found $v_0 = 870.7602 \pm 0.0002$ GHz. The experimental equipment typically used for this type of experiment is shown in Fig. 7.10. In the Knight and Bradley experiment the NPL 1 MHz standard frequency, which is derived from the caesium clock, was successively multiplied and synthesised to 18·57 GHz and this signal was mixed with the output from an 0-band klystron at 74·24 GHz. The phase-lock loop, fed from an external quartz-crystal controlled 30 MHz signal, served to keep the beat note between the fourth harmonic of the 18·6 GHz signal and the klystron output frequency at exactly

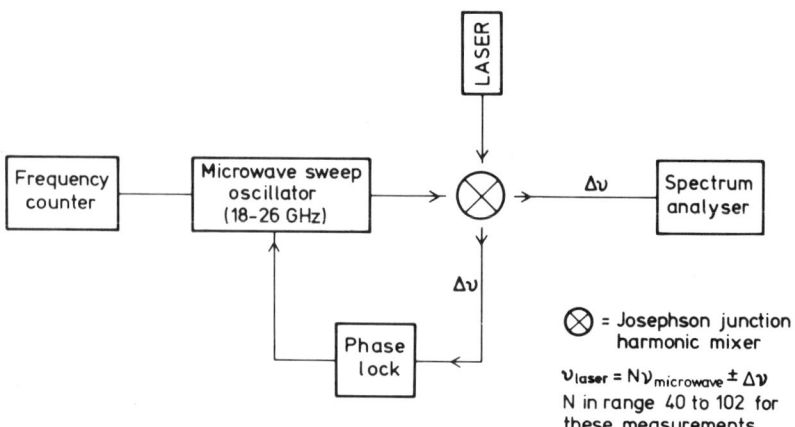

FIG. 7.10. Schematic of a direct frequency measurement arrangement (after Blaney, Knight and Murray-Lloyd [926]).

30 MHz. It did this by providing a correcting DC voltage to the klystron if the beat note drifted away from 30 MHz. One gets therefore a highly stable output of known frequency from the klystron. The 74.24 GHz signal was mixed with radiation from the laser in a tungsten on semiconductor (silicon or gallium-arsenide) "cats-whisker" diode. The twelfth harmonic of the klystron lies less than 100 MHz away from the laser and the beat notes can therefore be studied and measured on a spectrum analyser [926]. In this way the laser frequency can be directly counted relative to the fundamental standard. The HCN laser does not provide a convenient standard frequency source because it does not show any Lamb-dips on its own emission lines and there is not a nearby and very narrow line in some other gas to provide passive locking [927]. Thus even with good experimental care the output frequency variations can approach 1 MHz and this compares poorly with the instantaneous line width which is only a few kHz. The best precision that has so far been achieved with HCN lasers is about 0.2 MHz, i.e. 2 parts in 10^7. With the H_2O laser one can do better, mostly because the lines show a natural Lamb-dip. [128]. Frenkel *et al.* [928] were able to lock a 118·6 μm water vapour laser to its own Lamb-dip and then measured the frequency to be 2527·9528 ± 0·0001 GHz which represents an accuracy of 2.5 parts in 10^9. Subsequently Hocker and his colleagues [929] measured the 84 μm line of D_2O to have a frequency of 3557·143 ± 0·002 GHz but since the HCN laser was involved as a transfer standard in the frequency multiplication chain this measurement could not claim very high precision.

Extensions of direct frequency measurements into the mid- and near-infrared regions depend on the availability of non-linear devices which can respond to these enormously high frequencies and on the availability of suitable chains of high-precision transfer oscillators. The first of these requirements has been met by the discovery (accidental as rumour has it) of the metal on metal or MIM diode [355] which has been shown to be capable of video detection even at CO_2 laser frequencies. Evenson and his colleagues [930] have established a series of transfer oscillators which have enabled them to measure the frequencies of the 78 μm (3·821 775 THz ± 3 MHz) and the 28 μm (10·718 073 THz ± 2 MHz) water vapour lines and Daneu and his co-workers [931] have extended the chain upwards by using the 28 μm laser to reach the 10 μm region where they measured some frequencies of the CO_2 laser. The R(10) and R(12) lines of the 9·3 μm band lie below and above respectively the third harmonic of the 28 μm laser line. The difference frequencies, −19·950 and +21·866 GHz are too high for conventional spectrum analysers too handle but when a triple mixing experiment using, in addition, the radiation from a K-band klystron was performed, strong beat notes were observed when the klystron frequency coincided with either of the above frequencies.

The measurements yielded

R(10) = 32·134 269 THz ± 6 MHz
R(12) = 32·176 085 THz ± 6 MHz

These values agree with the calculated ones given in Table 5.2.

In 1973, Evenson and his colleagues at the National Bureau of Standards succeeded in measuring the frequency of the 3·39 μm line of the He/Ne laser stabilised to the saturated molecular absorption in methane. Their value was 88·376 181 627(50) THz. This measurement was subsequently repeated both at NBS and at other National Standards Laboratories. The most recent measurement was by Knight, Edwards, Pearce and Cross [932] in 1980 whose work was of interest since it centred on small stabilised lasers of similar design to those used as wavelength standards. This is clearly very relevant to the proposal to abandon separate fundamental standards of length and to have instead a defined constant speed of light in vacuum combined with time and frequency standards. Their final result was $v_0 = 88·376\ 181\ 616$ THz $\pm$ 3 kHz, i.e. they obtained an accuracy of 3 parts in 10^{11}. For this work they were awarded the Helmholtz Medal at the 1980 Conference on Precise Electromagnetic Measurement held at Braunschweig.

The highest frequency line so far measured [933] with an MIM diode is the 1·5 μm He/Ne line whose frequency is 196·780 271(25) THz but it will probably not prove possible to extend the use of the wideband MIM diodes to significantly higher frequencies and in particular it is very unlikely that they can be used to reach the visible region. This is a blow since a main target of frequency measurement by harmonic multiplication chains is to reach a suitably stabilised laser working in the visible region where lies the defined standard of length and where the most accurate interferometry can be carried out. Fortunately, however, one can probably use bulk non-linearities in non-centrosymmetric crystals to extend the chain into the visible by successive frequency doubling stages or else infer the frequency of a visible laser line from the use of the Ritz combination principle plus the frequencies of known infrared transitions connecting the upper and lower levels of the same system. Some values for the absolute frequencies of visible lasers have in fact been reported [934]. Using wavelength ratios with respect to the stabilised 3·39 μm line these have given accuracies of about 2 parts in 10^{10}. Useful direct frequency measurements are underway but only one result has been achieved so far and this has given rather less accuracy than the presently defined length standards. However in the future there is little doubt that this difficulty will be overcome and then with a precisely defined time and frequency scale extending up into the visible region; the time will have arrived when it will be possible to finally abandon an independent definition of the metre.

This said, there does remain the question of what value to adopt for the speed of light *in vacuo*. One could be inconoclastic and suggest the value 3 $\times 10^8$ m s^{-1} which would certainly be easy to remember but the difference between this value and that presently used, though small ($\sim 0·07\%$), is big enough to affect all accurate measurements and most metrologists prefer a value as closely identical as it is possible to make it to that which emerges as the best estimate based on the, at present, independently defined length and frequency standards even at the expense of having nine significant figures! Thus Barger and Hall give for the centre-of-gravity wavelength of the asymmetric ^{86}Kr standard the value $\lambda_{cg} = 3.392\,231\,376$ μm which when combined with the frequency given above results in the value

$$c_{cg} = 2.997\,924\,56_2 \times 10^8\,\text{m}\,\text{s}^{-1}.$$

This measurement has contributed to and in fact agrees with the adopted value given in Table 1.2.

The availability of moderately tunable, high power, frequency stabilisable lasers has greatly excited those fundamental metrologists who are concerned with establishing highly accurate values for the fundamental constants of nature, e, m, h etc. Usually they do not measure any of these independently but rather measure several different combinations, e/m for example, and then derive the separate values and their associated uncertainties by appropriate mathematical techniques. An obvious combination which is accessible to spectroscopic measurement is the Rydberg constant (equation 5.1.6) R_∞ whose presently agreed value is $1.097\,373\,177(83) \times 10^7$ m^{-1}. An absolute measurement of any hydrogen line will in principle give R_∞ at once via equation (5.1.2) but an immediate difficulty is that the simple Bohr treatment, which gives coincident energies for the S, P, D etc. states of hydrogen is inadequate for the precision available with laser spectroscopy. The Dirac theory is much better but even here one needs to make full allowance for the Lamb shift and other quantum electrodynamical effects before one has an acceptable treatment. The latest measurements [433] which take full advantage of Doppler suppression, via two-photon absorption, have given some very valuable data to put into the never-ending struggle to determine the fundamental constants of nature to higher and still higher precision.

7.6 Energy-generation and energy-saving aspects of solar radiation

It was mentioned in Chapter 2 that the greenhouse and solar-panel effects can be used very effectively to both store and to conserve energy from the sun. Up till very recently these uses represented the economic limits of solar energy activity because fossil fuels, and in particular petroleum, were so cheap. However in the 1970s the price of oil and other fuels rapidly escalated and even though there has been a marked price reduction in the early 1980s, it remains obvious that on the long time scale the price of fossil energy must continually increase. In the light of this scenario, solar energy starts to look very attractive, especially to countries such as the USA and Australia which have high energy demand and also large areas of high insolation.

The simplest application of solar energy is in space heating, and here the greenhouse effect is seeing a renewed lease of life with the development of coated float process glasses suitable for use in offices, homes and public buildings. Pilkington's market a plate-glass called "Kappa-Float" which has outstanding transparency in the near infrared, high reflectivity in the mid infrared, good mechanical properties and a very pleasing appearance. It is claimed that double-glazing with this glass is as good as triple-glazing with more conventional materials.

The next simplest approach is to use solar panels on the roofs of houses to generate hot water both for washing and for central heating systems. These suffer from the obvious drawback that the heat is mostly needed at night when the sun is not shining. The use, however, of a double system involving conventional heating in tandem with the solar panels and having the two interact via a large storage tank which also doubles as a heat exchanger provides an ideal solution. It is estimated that a system of this type will recover its capital costs, even in cloudy countries such as England, within three to four years.

Going on from these rather passive applications, we come next to consider how solar radiation can be used to generate electrical power. The traditional approach was to use large cylindrical mirrors, or concentrators, to focus large amounts of radiant power onto the receiving area of an otherwise quite conventional steam-raising boiler. The steam was then used to drive the turbines in a nearby power station. The problems with this system are that firstly the concentrators having large surface areas tend to be very expensive, and secondly they have to be steered so that the movement of the sun across the sky is compensated for and the solar image thus remains locked firmly on the boiler. The area problem can be solved by having each mirror made up from a very large number of much smaller individual mirrors, and the steering problem can be solved to a certain extent by having slightly afocal or "fuzzy" systems, but on both counts these installations prove to be very costly. Their operation also requires very skilled technicians, so one has a paradox that devices which could solve many of the energy problems of the Third World are too expensive and too technologically advanced to be used in those countries. In advanced nations, however, they can be very effective especially for relatively modest ($< 1\,MW$) power demands for installations or towns with guaranteed sunshine. Many of the smaller townships in Australia, the Western USA and South Africa qualify very well. In these situations hybrid power stations, in which the hot water coming out of the turbos of a conventional assembly is heated up again by the solar concentrators, can be very effective. Thus in Meekatharra, a remote town in Western Australia, a hybrid solar installation costing $A 3·6 million and constructed by the German Company MAN is producing 100 kW of electrical power and at the same time saving 50 kW of power from the town's conventional power-station. In this way there will be a saving of 10^5 litres of diesel oil per year which, bearing in mind the remoteness of the installation, would be worth $A 40 000 per year. Since it will therefore take nearly 100 years for the plant to recover its costs, its installation even there is economically marginal, but scaled up versions in other locations could nevertheless be attractive commercial propositions.

A possible alternative approach, in some ways more appropriate to dry hot sub-tropical countries, is the use of "solar ponds". In these a large expanse of water is required divided into a surface layer of pure water and a deeper layer of virtually saturated brine. The near-infrared radiation from the sun passes through the upper layer but is strongly absorbed in the lower, which then develops a strong thermal gradient, cool at the top, hot at the bottom. If all goes well the lower brine layers can become hotter than 100 °C but do not boil because of the elevation of the boiling point caused by the high salt content. The very hot brine can then be taken to a heat exchanger where ordinary water is turned to steam and thus drives turbines as before, while the cooler brine is returned to the pond. The attraction of solar ponds is the ease and low cost with which large collecting areas can be readily assembled. Ponds of this type are being tried out in several areas of the Middle East but it is too early to say whether they will prove technically and economically viable.

The most attractive way of all of generating electrical power is directly by means of silicon photovoltaic cells which in this connection are universally known as "solar cells". They were originally developed for the space program where large area, but low weight, solar panels enabled the operation of complex spacecraft for long periods without the need to incur the high weight penalty associated with batteries or nuclear power sources. Back on earth, the commercial use of solar cells as power-generators has been very slow to take off because these cells, requiring for their fabrication highly sophisticated operations in monocrystalline starting material, are very expensive. Thus typical cells have a capital cost of about £6 per watt of electricity generated and perform with an efficiency of only 10%. The lifetimes of solar cells in the field are so far unknown, but it seems likely that such figures would not be economic and that solar cell generation would have to be restricted to *in situ* generation in locales where there was no conventional electricity supply. There might be very many such applications in Third World countries, especially since batteries are becoming so costly. In the developed world though, research has to be directed towards making cheaper solar cells. The four approaches favoured are (1) the use of polycrystalline silicon, (2) the development of amorphous silicon cells, (3) the use of "wet" or electrolytic cells based on polycrystalline $GaAs_x P_{1-x}$, (4) the use of alternative photovoltaic technologies such as CdTe/CdSe. All of these have their problems as well as their merits. Thus polycrystalline or amorphous silicon cells are still relatively expensive and of low efficiency even though they can be mass-produced, the wet cells are efficient but involve expensive semiconductors and have corrosion difficulties, and the CdTe/Se system is not suitable at the moment for high-power applications. Nevertheless progress is steady and capital costs of £3 per watt are being talked of. This probably puts photovoltaic generation within a factor of two of conventional generation as far as price is concerned. So, although true photovoltaic power stations are some way off, the use of solar cells in the home, in display advertising, in powering road-signs, in driving irrigation pumps, in powering radio-relay stations, for outdoor clocks, and for countless other applications ensures this form of solar energy utilisation an expanding and secure commercial market.

APPENDIX 4

Line-shape Functions

To derive the line-shape function, that is $\alpha(\tilde{v})$, one needs to know the frequency dependence of the complex permittivity $\hat{\varepsilon}(\omega)$ since α follows immediately from $\varepsilon'(\omega)$ and $\varepsilon''(\omega)$ via the equations given in Chapter 3. There are two approaches: one can set up an equation of motion and solve it or else one can derive a correlation function $f(t)$ and insert this into the Kubo relation

$$\frac{\hat{\varepsilon} - \varepsilon_\infty}{\varepsilon_s - \varepsilon_\infty} = -\int_0^\infty \dot{f}(t) \exp(i\omega t) dt. \tag{A4.1}$$

As an example of the first approach one can consider a particle with charge e and mass m bound to its equilibrium position by a force which would produce a natural oscillation frequency ω_0. If one now applies an electric field

$$E = E_0 \exp(i\omega t), \tag{A4.2}$$

the particle will undergo forced oscillations and the equation of motion will be

$$m\frac{d^2\hat{x}}{dt^2} + \left[\frac{m}{\tau}\right]\frac{d\hat{x}}{dt} + m\omega_0^2\hat{x} = eE_0 \exp(i\omega t), \tag{A4.3}$$

where the second term has been introduced to take account of damping of the motion with natural time constant τ. The solution of this equation is

$$\hat{x} = \left[\frac{e}{m}\right]\frac{E_0 \exp(i\omega t)}{\omega_0^2 - \omega^2 + i\omega\tau^{-1}}. \tag{A4.4}$$

The polarisation is thus complex and, noting that the induced dipole moment is related to the applied field by

$$\mu = ex = \chi E \tag{A4.5}$$

where χ is the polarisability, it follows that the polarisability is complex and given by

697

$$\hat{\chi} = \left[\frac{e^2}{m}\right]\frac{1}{\omega_0^2 - \omega^2 + i\omega\tau^{-1}}. \tag{A4.6}$$

Finally one notes that

$$\hat{\varepsilon} = 1 + 4\pi N\chi, \tag{A4.7}$$

where N is the number of such oscillators per unit volume and one has

$$\varepsilon' = 1 + \left[\frac{4\pi Ne^2}{m}\right]\left[\frac{\omega_0^2 - \omega^2}{(\omega_0^2 - \omega^2)^2 + \omega^2\tau^{-2}}\right], \tag{A4.8a}$$

$$\varepsilon'' = \left[\frac{4\pi Ne^2}{m}\right]\left[\frac{\omega\tau^{-1}}{(\omega_0^2 - \omega^2)^2 + \omega^2\tau^{-2}}\right]. \tag{A4.8b}$$

These equations are derived purely classically but as it turns out the corresponding quantum mechanical equations are very similar. The basic reason for this is the identity of the classical and the quantum mechanical amplitude for a simple harmonic oscillator. However, since at the moment we are entirely concerned with the *form* of the dispersion we will merely remark that the quantum version can be obtained by replacing (e^2/m) by $8\pi^2 v_{ij}|\mu_{ij}|^2/3h$ where v_{ij} is the transition frequency and μ_{ij} is the transition moment, and by replacing N by the population difference ΔN. Returning now to (A4.8) and making the weak dispersion approximation (i.e. n constant) one gets for the form of α.

$$\alpha \sim \frac{\omega^2\tau^{-1}}{(\omega_0^2 - \omega^2)^2 + \omega^2\tau^{-2}}. \tag{A4.9}$$

This function has a minimum of absolute value zero at the origin and has maxima at $\omega = \pm\omega_0$. For frequencies $\omega \gg \omega_0$, α falls to zero s. this line-shape is very acceptable. It is usually known as the Lorentz line-shape after H. A. Lorentz who carried out monumental studies of the classical theory of electromagnetic interactions [133]. Equation (A4.9) can be manipulated by means of the substitutions

$$\omega_0^2 = \omega_L^2 + \tau_L^{-2}, \tag{A4.10a}$$

$$\tau = \tfrac{1}{2}\tau_L, \tag{A4.10b}$$

to produce a factorisable form which can be split into two partial fractions

$$\alpha \sim \frac{1}{2\omega_L}\left[\frac{\omega\tau^{-1}}{(\omega_L - \omega)^2 + \tau_L^{-2}} - \frac{\omega\tau^{-1}}{(\omega_L + \omega)^2 + \tau_L^{-2}}\right]. \tag{A4.11}$$

The absorption in the physical region ($\omega > 0$) therefore contains a contribution from a component which peaks in the non-physical region ($\omega < 0$). This contribution, however, varies slowly in the region of the peak of the "real" component (at $\omega = +\sqrt{(\omega_L^2 + \tau_L^2)}$) so it can usually be ignored. For this reason it is very common practice to write the Lorentz line-shape function in the approximate form

$$\alpha = \frac{A\omega}{(\omega_0 - \omega)^2 + (\Delta\omega)^2}. \tag{A4.12}$$

This is very convenient for practical spectroscopy but it must be stressed that the quantity ω_0 which appears in it is *not* the true resonant frequency. For mid- and near infrared spectroscopy where $(\Delta\omega/\omega)$ may be of the order 10^{-2} or less it is an excellent approximation but care would be needed in its use to describe a very broad band in the far infrared.

To illustrate the second approach, one notes that if one has an oscillator of frequency ω_0 which is gradually being dephased with a time constant τ, then the correlation function would be

$$f(t) = \cos \omega_0 t \, \exp(-t/\tau). \tag{A4.13}$$

Substituting the derivative of this into (A4.1) after replacing the cosine function by its complex exponential equivalent leads to

$$\frac{\hat{\varepsilon} - \varepsilon_\infty}{\varepsilon_s - \varepsilon_\infty} = \frac{1}{2}\left[\frac{1 - i\omega_0\tau}{1 - i(\omega + \omega_0)\tau} + \frac{1 + i\omega_0\tau}{1 - i(\omega - \omega_0)\tau} \right]. \tag{A4.14}$$

The imaginary component of this is

$$\varepsilon'' = \left[\frac{\varepsilon_s - \varepsilon_\infty}{2} \right]\left[\frac{\omega\tau}{1 + (\omega - \omega_0)^2\tau^2} + \frac{\omega\tau}{1 + (\omega + \omega_0)^2\tau^2} \right]. \tag{A4.15}$$

This line-shape function is usually called the Van-Vleck–Weisskopf formula because its quantum mechanical equivalent was introduced by Van-Vleck and Weisskopf [134] to explain the shape of pressure-broadened microwave pure rotation lines. Again just as was done for the Lorentz expression it is normal to retain only the "physical" term and to write

$$\alpha = \frac{A\omega^2}{(\omega_0 - \omega)^2 + (\Delta\omega)^2}. \tag{A4.16}$$

This time, however, the ω_0 which appears *is* the true resonant frequency and one can rewrite (A4.16) in the alternative form

$$\alpha = \alpha_{max}\left[\frac{\omega}{\omega_0} \right]^2 \frac{(\Delta\omega)^2}{(\omega_0 - \omega)^2 + (\Delta\omega)^2}, \tag{A4.17}$$

which is frequently a useful formulation. The peak absorption coefficient α_{max} under microwave conditions (i.e. $h\nu_0 \ll kT$) can be written

$$\alpha_{max} = \left[\frac{4\pi^2 h N_0 c \tilde{\nu}_0^3 \mu^2}{3(kT)^2 \Delta\tilde{\nu}} \right] \exp\left[-\frac{hcBJ(J+1)}{kT} \right] \tag{A4.18}$$

This is an interesting expression since both N_0, the number of molecules per unit volume, and $\Delta\tilde{v}$, the half-width in wave number terms ($\Delta\omega/2\pi c$), depend on the first power of the pressure, so α_{max} is pressure independent! This strange behaviour has been confirmed by microwave studies but it is not strictly a paradox since the integrated area under the line (which depends on $\Delta\tilde{v}$) does fall linearly with the pressure. Of course at a low enough pressure, Doppler broadening will become apparent and then α will start to fall as the pressure is lowered still further. The Van-Vleck–Weisskopf line-shape function transforms smoothly into the Debye equation as $\omega_0 \to 0$, but note of course that equations (A4.16), (A4.17) and (A4.18) are now no longer appropriate since the approximation of constant n is invalid. Like the Debye equation, the Van-Vleck–Weisskopf has the physically unacceptable property that it predicts a "plateau" of finite absorption at infinite frequency. The fault lies in the use of a correlation function which does not behave properly at $t = 0$ (see section 5.5 for details). The line-shape function could be improved by incorporating inertial "roll-off" but so far nobody has thought it worthwhile to do this. Within its strictly limited role of dealing with rotational lines at low frequencies the Van-Vleck–Weisskopf equation is eminently successful but it is inappropriate for anything else. The Lorentz equation should not be used for very low frequencies and clearly it cannot account for Debye dielectric dispersion but within its limited role of dealing with vibrational bands without resolved line structure in the mid infrared it too is successful and of course widely used. Brot has discussed various line-shape functions in some detail [135].

An interesting point which arises in line-shape theory is the question of whether there is a difference between a resonance and a relaxation. Theoretically this question means whether there is a well-defined oscillation frequency ω_0 or not. If one has a sound theoretical reason for expecting this to be the case, then that is all there is to it, but it is also frequently asserted that one can deduce whether a given feature is a resonance or not purely from experimental considerations. This is not always true. In the mid infrared, a sharply peaked feature must be a resonance but in the far infrared it is possible to get a peaked feature when there is no well-defined resonant frequency. An obvious example occurs when one introduces inertial terms into the Debye equation for dielectric dispersion (see section 5.5). The first approximation, the Rocard–Powles formula, is then

$$\frac{\hat{\varepsilon} - \varepsilon_\infty}{\varepsilon_s - \varepsilon_\infty} = \frac{1}{(1 - i\omega\tau_D)(1 - i\omega\tau_D\gamma)}. \tag{A4.19}$$

The imaginary component of which is

$$\varepsilon'' = \frac{\omega\tau_D(1 + \gamma)}{(1 - \omega^2\tau_D^2\gamma)^2 + \omega^2\tau_D^2(1 + \gamma)^2}, \tag{A4.20}$$

and the absorption coefficient α will show a peak since it is zero at $\omega = 0$ and

$\omega = \infty$. There is clearly a sort of "pseudo" resonance frequency given by

$$\omega = \sqrt{\left(\frac{1}{\tau_D^2 \gamma}\right)},$$

but whether this deserves the title or not is a matter for debate. It is also sometimes stated that one can infer the presence of a resonance from the behaviour of ε' or equivalently n even if there is no evidence for it in ε'' or α. Clearly this cannot be absolutely true since by the Kramers–Kronig theorem the variation of ε' with frequency is completely defined once one knows that of ε''. It may be true that a given kind of behaviour may be more *obvious* in one presentation than in another but that is all one can say. In the present case if there is a peak in α, then n will approach its limiting value n_∞ from lower values, whereas if there is not n will approach n_∞ monotonically from higher values. One is clearly not justified in deducing the existence of a real resonance from the fact that $n(\tilde{\nu})$ shows a minimum. Likewise the practice of resolving broadened far-infrared absorption bands of liquids into a series of "resonances" may not be physically very meaningful in the absence of a good theory which tells you *a priori* that it is!

APPENDIX 5

Fourier Transform Operations

Introduction

It is an increasingly common occurrence in experimental physics that one observes a function of say time, $f(t)$, but what one really wants is the function of (in this case) frequency $\phi(v)$ which is the Fourier transform of $f(t)$, i.e.

$$\hat{\phi}(v) = \int_{-\infty}^{+\infty} f(t) \exp(2\pi i v t) \, dt. \tag{A5.1}$$

Commonly $f(t)$ will be a symmetric function of time, i.e. $f(t) = f(-t)$, in which case $\phi(v)$ is pure real and given by

$$\phi(v) = \int_{-\infty}^{+\infty} f(t) \cos 2\pi v t \, dt. \tag{A5.2}$$

One cannot actually evaluate integrals such as (A5.1) or (A5.2) because the range of time over which observation can be taken is limited and because, since one is working with non-analytical (i.e. experimental) functions, one can only work with a finite number of samples. In practice one is forced therefore to replace (A5.2) by the summations

$$\phi(v) = \Delta t \sum_{-M}^{+M} f(m\Delta t) \cos 2\pi v m \Delta t, \tag{A5.3}$$

where the total number of samples taken is $2M + 1$ and the distance between samples is Δt. This replacement has two effects:

a. The resolution, i.e. the faithfulness of the reconstruction of $\phi(v)$, is reduced. This is a natural consequence of the failure to include the data beyond $\pm M\Delta t$.

b. The reconstructed function, instead of being uniquely defined throughout the whole range of v, now shows mirror and translational symmetry

702

about a "folding" frequency given by $v_f = 1/2\Delta t$. The function in the fundamental range (i.e. $v = 0$ to $v = v_f$) is endlessly replicated throughout the frequency domain—see Fig. A5.1. This "aliasing" arises from our having replaced a Fourier integral (A5.2) by a Fourier series (A5.3) and, as is well known, Fourier series are periodic functions.

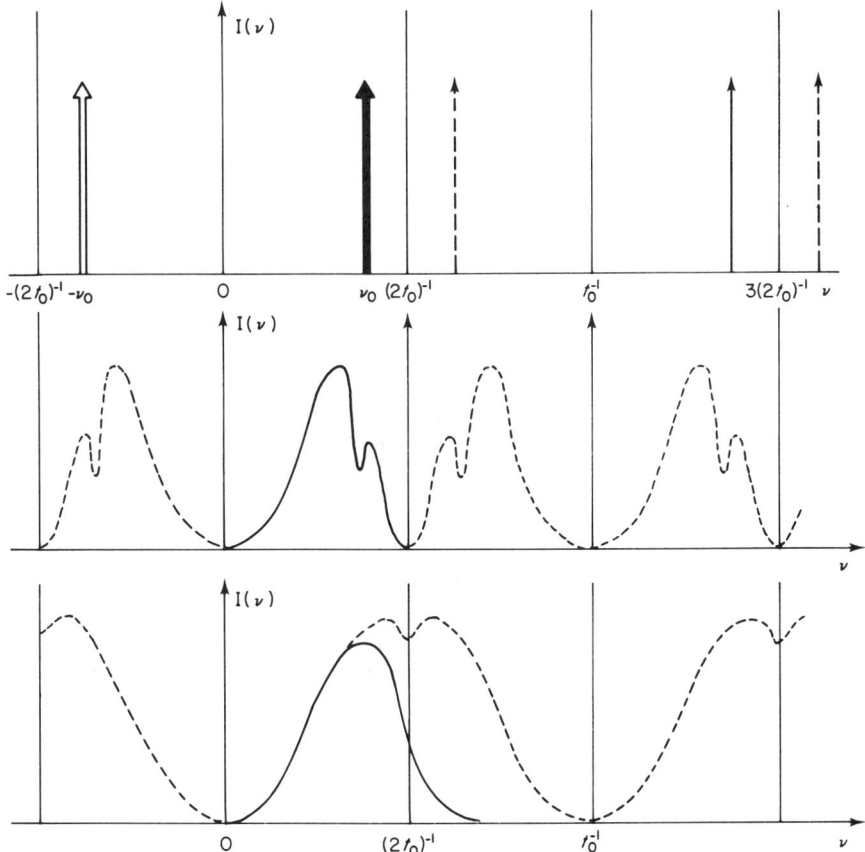

FIG. A5.1. The phenomenon of aliasing or replication. *Upper inset*: A real spectral feature at v_0 is, because of the nature of the transform process, accompanied in the transformed spectrum by an image at $-v_0$. In all practical spectroscopy, the spectrum will be calculated from a sampled interferogram, with sample spacing t_0, and both the element and its image will be endlessly replicated at frequencies $(v_0 \pm m/t_0)$ and $(-v_0 \pm m/t_0)$ respectively. The spectral domain therefore has mirror symmetry at frequencies $v_f = \pm m/2t_0$. *Middle inset*: For broad-band spectroscopy, it is essential to restrict the spectral band so that no real power lies above the "folding" frequency $(2t_0)^{-1}$. *Lower inset*: If this is not done, then real power (solid line) lying above $(2t_0)^{-1}$ will be folded back or "aliased" into the fundamental region $(0 - 1/2t_0)$ and give erroneous results.

The first of these effects just has to be endured—after all finite resolution is unavoidable. If the resolution is inadequate then M has to be increased and that is all there is to it. The second effect, strangely enough, need not necessarily be onerous at all. If it has been arranged, as it usually can be, that the real frequency spectrum has no components lying above the folding frequency, then the reconstruction will be exact since there will be nothing to "fold" back or alias into the fundamental regions.

The slow Fourier transform

To evaluate the summation (A5.3), for all save the most trivial cases, involves the use of a digital computer. The computer has to be programmed to carry through all the operations in the correct sequence and this programming involves some deft operations if the computer is to carry through the calculation in the shortest possible time. The first matter to arise is the evaluation of the set of cosines and sines that will be necessary in the general case. A computer cannot look up the values of functions in a table and so every time a transcendental function is required it has to be evaluated from its equivalent power series. For $\cos x$ and $\sin x$ we have

$$\cos x = 1 - 1/2\, x^2 + \frac{1}{24} x^4 - \ldots\ldots (-1)^n \frac{1}{(2n)!} (x)^{2n}, \qquad \text{(A5.4)}$$

$$\sin x = x - \frac{1}{6} x^3 + \frac{1}{120} x^5 \ldots\ldots (-1)^n \frac{1}{(2n+1)!} x^{2n+1}. \qquad \text{(A5.5)}$$

To evaluate the series to sufficient precision takes appreciable time and if one were attempting to carry through a large transform (i.e. M large) for a large number of output points the computing time just for calculating the trigonometrical functions would be prohibitive. Fortunately, all the arguments of the functions are integral multiples of the smallest (namely $2\pi v \Delta t$) and one can take recourse to the formulae for the addition of two cosines, or sines, in the form

$$\cos (N+1)x + \cos (N-1)x = 2 \cos Nx \cos x \qquad \text{(A5.6)}$$

and $\qquad \sin (N+1)x + \sin (N-1)x = 2 \sin Nx \cos x \qquad \text{(A5.7)}$

to derive the Chebyshev recurrence relations which are used in practice. These relations permit one to rapidly calculate a series of cosines or sines in terms of the smallest and the computing time is drastically reduced.

With this step built into the programme one sets out to evaluate the summation. It is readily seen that for each frequency value, $(2M+1)$ multiplications are necessary and, if one is calculating N output points, the total number will be $N(2M+1)$. For the simple case given by (A6.3) this number can be reduced somewhat since by definition $f(t)$ is then symmetric and one need only run over half the range, say from $m = 0$ to $m = M$, but in all cases one will be faced with a number of the order of M^2 since N will commonly equal M. The total time taken will be roughly proportional to the total number of multiplications, so in this way of doing the computation one can say that the total time is proportional to the square of the number of points to be transformed. If M is large the transformation can be very slow and this approach is therefore usually called the slow Fourier transform.

The fast Fourier transform

The main advantages of the slow Fourier transform are that one can calculate at arbitrarily fine intervals of v and that one can calculate over a restricted band of frequencies if that band contains all the features of interest. On close examination both these advantages can be seen to be illusory. For example, since the resolution will be restricted to no better than $\Delta v = 1/(2M\Delta t)$ there is no point in calculating to a finer apparent resolution. If one is prepared to accept computation at frequency intervals given by the resolution limit and over the entire band up to the folding frequency then very great savings in computing time can be realised. One would have analogously to (A5.3) in the general case

$$\hat{\phi}(v) = \Delta t \sum_{-M+1}^{+M} f(m\Delta t) \exp(2\pi i v m \Delta t), \qquad (A5.8)$$

where we have for future convenience slightly modified the previous treatment so that the total number of points will be even. Now, since the correct resolution limit sampling in the frequency domain Δv is $(2M\Delta t)^{-1}$, we have

$$\hat{\phi}(n\Delta v) = \Delta t \sum_{-M+1}^{M} f(m\Delta t) \exp\left(\frac{i\pi mn}{M}\right). \qquad (A5.9)$$

This equation merits careful study. The cosines and sines involved are of angles simply related to π radians. The samples are therefore being taken off the cosine (or sine) wave in a highly symmetrical way and the set of cosines and sines will therefore repeat after $2M$ samples. This symmetry lies at the heart of the method—of dramatically reducing the time required for computing a Fourier transform—the so called Fast Fourier Transform. In practice the negative indices in (A5.9) are inconvenient and it may readily be shown that if one replicates the $f(t)$ function in the same way that $\phi(v)$ is replicated, then one may write

$$\hat{\phi}(n\Delta v) = \hat{\phi}_n = \Delta t \sum_{0}^{2M-1} f_m \exp\left(i\frac{\pi mn}{M}\right). \qquad (A5.10)$$

The replication of $f(t)$ brings a complete symmetry to the problem for now we have two *periodic* functions $f(t)$ and $\phi(v)$ which form a Fourier pair and are completely equivalent to each other. The Fourier transform of a correctly sampled version of either gives the other. The replication of $f(t)$, in this light, is a consequence of the finite sampling of $\phi(v)$. The double replication is indicated in Fig. A5.1.

The next crucial step in the operation is to make $2M$ a power of two, that is $2M = 2^p$ where p is an integer. The sampling of the cosine (or sine) wave now becomes even more symmetrical. One can at once divide the f ordinates into two sets, those with m even and those with m odd. The basic transform equation (A5.10) then breaks up into two separate Fourier transformations,

$$\phi_n = \Delta t \left\{ \sum_0^{M-1} f_{2m} \exp\left(\frac{i\pi 2mn}{M}\right) + \exp\left(\frac{i\pi n}{M}\right) \sum_1^{M-1} f_{2m+1} \exp\left(\frac{i\pi 2mn}{M}\right) \right\},$$
(A5.11)

each of which involves a data set only half the original size. Each of these sets in their turn may be relabelled and their ordinates again divided into even and odd sets. Clearly the process can be continued until we have $M/2$ sets each of which contains only two members. The Fourier transform operation therefore reduces to an ordered sequence of operations each of which is trivial. There is thus an enormous saving of time.

What is required now is a routine, that is an algorithm, to be used in the programming of a high-speed digital computer so that it can efficiently carry out the transformation. The required algorithm was apparently hit upon (as is so often the case) by several people independently and at about the same time (*circa* 1960) but since the clearest description of it was given by Cooley and Tukey [935] it is nowadays known usually as the Cooley–Tukey (or sometimes simply as the FFT) algorithm. The basic operation is the division into even and odd sets and it will be realised, on reflection, that this repeated division will result in the $f(t)$ ordinates being sorted into *inverted binary order*. To see what this means, imagine that we are carrying out a transformation of a mere 8 ordinates. These will be

$$f_0 \quad f_1 \quad f_2 \quad f_3 \quad f_4 \quad f_5 \quad f_6 \quad f_7.$$

Numbered in binary these would occupy locations (in the memory store of the machine):

$$000 \quad 001 \quad 010 \quad 011 \quad 100 \quad 101 \quad 110 \quad 111.$$

After the first sort, we would have

$$000 \quad 010 \quad 100 \quad 110 \quad 001 \quad 011 \quad 101 \quad 111.$$

And after the second

$$000 \quad 100 \quad 010 \quad 110 \quad 001 \quad 101 \quad 011 \quad 111,$$

that is

$$f_0 \quad f_4 \quad f_2 \quad f_6 \quad f_1 \quad f_5 \quad f_3 \quad f_7.$$

This is the inverted binary (or bit) order. If the incoming data are held in a buffer store and then read into the machine's operating store in this inverted bit order, the principal requirement of the algorithm will have been satisfied.

The second major requirement is to use the available memory store as efficiently as possible. Spectroscopists are always wanting to transform more and more points and computer store becomes a limitation. To get as far round this difficulty as is possible the algorithm is usually written so that the results of each step are "overwritten" (that is replace) the results of the previous step. The algorithm is general and transforms one set of possibly complex data into another set of possibly complex data so in this connection a memory location is actually a double location one for the real part and the other for the imaginary part.

With the preliminaries out of the way we can now see how the algorithm works in practice and to crystallise the ideas we will work with a specific example, namely the eight point transform. Equation (A5.10) then gives

$$\phi_0 = (f_0 + f_4) + (f_2 + f_6)\, E(0) + E(0)\left[(f_1 + f_5) + (f_3 + f_7)\, E(0)\right]$$

$$\phi_1 = (f_0 - f_4) + (f_2 - f_6)\, E\left(\frac{\pi}{2}\right) + E\left(\frac{\pi}{4}\right)\left[(f_1 - f_5) + (f_3 - f_7)\, E\left(\frac{\pi}{2}\right)\right]$$

$$\phi_2 = (f_0 + f_4) + (f_2 + f_6)\, E(\pi) + E\left(\frac{\pi}{2}\right)\left[(f_1 + f_5) + (f_3 + f_7)\, E(\pi)\right]$$

$$\phi_3 = (f_0 - f_4) + (f_2 - f_6)\, E\left(\frac{3\pi}{2}\right) + E\left(\frac{3\pi}{4}\right)\left[(f_1 - f_5) + (f_3 - f_7)\, E\left(\frac{3\pi}{2}\right)\right]$$

$$\phi_4 = (f_0 + f_4) + (f_2 + f_6)\, E(2\pi) + E(\pi)\left[(f_1 + f_5) + (f_3 + f_7)\, E(2\pi)\right]$$

$$\phi_5 = (f_0 - f_4) + (f_2 - f_6)\, E\left(\frac{\pi}{2}\right) - E\left(\frac{\pi}{4}\right)\left[(f_1 - f_5) + (f_3 - f_7)\, E\left(\frac{\pi}{2}\right)\right]$$

$$\phi_6 = (f_0 + f_4) + (f_2 + f_6)\, E(\pi) - E\left(\frac{\pi}{2}\right)\left[(f_1 + f_5) + (f_3 + f_7)\, E(\pi)\right]$$

$$\phi_7 = (f_0 - f_4) + (f_2 - f_6)\, E\left(\frac{3\pi}{2}\right) - E\left(\frac{3\pi}{4}\right)\left[(f_1 - f_5) + (f_3 - f_7)\, E\left(\frac{3\pi}{2}\right)\right]$$

(A5.12)

Where as a shorthand we have written $E(\theta)$ instead of $\exp(i\theta)$. The symmetry of these equations is obvious: thus the arguments of the various exponential functions steadily increase but since $E(\pi + \theta) = -E(\theta)$, the ordinates beyond E_4 are simply found from the corresponding ones on the near side. This is a general result and one crucial to the success of the FFT. Each operation generates two (generally complex) quantities say $\hat{A}$ and $\hat{B}$ but one merely needs to take the sum and the difference of $\hat{A}$ and $\hat{B}$ to get *two* of the required quantities. The vital point is that multiplications are relatively slow in the computer but additions are virtually instantaneous. The great saving of time in the FFT comes about through replacing most of the (slow) multiplications by (quick) additions and subtractions. This process is carried out likewise at each step of the algorithm. Thus the terms in square brackets for ϕ_0 and ϕ_2 or in fact *any* pair of the form ϕ_n and ϕ_{n+2} are simply related to one another by the sum and difference of two terms. The equivalent terms not in square brackets are related in the same way. It will readily be seen therefore how the steps of proceding from the f_m to the ϕ_n can be set out in an orderly sequence. This is shown in Table A5.1. The incoming data are read into the memory stores in reverse binary order using a small subroutine. Starting at location zero the sum is formed and the difference of adjacent ordinates and these are over written onto the original locations (first step). This step is then repeated except that every other ordinate is now combined and the multiplying factor $E[j(\pi/2)]$, where j is the number of the location is introduced. Thus two quantities are formed:

$$\hat{A}_n = (n) + E\left(\frac{n\pi}{2}\right)(n+2). \tag{A5.13a}$$

$$\hat{A}_{n+2} = (n) - E\left(\frac{n\pi}{2}\right)(n+2), \tag{A5.13b}$$

where (n) represents the contents of location n. The machine proceeds along the array, skipping each location which has already been overwritten. The final step, in this example, consists of taking the contents of data registers (n) and $(n+4)$ and using multiplying factors $E[(n\pi/4)]$. In this particular example, the machine would then stop since the next operation would involve the pairs (n) and $(n+8)$ and the latter does not exist! However, if one had started with sixteen points (i.e. $M = 8$) then this next step could go ahead with multiplying factors $E[(n\pi/8)]$. Clearly each step is basically the same, the variations being regular and rigidly ordered. Thus we have the basis for programming an automatic calculating machine to start at the beginning and to proceed mechanically until it reaches the end.

The number of steps is clearly p where $2^p = 2M$ is the total number of points to be transformed. At each step there will be M multiplications and M additions. The total number of operations will therefore be

$$M \log_2 M.$$

The time required will depend on the relative speed of multiplication and additions and on the transfer time between registers, amongst other factors, but clearly one will have

$$T_{\text{FFT}} \sim M \log_2 M. \tag{A5.14}$$

and, from the analysis given earlier,

$$T_{\text{SFT}} \sim M^2. \tag{A5.15}$$

TABLE A5.1
Flow chart of the FFT algorithm for eight input points

Location	Initial contents	First step	Second step	Third step
000	f_0	f_0+f_4	$(f_0+f_4)+(f_2+f_6)$	$(f_0+f_4)+(f_2+f_6)+(f_1+f_5)+(f_3+f_7)$
001	f_4	f_0-f_4	$(f_0-f_4)+(f_2-f_6)E\left(\dfrac{\pi}{2}\right)$	$(f_0-f_4)+(f_2-f_6)E\left(\dfrac{\pi}{2}\right)+E\left(\dfrac{\pi}{4}\right)\left[(f_1-f_5)+(f_3-f_7)E\left(\dfrac{\pi}{2}\right)\right]$
010	f_2	f_2+f_6	$(f_0+f_4)-(f_2+f_6)$	$(f_0+f_4)-(f_2+f_6)+E\left(\dfrac{\pi}{2}\right)\left[(f_1+f_5)-(f_3+f_7)\right]$
011	f_6	f_2-f_6	$(f_0-f_4)-(f_2-f_6)E\left(\dfrac{\pi}{2}\right)$	$(f_0-f_4)-(f_2-f_6)E\left(\dfrac{\pi}{2}\right)+E\left(\dfrac{3\pi}{4}\right)\left[(f_1-f_5)-(f_3-f_7)E\left(\dfrac{\pi}{2}\right)\right]$
100	f_1	f_1+f_5	$(f_1+f_5)+(f_3+f_7)$	$(f_0+f_4)+(f_2+f_6)-(f_1+f_5)-(f_3+f_7)$
101	f_5	f_1-f_5	$(f_1-f_5)+(f_3-f_7)E\left(\dfrac{\pi}{2}\right)$	$(f_0-f_4)+(f_2-f_6)E\left(\dfrac{\pi}{2}\right)-E\left(\dfrac{\pi}{4}\right)\left[(f_1-f_5)+(f_3-f_7)E\left(\dfrac{\pi}{2}\right)\right]$
110	f_3	f_3+f_7	$(f_1+f_5)-(f_3+f_7)$	$(f_0+f_4)-(f_2+f_6)-E\left(\dfrac{\pi}{2}\right)\left[(f_1+f_5)-(f_3+f_7)\right]$
111	f_7	f_3-f_7	$(f_1-f_5)-(f_3-f_7)E\left(\dfrac{\pi}{2}\right)$	$(f_0-f_4)-(f_2-f_6)E\left(\dfrac{\pi}{2}\right)-E\left(\dfrac{3\pi}{4}\right)\left[(f_1-f_5)-(f_3-f_7)E\left(\dfrac{\pi}{2}\right)\right]$

So the saving in time using the FFT is a factor of the order $M/\log_2 M$. When M is large this factor can be considerable—thus if M is 10^6, it becomes roughly 50 000! In practical terms this means that the FFT makes possible Fourier transformations that would take an impossibly long time by the straightforward slow transform.

The outline given above assumes the data at the beginning to be in binary inverted order. However the algorithm is perfectly symmetrical and one may begin with the data in the correct order but then one will get the spectral ordinates in reversed binary order! The basic operation of taking every $(2^j)^{\text{th}}$ element, where j is the number of the step, is reminiscent of the ancient Roman punishment of decimation, i.e. killing every tenth person in a line, and this approach to the algorithm is therefore sometimes known as "decimation in time". The phrase is not very helpful since the loose English usage of decimation, to mean to reduce drastically, has led spectroscopists to think that what is intended is a reference to the great reduction in the time required for the computation. That this is not so can be seen by the existence of an alternative way of writing the algorithm which is referred to as "decimation in frequency". Clearly there is no simple meaning to be attached to this phrase!

Refinements to the FFT

In the general case where all the f_i are complex, the only symmetry in the ϕ_j is that represented by equation (A5.13), namely that the ordinates occur in pairs, one being the sum and the other the difference of two terms. However, in practical spectroscopy the f_i are nearly always real and, furthermore, they are either symmetric (i.e. $f_i = f_i$) or else antisymmetric (i.e. $f_i = -f_{-i}$) about the origin. When the data are real, further symmetry is involved for then the set of complex $\hat{\phi}(v)$ must be Hermitian since it has a real Fourier transform. This being so there is not only a waste of computer store (since half the original locations will contain zeros) but a waste of computer time since one will be calculating the full set of spectral ordinates when in fact half of them can be derived from the other half by the simple symmetry operation

$$\hat{\phi}_{M-n} = \hat{\phi}^*_{M+n},\qquad\qquad (A5.16)$$

which follows from the Hermitian character of ϕ. Connes [936], Bell [937] and Fleming [938] have discussed how this redundancy may be eliminated. Their solution is to take the set of purely real data f_m and to divide it into a set for which m is even and one for which m is odd. One then assembles the complex set

$$\hat{F}_k = f_{2k} + if_{2k+1}.\qquad\qquad (A5.17)$$

This set is one half the size of the original set. The Fourier transform of $\hat{F}_k$,

$$\hat{E}_l = \sum_0^{M-1} \hat{F}_k \exp\left(\frac{i2\pi lk}{M}\right),\qquad\qquad (A5.18)$$

may be written

$$\hat{E}_l = \hat{C}_l + i\hat{D}_l, \tag{A5.19}$$

where $\hat{C}_l$ is the Fourier transform of the even set and $\hat{D}_l$ that of the odd set. These quantities are Hermitian and may therefore be constructed from $\hat{E}_l$ by means of the relations

$$\hat{C}_l = 1/2 \left[\hat{E}_l + \hat{E}^*_{-l} \right], \tag{A5.20a}$$

$$\hat{D}_l = 1/2i \left[\hat{E}_l - \hat{E}^*_{-l} \right]. \tag{A5.20b}$$

The values of E_l^* at the negative values of l may themselves be found from the relation

$$\hat{E}^*_{-l} = \hat{E}^*_{M-l}, \tag{A5.21}$$

which follows from the periodicity of E_l. The desired half range spectrum may then be found from

$$\hat{\phi}_n = \hat{C}_l + \hat{D}_l \exp\left(\frac{i\pi n}{M}\right). \tag{A5.22}$$

The extra step involved in this approach (equation A5.22) makes this form of the FFT marginally slower, but the great saving of memory store makes this well worth while.

The experimental data set produced in spectroscopy is not only usually real, it is also usually symmetric or antisymmetric about the origin. When this is the case the transform instead of being complex will be either pure real or else pure imaginary and will show perfect symmetry about the ordinate at $n = M$. Still further redundancy therefore exists. It is worth pointing out here that if, because of electrical or mechanical imperfections, the experimental data is not actually perfectly symmetric or antisymmetric, it can always be made so by convolution with a suitable phase factor. This approach is very attractive because one need only observe half a data run. One may therefore use the whole range of the variable to get meaningful results and thus obtain the maximum permissible resolution. A method by which the computer store requirement can be reduced by a factor of four and all redundancy be removed has been described by Connes [936]; Bell [937] and Fleming [938]. This method gives the fastest possible and most economical Fourier transform of a given large data set.

The Classical Theory of the Vibration of Polyatomic Molecules

It is a remarkable fact that the transition frequencies calculated by quantum mechanics for the N-dimensional harmonic oscillator are exactly the same as the corresponding classical vibration frequencies. Of course, the fine detail of the quantum result, zero-point energy, tunneling for example, are not given by the classical treatment, but if one is interested solely in the frequencies and the amplitudes of vibration, as one usually is in most aspects of infrared spectroscopy, then one may go ahead and use purely classical methods of calculation.

In classical mechanics it is not usual to start with fundamentals, that is the application of Newton's laws to the system followed by explicit solution of the resulting differential equations; rather expressions for the kinetic and the potential energy are set up and then the formalisms of Lagrange or of Hamilton are applied. For the particular case of the harmonic oscillator this gives some very simple results. Thus if one defines a set of cartesian coordinates $X = x_1, \ldots x_i, \ldots x_N$, it follows that the kinetic energy is

$$2T = m_i \dot{x}_i^2, \tag{A6.1}$$

which can be written in matrix-vector form as

$$2T = \tilde{\dot{X}} . M . \dot{X}, \tag{A6.2}$$

where the tilde ($\sim$) signifies transposition. The elements of the matrix M will be identical in groups of three since each atom is defined by its three cartesian coordinates. One can similarly define the potential energy via the relations

$$2V = f_{ij} x_i x_j, \tag{A6.3}$$

which can equivalently be written

$$2V = \tilde{X} . F . X. \tag{A6.4}$$

For the harmonic oscillator one now requires that there be conservation of energy and this, with the assumed solution $x = x_0 \cos \omega_0 t$, gives immediately

$$x_0^2[m\omega_0^2 + (f - m\omega_0^2)\cos^2 \omega_0 t] = \text{constant} \qquad (A6.5)$$

for the one-dimensional case. It will be seen that, apart from the trivial solution $x_0 = 0$, equation (A6.5) demands that

$$\omega_0 = \sqrt{\left(\frac{f}{m}\right)}. \qquad (A6.6)$$

This is the core of the argument. Equations of the form (A6.5) are called secular equations and their existence demands that the system have only a set of discrete frequencies, in the present case just the one given by (A6.6). In the general N-dimensional case the secular equation would restrict the system to just N discrete vibration frequencies.

The treatment continues by tidying up the algebra with the introduction of mass weighted coordinates,

$$w_i = \sqrt{m_i x_i}, \qquad (A6.7)$$

in terms of which (A6.2) and (A6.4) can be written

$$2T = \tilde{W} \cdot \dot{W}, \qquad (A6.8)$$

and

$$2V = \tilde{W} \cdot F_m \cdot W. \qquad (A6.9)$$

F_m is a square matrix with elements $f_{ij}/m_i m_j$. In this notation, (A6.8) represents a hypersphere in N-dimensional space and (A6.9) represents a hyperellipsoid. Clearly if one rotates the N-dimensional cartesian system without displacing its origin, (A6.8) will merely go into another hypersphere, but by a judicious choice of angles, (A6.9) will become diagonalised. In other words its principal axis system and the reference coordinate system will coincide. This is a particular illustration of a general theorem that if one has two arbitrary but positive definite quadratic forms one can always find a coordinate transformation which will diagonalise both. If the coordinate transformation is given by

$$Q = U \cdot W, \qquad (A6.10)$$

where U is an orthonormal matrix (i.e. $U \cdot \tilde{U} = E$), then the set Q is called the set of normal coordinates. It follows from the definitions that

$$2T = \tilde{\dot{Q}} \cdot \dot{Q}, \qquad (A6.11)$$

and

$$2V = \tilde{Q} \cdot \Lambda \cdot Q, \qquad (A6.12)$$

where Λ is a diagonal matrix whose elements are the squares of the eigenfrequencies ω_i. The concept of the normal coordinates of vibration is a very fundamental one and it may be shown, for example, that any arbitrary motion of the system can always be resolved into suitably weighted linear combinations of the normal coordinates. There are obvious parallels with the wave functions of quantum mechanics.

The simple treatment, outlined above in terms of the cartesian coordinates, is a valuable way of bringing out all the fundamental points of the topic, but it is not commonly used as the basis for practical calculation. There are two reasons for this. Firstly, it will produce six zero frequencies—the three translations plus the three rotations—and this is an unnecessary complication when one is considering purely vibrational motion. Secondly the description of the restoring forces within the molecule solely in terms of interactions between pairs of atoms tends to obscure the physics of the situation since the actual forces are strongly concentrated in holding the bonds and the interbond angles at their equilibrium values. A coordinate system made up of the changes in the bond lengths and in the interbond angles would be preferable on both counts and such an internal coordinate system is used almost invariably in molecular normal coordinate analysis. One point needs to be made, however, and this is that the use of equations such as (A6.3), which is essentially a truncated Taylor's series, requires that the coordinate system be independent, for only if this is so is it justified to omit the linear term. Now one frequently finds that if one includes all the angles of a symmetrically equivalent set, the HCH angles in methane for example, one will finish up with one coordinate too many. There will then be a necessary relationship between the members of the set and the coordinates will not be independent. One is said to have a redundant coordinate in the system. This introduces complications but most workers in the field prefer to face up to these rather than to arbitrarily omit one member of a symmetrically equivalent set. One therefore sets up the complete set of internal coordinates R and has

$$X = AR. \tag{A6.13}$$

where A is the transformation matrix. It follows by augmentation and expansion that the vibrational part of the kinetic energy is now given by

$$2T = \tilde{R}\tilde{A}MA\dot{R}. \tag{A6.14}$$

This equation unfortunately is not easy to use in practice because A can only be set up by the use of explicit "no rotation and translation" conditions. Fortunately if one writes down the inverse relation to (A6.13),

$$R = BX, \tag{A6.15}$$

then it may readily be shown that $BM^{-1}\tilde{B}$ is the inverse of $\tilde{A}MA$ and B can be written down immediately. It follows that one can now write

$$2T = \tilde{R}G^{-1}R, \tag{A6.16}$$

where G is defined by the relation

$$G = BM^{-1}\tilde{B}. \tag{A6.17}$$

The next step is to introduce the matrix of the internal coordinate force constants F_i and to write

$$2V = \tilde{R}F_iR. \tag{A6.18}$$

From what was said above, it will be realised that the large elements of F_i tend to be distributed along its leading diagonal. One now continues by introducing the normal coordinates defined by

$$Q = L^{-1}R \tag{A6.19}$$

and finds relations formally equivalent to (A6.11) and (A6.12). The matrix L satisfies the two relations

$$L\tilde{L} = G$$

and

$$\tilde{L}F_iL = \Lambda \tag{A6.20}$$

which can be combined to give the secular equation

$$GF_iL = L\Lambda. \tag{A6.21}$$

As before this has, apart from the trivial solution $L = 0$, the solution

$$|GF_i - \lambda E| = 0. \tag{A6.22}$$

The squares of the vibration frequencies are therefore given by the eigenvalues (that is the latent roots) of the matrix product GF.

The matrices G and F and their product will be square and of size $(3N - 6)$ by $(3N - 6)$ where N is the number of atoms in the molecule. In the early days of the art, calculating facilities were rather limited and the problem of diagonalising even a 9×9 matrix, as would arise for example with an XY_4 molecule, were formidable. It was universal therefore to use any symmetry which the molecule had to factor the big matrix down into a collection of more tractable smaller matrices. The symmetry elements which the molecule possesses will form a group and the normal coordinates will form a representation of this group. It will be a reducible representation and can therefore be reduced into a combination of the irreducible representations by the usual methods of group theory [486]. As an example, the nine normal coordinates for an XY_4 molecule form the representation

$$\Gamma_Q = A_1 + E + 2F_2 \tag{A6.23}$$

of the molecular point group T_d. There are therefore only three distinct matrices in the resolution and only one of these, the 2×2 F_2 matrix actually requires any solution! The problem is thus enormously simplified. The factoring is done by finding a set of symmetry coordinates, using standard group theoretical methods, assembling these into an orthonormal matrix U so

$$S = UR, \tag{A6.24}$$

and then using U to factor both G and F thus:

$$\mathscr{G} = UG\tilde{U}, \tag{A6.25a}$$

$$\mathscr{F} = UF\tilde{U}. \tag{A6.25b}$$

The $\mathscr{G}$ and $\mathscr{F}$ matrices consist of primitive blocks distributed down the leading diagonals.[†] It is important to note that the elements of U are determined solely by the symmetry of the molecule. They will be the same for all XY_4 molecules, for example, whereas the elements of either L or its symmetry coordinate equivalent $\mathscr{L}$ given by

$$\mathscr{L} = UL, \tag{A6.26}$$

which will completely factorise GF, down to diagonal form, depend on the masses and force constants of the particular molecule under study. The process of factorisation using the U matrix can be illustrated by once more considering the XY_4 molecule. If one numbers the bond lengths (r) by 1, 2, 3, 4, then there will be ten (one redundancy) internal coordinates r_1, r_2, r_3, r_4 and $\theta_{12}, \theta_{13}, \theta_{14}$, θ_{23}, θ_{24} and θ_{34}. In these, the initial Δs have been dropped to keep the notation uncluttered. With this set, there are seven symmetrically distinct force constants and in terms of these the potential energy may be written

$$2V = f_r \sum_i^4 r_i^2 + f_{rr} \sum_{i \neq j} r_i r_j + f_{r\theta} \sum_{i \neq j} r_i(r_0 \theta_{ij}) + f'_{r\theta} \sum_{i \neq j \neq k} r_i(r_0 \theta_{jk}) \tag{A6.27}$$

$$+ \tfrac{1}{2} f_\theta \sum_{i \neq j} (r_0 \theta_{ij})^2 + f_{\theta\theta} \sum_{i \neq j \neq k} (r_0 \theta_{ij})(r_0 \theta_{ik}) + f'_{\theta\theta} \sum_{i \neq j \neq k \neq l} (r_0 \theta_{ij})(r_0 \theta_{kl}),$$

which may then be tabulated in matrix vector form as in (A6.18). The effects of the redundant coordinate are now immediately obvious since the factored F matrix, from (A6.23), contains only five independent force constants. What happens is that only certain combinations of the force constants in (A6.27) have physical meaning and may therefore be determined; these are

$$\phi_r = f_r$$
$$\phi_{rr} = f_{rr}$$
$$\phi_{r\theta} = f_{r\theta} - f'_{r\theta}$$
$$\phi_\theta = f_\theta - f'_{\theta\theta}$$
$$\phi_{\theta\theta} = f_{\theta\theta} - f'_{\theta\theta} \tag{A6.28}$$

The symmetry coordinates are derived, as was mentioned above, using standard group theoretical methods. If the molecular point group has no degenerate representations, this is a very simple procedure and one merely needs to have, in addition to the group character table, a list of the transformations which any chosen internal coordinate undergoes under the application of the group symmetry elements. If the operations of the group are denoted by O_i and the characters for an irreducible representation Γ are denoted by χ_i^{Γ} then the symmetry coordinates are generated by

$$S^{\Gamma}(R) = \eta \sum_i \chi_i^{\Gamma}(O_i R) \tag{A6.29}$$

where η is a normalising coefficient. The routine then is to take one of each symmetrically equivalent set and to apply equation (A6.29) in turn and for each irreducible representation known to contain a genuine vibration. If this is done for the XY_4 molecule, whose point group T_d *does* contain degenerate representations, then only the following set is obtained:

$$S^{A_1}(r) = \tfrac{1}{2}[r_1 + r_2 + r_3 + r_4]$$

$$S^E(\theta) = \frac{1}{\sqrt{12}}[2\theta_{12} + 2\theta_{34} - \theta_{13} - \theta_{14} - \theta_{23} - \theta_{24}]$$

$$S^{F_2}(r) = \frac{1}{\sqrt{12}}[3r_1 - r_2 - r_3 - r_4]$$

$$S^{F_2}(\theta) = \frac{1}{\sqrt{2}}[\theta_{12} - \theta_{34}] \tag{A6.30}$$

in addition to the redundant coordinate $S^{A_1}(\theta) = 1/\sqrt{6}[\theta_{12} + \theta_{13} + \theta_{14} + \theta_{23} + \theta_{24} + \theta_{34}]$. Finding the remaining five coordinates is a rather fiddly operation. Fortunately means are available (see Wilson, Decius and Cross [486] for details) for arriving at the fully factored matrices† knowing only the primitive set (A6.30). The results for XY_4 are

A_1	$\mathscr{G}_{11} = \mu_Y$	$\mathscr{F}_{11} = \phi_r + 3\phi_{rr}$ E	$\mathscr{G}_{22} = 3\mu_Y \; \mathscr{F}_{22} = \phi_\theta - 2\phi_{\theta\theta}$
F_2	$\mathscr{G}_{33} = \mu_Y + (4/3)\mu_X$	$\mathscr{G}_{34} = -(8/3)\mu_X$	$\mathscr{G}_{44} = 2\mu_Y + (16/3)\mu_X$
	$\mathscr{F}_{33} = \phi_r - \phi_{rr}$	$\mathscr{F}_{34} = \phi_{r\theta}$	$\mathscr{F}_{44} = \phi_\theta \qquad \text{(A6.31)}$

Taking as an example the $C^{35}Cl_4$ molecule which has $\phi_r = 3\cdot157$, $\phi_{rr} = 0\cdot415$, $\phi_\theta = 0\cdot420$, $\phi_{r\theta} = 0\cdot406$, $\phi_{\theta\theta} = 0\cdot040$, all in units of 10^2 N m^{-1}, the calculated frequencies are

A_1	$\nu_1 = 14\cdot11$ THz,	i.e. 462 cm^{-1}
E	$\nu_2 = 6\cdot67$ THz,	i.e. 222 cm^{-1}
F_2	$\nu_3 = 23\cdot30$ THz,	i.e. 777 cm^{-1}
F_2	$\nu_4 = 9\cdot35$ THz,	i.e. 312 cm^{-1}

These results agree exactly with the experimental values (Fig. 5.12) corrected for anharmonicity. Unfortunately, this is a necessity since the only way we have for deriving the force constants is to use the observed frequencies as input! Even so one does not usually have enough information and it is necessary to find additional data such as the vibrational frequencies of

---

† The rows and columns of G corresponding to the redundant coordinate will contain only zero entries. This is a useful check that one has carried out the algebraic manipulations correctly.

isotopically substituted variants, vibrational amplitudes, Coriolis splitting coefficients, vibrational intensities etc. before one can completely define the force field. Despite this complication, there is enough internal consistency in the theory and its practical applications for us to believe that the simple harmonic approach, as detailed above, is an excellent first approximation. As just one example one might consider the vibration frequencies of isotopic molecules. It follows immediately from equation (A6.21), that

$$\Pi\omega_i^2 = |GF| = |G|\,|F| \tag{A6.32}$$

and from the symmetry arguments developed above it will be realised that this equation, in fact, applies to each symmetry species separately. Now, the F matrix, depending as it does solely on electronic properties, will be an isotopic invariant, so if one were to assemble the products of the squares of the frequencies for two different isotopic species, but for the same symmetry type and take their ratio, this ratio would depend only on the masses and would therefore be a known quantity. This result is usually known as the product rule. If it is applied to the raw frequencies it gives fair agreement, but for those cases where, because sufficient overtone and combination bands are known, the equilibrium frequencies themselves can be used, the results are always excellent. The product rule therefore gives solid support to the conceptual framework used to treat molecular vibrations, but the other side of the coin is that because of it the amount of extra information available from isotopic studies is limited. Thus if one had determined the eight frequencies of two different isotopic variants of carbon tetrachloride, one would in fact have only five independent pieces of information—only just enough to determine the force constants.

It was mentioned above that the use of symmetry coordinates was originally introduced as an essential aid to computation. This is nowadays no longer a pressing requirement since modern high-speed digital computers can not only assemble G and F from the raw data, they can also diagonalise the product to give the vibration frequencies and all this in an almost unbelievably short time. Even for large molecules of high symmetry, the straightforward "brute-force" method will give the answers considerably faster. However, the symmetry coordinate method, though now slower, does give far more insight into the physics of the situation and it does give the important polarisation information which is very valuable in practical spectroscopy. There will probably be room for both approaches for some time to come. The symmetry coordinate method will be used for in-depth studies of small highly symmetrical molecules. The internal coordinate (or for that matter even the cartesian coordinate) method will be used in conjunction with roughly transferrable force constants to calculate a spectrum whose comparison with the actual spectrum will give much help in the assignment of the latter. This approach may well turn out to be indispensable when vibrational spectroscopy is used to investigate large biological molecules.

References

[1] SI: the International System of Units (translation of the International Bureau of Weights and Measures publication "Le Système international d'unites") (Eds C. H. Page and P. Vigoureux), HMSO, 3rd Edition (1977), ISBN 0.11.480056.

[2] D. J. E. Knight and W. R. C. Rowley, Survey Review XXIV, 185 (1977).

[3] E. R. Cohen and B. N. Taylor, CODATA task group on Fundamental Constants, CODATA Bulletin No. 11 (1973), *J. Phys. Chem. Ref. Data* **2** (4), 663 (1973).

[4] R. A. Smith, F. E. Jones and R. P. Chasmar, "The Detection and Measurement of Infrared Radiation", First Edition (1957) Revised Second Edition (1968), Oxford University Press.

[5] L. J. Bellamy "The Infrared Spectra of Complex Molecules", Methuen, London (1958).

[6] C. G. Granquist, *Appl. Opt.* **20**, 2606 (1981).

[7] John Chamberlain, "The Principles of Interferometric Spectroscopy" (completed, collated and edited by G. W. Chantry and N. W. B. Stone), John Wiley, Chichester, UK (1979).

[8] G. W. Chantry, "Submillimetre Spectroscopy", Academic Press, London and New York (1971).

[9] For an account see Reference 15 pp. 413–415.

[10] R. F. Hoskins, "Generalised Functions", Ellis Horwood, Chichester, UK (1979).

[11] R. Bracewell, "The Fourier Transform and Its Applications", McGraw Hill, New York (1965).

[12] J. Connes, Thesis, University of Paris, 7 October 1960, subsequently published as a special issue of *Rev. Opt. Théor. Instrum.* serie, A No. 3579, No. d'Ordre 4451 and in *Rev. Opt. Théor. Instrum.* **40**, 45, 116, 171 and 231 (1961). Mme Connes' thesis was also issued as a US Air Force Special Report, AFCRL Rep No. 11471–0019.

[13] A. H. Filler, *J. Opt. Soc. Am.* **54**, 762 (1964).

[14] R. H. Norton and R. Beer, *J. Opt. Soc. Am.* **66**, 259 (1976).

[15] G. Arfken, "Mathematical Methods for Physicists", Academic Press, 2nd Edition, New York and London (1970).

[16] R. B. Blackman and J. W. Tukey, "The Measurement of Power Spectra from the Point of View of Communication Engineering" Dover, New York (1959).

[17] For good introductions to information theory see L. Brillouin, "Science and Information Theory", 2nd Edition, Academic Press, New York (1962), and Shannon and Weaver "The Mathematical Theory of Information", University

of Illinois Press, Urbana, Illinois (1949). A good modern article is by A. D. Wyner, *Proc. IEEE*, **69**(2), 239 (1981).

[18] M. S. Bartlett, "An Introduction to Stochastic Processes", 2nd Edition, Cambridge University Press (1966).

[19] J. K. Kauppinen, D. J. Moffatt, H. H. Hantsch and D. G. Cameron, *Appl. Spectrosc.* **35**, 211 (1980); J. K. Kauppinen, D. J. Moffatt, D. G. Cameron and H. H. Hantsch, *Appl. Opt.* **20**, 1866 (1981).

[20] For good general accounts of maximum entropy methods see D. G. Childers, "Modern Spectrum Analysis", IEEE Press, New York (1978); T. E. Barnard, Rep. TR−75−01, Texas Instruments, June (1975); L. G. Griffiths and R. Prieto−Diaz, *Trans. GeoSci. Electron.* **GE-15**, 13 (1977); T. E. Landers and R. T. Lacoss, *IEEE Trans. GeoSci. Electron.* **GE-15**, 26 (1977).

[21] "Information Theory and Statistical Mechanics" E. T. Jaynes, Papers I and II, *Phys. Rev.* **106**, 620 (1957) and **108**, 171 (1957); see also E. T. Jaynes New engineering applications of information theory, *in* "Proceedings of the First Symposium on Engineering Applications of Random Function Theory and Probability" (Eds J. R. Bogdanoff and F. Kozin), pp. 163−203, John Wiley, New York (1963).

[22] R. T. Lacoss, *Geophysics*, **36** (1971).

[23] G. White, M.Sc. Thesis, Jodrell Bank, University of Manchester (1973); an account of the non-linear theory can be found in C. H. Chen, *Proc. IEEE*, **69**, 839 (1981); for a criticism of the MEM see M. A. Fiddy and T. J. Hall, *J. Opt. Soc. Am.* **71**, 1406 (1981).

[24] V. Cappellini, A. G. Constantides and P. Emiliani, "Digital Filters and their Applications", Academic Press, New York and London (1978), ISBN 0−12−159250−2.

[25] G. Winnewisser, E. Churchwell and C. M. Walmsley, Astrophysics of inter-stellar molecules, *published in* "Modern Aspects of Microwave Spectroscopy" (Ed. G. W. Chantry), Academic Press, New York and London (1979).

[26] H. P. Baltes, *Infrared Phys.* **16**, 1 (1976).

[27] J. R. Benoit, C. Fabry and A. Perot, *Trav. Mém. Bur. Int. Poids Mes.* **15**, 1 (1913); see also, H. Barrell *Proc. R. Soc.* **A186**, 164 (1946).

[28] R. Y. Chiao and B. P. Stoicheff, *J. Opt. Soc. Am.* **54**, 1286 (1964).

[29] E. A. Baker and B. Walker, *J. Phys. E.* (*Sci. Instrum.*) (1982).

[30] A. G. Fox and T. Li, *Bell. Syst. Tech. J.* **40**, 453 (1961); G. D. Boyd and J. P. Gordon, *Bell. Syst. Tech. J.* **40**, 489 (1961).

[31] H. Kogelnik and T. Li. *Proc. IEEE*, **54**, 1312 (1966); see also P. W. Smith, Single-frequency lasers *in* "Lasers" (Eds A. K. Levine and A. J. De Maria), Vol. 4, Marcel Dekker, New York (1976).

[32] A. F. Harvey, "Coherent Light", Wiley Interscience, London and New York (1970).

[33] L. A. Weinstein, "Open Resonators and Open Waveguides", Golan Press Boulder Colorado (1969).

[34] G. Goubau and F. Schwering, *IRE Trans.* **AP-9**, 248 (1961).

[35] D. H. Martin and J. Le Surf, *Infrared Phys.* **18**, 405 (1978).

[36] A. Abraham and S. D. Smith, *J. Phys. E.* (*Sci. Instrum.*), **15**, (1982).

[37] A. L. Cullen and P. Nagenthiram, Millimetre open resonators, *in* "High Frequency Dielectric Measurement" (Eds J. Chamberlain and G. W. Chantry), IPC Science and Technology Press, Guildford, England (1973).

[38] R. N. Clarke and C. B. Rosenberg, *J. Phys. E.* (*Sci. Instrum.*), **15**, (1982).

[39] A. Crocker, H. A. Gebbie, M. F. Kimmitt and L. E. S. Mathias, *Nature* (*London*) **201**, 250 (1964); H. A. Gebbie, N. W. Stone and F. D. Findlay, *Nature* (*London*) **202**, 685 (1964).

[40] See for example F. K. Kneubuhl and E. Affolter, Infrared and submillimetre-wave waveguides, *in* "Infrared and Millimeter Waves", Vol. 1. "Sources of Radiation" (Ed. K. J. Button), Chapter 6, Academic Press, New York and London (1979).

[41] P. Belland, D. Veron and L. B. Whitbourn, *J. Phys. D: Appl. Phys.* **8**, 2113–2122 (1975).

[42] D. T. Hodges and T. S. Hartwick, *Appl. Phys. Lett.* **23**, 252–253 (1973).

[43] R. H. Dicke, P. J. E. Peebles, P. G. Roll and D. T. Wilkinson, *Astrophys. J.* **142**, 414 (1965).

[44] S. Weinberg "The First Three Minutes", Andre Deutch, London (1977).

[45] A. A. Penzias and R. W. Wilson, *Astrophys. J.* **142**, 419 (1965).

[46] D. P. Woody, J. C. Mather, N. S. Nishioka and P. L. Richards *Phys. Rev. Lett.* **34**, 1036 (1975); see also E. I. Robson, D. G. Vickers, J. S. Huizinger, J. E. Beckman and P. E. Clegg, *Nature* **251**, 591 (1974); also article by P. E. Clegg *in* "Infrared Astronomy" (Eds G. Selti and G. G. Fazio), D. Reidel, North Holland (1978).

[47] The phenomenon was apparently first reported by C. H. Townes and A. C. Cheung for cosmic formaldehyde in *Astrophys. J. Lett.* **157**, L103 (1969).

[48] This work is lucidly reviewed by P. C. W. Davies *in Rep. Prog. Phys.* **41** (8), 1313 (1978).

[49] S. W. Hawking, *Nature* (London) **248**, 30 (1975); *Commun. Math. Phys.* **43**, 199 (1975); but also see reference [48] for an illuminating discussion.

[50] J. D. Barrow and F. J. Tipler, *Nature* (*London*) **276**, 453 (1978).

[51] J. E. Beckman and A. F. M. Moorwood, Infrared astronomy *published in* Rep. Prog. Phys. **42**, 87 (1979).

[52] J. W. Fleming and G. W. Chantry, *IEEE Trans. Instrum. Meas.* **IM-23**, 473 (1974).

[53] H. A. Willis and M. E. A. Cudby, private communication.

[54] J. E. Harries *J. Opt. Soc. Am.* **67**, 880 (1977).

[55] E. B. Wilson Jr. and A. J. Wells, *J. Chem. Phys.* **14**, 578 (1946).

[56] Th. Encrenaz, *Infrared Phys.* **19**, 353–373 (1980).

[57] J. Hansen, D. Johnson, A. Lacis, S. Lebedeff, P. Lee, D. Rind and G. Russell, *Science* **213**, 957 (1981).

[58] A. Shkolnik, C. R. Taylor, V. Finch and A. Borat, *Nature* (*London*) **283**, 373 (1980); R. Dmiel, A. Prevulotzky and A. Shkolnik, *Nature* (*London*) **283**, 761 (1980).

[59] H. M. Nuzzenzveig, "Introduction to Quantum Optics", *in* the series "Documents on Modern Physics", Gordon and Breach, London and New York (1975); G. J. Troup and R. G. Turner, Optical coherence theory, *Rep. Prog. Phys.* **37**, 771 (1974).

[60] M. Born and E. Wolf, "Principles of Optics", Pergamon Press, Oxford (1959).

[61] M. J. Halmos and J. Shamir, *Appl. Opt.* **21**, 265 (1982).

[62] G. W. Chantry and John Chamberlain, Far infrared spectra of polymers, *in* "Polymer Science", a Materials Science Handbook (Ed. A. D. Jenkins), Vol. 2, North-Holland, Amsterdam (1972).

[63] L. Jannossy, *Nuovo Cim.* **6**, 111 (1957); ibid. **12**, 369 (1959).

[64] A. A. Michelson, "Light Waves and their Uses", University of Chicago Press (1902); "Studies in Optics", Phoenix Edition, University of Chicago Press (1962).

[65] L. Mandel and E. Wolf, *J. Opt. Soc. Am.* **51**, 815 (1961).

[66] R. H. Brown and R. Q. Twiss, *Proc. R. Soc.* A242, 300 (1957); ibid. A243, 291 (1957); ibid. A248, 199 (1958).

[67] P. W. Smith, *IEEE J. Quant. Elect.* **QE-1** (8), 343 (1965).

[68] M. Yamanaka, H. Yoshinaga and S. Kon. *Jap. J. Appl. Phys.* **7**, 827 (1967).

[69] W. E. Lamb, *Phys. Rev.* **134**, A1429 (1964); the original "hole-burning" concept is due to W. R. Bennett, *Phys. Rev.* **126**, 580 (1962).

[70] C. K. N. Patel and E. D. Shaw, *Phys. Rev. Lett.* **24**, 451 (1970).

[71] O. R. Wood and S. E. Schwarz, *Appl. Phys. Lett.* **11**, 88 (1967).

[72] R. G. Jones, C. C. Bradley, J. Chamberlain, H. A. Gebbie, N. W. B. Stone and H. Sixsmith, *Appl. Opt.* **8**, 701 (1969).

[73] C. K. N. Patel, Gaseous optical masers, *in* "Lasers and Applications", Chapter 2, Ohio State University (1963); Gas Lasers *in* "Lasers" (Eds A. K. Levine and A. J. De Maria), Vol. 2, Marcel Dekker, New York (1971).

[74] T. H. Maiman, *Phys. Rev. Lett.* **4**, 564 (1960).

[75] J. C. Polanyi, *J. Chem. Phys.* **34**, 347 (1961); J. V. V. Kasper and G. C. Pimentel, *Phys. Rev. Lett.* **14**, 352 (1965); P. J. Kuntz, E. M. Nemeth and J. C. Polanyi, *J. Chem. Phys.* **50**, 4607 (1969); P. J. Kuntz, M. H. Mok and J. C. Polanyi, *J. Chem. Phys.* **50**, 4623 (1969).

[76] K. L. Kompa and G. C. Pimentel, *J. Chem. Phys.* **47**, 857 (1967); see also G. C. Pimentel, Infrared study of transient molecules in chemical lasers, *in* "Molecular Spectroscopy IX", General Lectures presented at the Ninth European Molecular Spectroscopy Congress, Butterworth's (London) (1969).

[77] A. J. De Maria and C. J. Ultee, *Appl. Phys. Lett.* **9**, 67 (1966).

[78] B. Lax, Progress in semi conductor lasers, *in* IEEE Spectrum, p. 62 (1965); see also J. F. Butler, Semiconductor diode lasers *in* "Applied Optics and Optical Engineering" (Eds R. Kingslake and B. J. Thompson), Vol. 6, Academic Press, New York and London (1980); E. D. Hinkley and P. L. Kelley, *Science* **171**, 635 (1971); E. D. Hinkley, *Appl. Phys. Lett.* **13**, 49 (1968).

[79] G. W. Chantry and G. Duxbury, Molecular lasing systems, *in* "Methods of Experimental Physics", Vol. III, "Molecular Spectroscopy" (Ed. D. Williams), Academic Press, New York and London (1974).

[80] D. R. Lide and A. G. Maki, *Appl. Phys. Lett.* **11**, 2 (1967).

[81] L. O. Hocker and A. Javan, *Phys. Lett.* **25A**, 489 (1967.

[82] B. Hartman and B. Kleman, *Appl. Phys. Lett.* **12**, 168 (1968); W. S. Benedict, ibid. p. 170; M. A. Pollack and W. J. Tomlinson, ibid. p. 173; W. S. Benedict, M. A. Pollack and W. J. Tomlinson, *IEEE J. Quant. Electron.* **QE5**, 108 (1969).

[83] M. A. Pollack, *IEEE. J. Quant. Electron.* **QE5**, 558 (1969).

[84] J. Reid and K. Siemsen, *Appl. Phys. Lett.* **29**, 250 (1976); K. J. Siemsen and J. Reid, *Opt. Commun.* **20**, 284 (1977).

[85] M. Yamanaka, *Rev. Laser Engng* **3**, 253 (1976); M. Rosenbluh, R. J. Temkin and K. J. Button, *Appl. Opt.* **15**, 2635 (1976); J. J. Gallagher, M. D. Blue, B. Bean and S. Perkowitz, *Infrared Phys.* **17**, 43 (1977); D. T. Hodges and D. J. E. Knight, "Laser Handbook", to be published.

[86] T. Y. Chang and T. J. Bridges, *Opt. Commun.* **1**, 423 (1970); T. Y. Chang, T. J. Bridges and E. G. Burkhardt, *Appl. Phys. Lett.* **17**, 249 (1970).

[87] T. Oka, *Adv. Atom. Mol. Phys.* **9**, 127 (1974).

[88] T. Y. Chang and J. D. McGee, *Appl. Phys. Lett.* **29**, 725 (1976); D. G. Biron, R. J. Temkin, B. Lax and B. G. Danly, *Opt. Lett.* **4**, 381 (1979).

[89] J. C. Slater, "Microwave Electronics", Van Nostrand, New York (1950).

[90] J. M. Manley and H. E. Rowe, *Proc. IRE* **47**, 2115–2116 (1959).

[91] J. R. Birch, *Electron. Lett.* **16**, 799 (1980).

[92] C. H. Townes and A. L. Schawlow, "Microwave Spectroscopy", McGraw Hill, New York (1955); W. Gordy, Microwave spectroscopy, *in* "Handbook of Physics", 2nd Edition (Eds E. U. Condon and H. Odishaw), Chapter 6, McGraw Hill, New York (1967); W. Gordy and R. L. Cook, "Microwave Molecular Spectra", Wiley Interscience, New York (1970); for further re-

ferences see G. Roussy and G. W. Chantry, Microwave spectrometers, *in* "Modern Aspects of Microwave Spectroscopy" (Ed. G. W. Chantry), Academic Press, London and New York (1979).

[93] See for example T. Oka, Infrared and radio frequency spectroscopy in the laser cavity, *in* "Frontiers in Laser Spectroscopy" (Eds Balian *et al.*). North Holland, Amsterdam (1977).

[94] H. Hellwig, "Frequency Standards and Clocks", NBS Technical Note 616 (2nd Revision); Atomic Frequency Standards Survey, *Proc. IEEE* **63**, 212 (1975); C. Audoin and J. Vanier, Atomic frequency standards and clocks. *J. Phys. E. (Sci. Instrum.)* **9**, 697 (1976).

[95] A. F. Pearce and D. J. Wootton, Reflex Klystrons, *in* "Millimetre and Submillimetre Waves" (Ed. F. A. Benson), Iliffe, London (1969).

[96] W. Gordy, *Chim. Pure Appliquee* **11**, 403 (1965); see also as an example, F. C. De Lucia, D. Helminger and W. Gordy, *Phys. Rev.* **A3**, 1849 (1971).

[97] H. Kilp, *J. Phys. E (Sci. Instrum.)*, **10**, 985 (1977).

[98] Ph. Helminger, F. C. De Lucia and W. Gordy, *Bull. Am. Phys. Soc.* **16**, 531 (1971).

[99] G. Kantorowicz and P. Palluel, Backward-wave oscillators, *in* "Infrared and Millimetre Waves" (Ed. K. J. Button), Vol. 1, Academic Press, New York and London (1979).

[100] A. F. Krupnov and A. V. Burenin, *in* "Molecular Spectroscopy: Modern Research" (Ed. K. N. Rao), Vol. II, pp. 93–126, Academic Press, New York and London (1976).

[101] A. F. Krupnov, Modern submillimetre microwave-scanning spectroscopy, *in* "Modern Aspects of Microwave Spectroscopy" (Ed. G. W. Chantry), Academic Press, London, New York (1979).

[102] C. E. Cleeton and N. H. Williams, *Phys. Rev.* **45**, 234 (1934).

[103] R. Q. Twiss, *Aust. J. Phys.* **11**, 424 and 564 (1958).

[104] J. Schneider, *Phys. Rev. Lett.* **2**, 504 (1959).

[105] A. V. Gaponov, Addendum, *Izv. Vyssh. Uchebn. Zaved. Radiofiz.* **2**, 450, 836 (1959).

[106] See for example A. A. Andronov, V. A. Flyagin, A. V. Gapanov, A. L. Goldenberg, M. I. Petelin, V. G. Usov and V. K. Yulpatov, *Infrared Phys.* **18**, 385 (1978).

[107] D. V. Kisel, G. S. Korablev, V. G. Navel'yev, M. I. Petelin and Sh. Ye Tsimring, *Radiotekh. Elektron.* **19**, 782 (1974). (English translation, *Radio Eng. Electron. Phys.* **19**, 781 (1974).

[108] R. M. Gilgenbach, M. E. Read, K. E. Hackett, R. Lucey, B. Hui, V. L. Granatstein, K. R. Chu, A. C. England, C. M. Loring, O. C. Eldridge, H. C. Howe, A. G. Kulchar, E. Lazarus, M. Marakami and J. B. Wilgen, *Phys. Rev. Lett.* **44**, 647 (1980).

[109] V. A. Flyagin, A. V. Gaponov, M. I. Petelin and V. K. Yulpatov, *IEEE Trans. Microwave Theory Tech*, **MTT-25**, 514 (1977).

[110] J. L. Hirshfield and V. L. Granatstein, The electron cyclotron maser—an historical survey, *IEEE Trans. Microwave Theory Tech.* **MTT-25**, 522 (1977).

[111] J. L. Hirshfield, Gyrotrons, *in* "Infrared and Millimeter Waves" (Ed. K. J. Button), vol. I, pp. 1–54, Academic Press, New York and London (1979).

[112] K. R. Chu, Y. Y. Lau, L. R. Barnett and V. L. Granatstein, *IEEE, Trans. Electron Devices*, **ED-28**, 866 (1981); L. R. Barnett, Y. Y. Lau, K. R. Chu and V. L. Granatstein, ibid. p. 872.

[113] P. Lagarde, *Infrared Phys.* **18**, 395 (1978): see also "Handbook on Synchrotron Radiation", in four volumes (series editors D. E. Eastman and Y. Farge), North-Holland, Amsterdam and New York (1982).

[114] F. E. Close, The quark-parton model, *Rep. Prog. Phys.* **42**, 1285 (1979).

[115] J. H. Poole, Daresbury Technical Memorandum DL/SRF/TM4 (Revised). Science Research Council, Daresbury Laboratory, Warrington, UK (1978); P. Meyer and P. Lagarde, *J. Phys.* **37**, 1387 (1976).

[116] P. Sprangle, R. A. Smith and V. L. Granatstein, Free electron lasers and stimulated scattering from relativistic electron beams, *in* "Infrared and Millimeter Waves", (Ed. K. J. Button), Vol. 1, Academic Press, New York and London (1979).

[117] P. L. Kapitza and P. A. M. Dirac, *Proc. Camb. Phil. Soc.* **29**, 297 (1933).

[118] L. R. Elias, W. M. Fairbank, J. M. J. Madey, H. A. Schwettman and T. I. Smith, *Phys. Rev. Lett.* **36**, 717 (1976).

[119] D. A. G. Deacon, L. R. Elias, J. M. J. Madey, G. J. Ramian, H. A. Schwettman and T. I. Smith, *Phys. Rev. Lett.* **38**, 892 (1977).

[120] V. L. Granatstein, S. P. Schlesinger, M. Herndon, R. K. Parker and J. A. Pasour, *Appl. Phys. Lett.* **30**, 384 (1977).

[121] S. J. Smith and E. M. Purcell, *Phys. Rev.* **92**, 1069 (1953).

[122] R. S. Rusin and G. D. Bogomolov, *JETP Lett*, **4**, 160 (1966).

[123] R. P. Leavitt, D. E. Wortman and C. A. Morrison, *Appl. Phys. Lett.* **35**, 363 (1979).

[124] K. Mizuno, S. Ono and Y. Shibata, *IEEE Trans. Electron Devices* **ED-20**, 749 (1973).

[125] R. P. Leavitt, D. E. Wortman and H. Dropkin, *IEEE J. Quant. Electron.* **QE-17**, 1333 (1981); D. E. Wortman, H. Dropkin and R. P. Leavitt, ibid. 1341 (1981).

[126] J. M. Wachtel, *J. Appl. Phys.* **50**, 49 (1979).

[127] D. H. Martin and K. Mizuno, *Adv. Phys.* **25**, 211, (1976): this review also covers many other "electronic" sources of infrared and submillimetre-wave radiation.

[128] M. P. Shaw, H. L. Grubin and P. R. Solomon, "The Gunn-Hilsum Effect", Academic Press, London and New York (1979).

[129] H. J. Kuno, IMPATT devices for the generation of millimeter waves *in* "Infrared and Millimeter Waves", (Ed. K. J. Button), Vol. 1, Academic Press, New York and London (1979).

[130] J. Nishizawa, 5th International Conference on Infrared and MM-Waves, FRG, Digest, p. 201 (1980).

[131] S. Ahmad and J. Freyer, *IEEE Trans. Electron Devices* **ED-26**, 1370 (1979).

[132] T. Oka, "Molecular spectroscopy using infrared lasers; a study of radiative and collisional processes", published in Horizons of Quantum Chemistry (Eds K. Fukui and B. Pullman), pp. 151–167, D. Reidel (1980).

[133] J. M. Stone, "Radiation and Optics", pp. 252ff, McGraw-Hill Book Company, New York (1963).

[134] J. H. Van Vleck and V. F. Weisskopf, *Rev. Mod. Phys.* **17**, 227 (1945).

[135] C. Brot, *Phys. Lett.* **30A**, 101 (1969).

[136] J. Cuthbert and E. J. Denney, "Application of microwave spectroscopy to chemical analysis", published in Molecular Spectroscopy, Institute of Petroleum (London), (1971).

[137] R. Karplus and J. Schwinger, *Phys. Rev.* **73**, 1020 (1948); see also H. S. Snyder and P. I. Richards, *Phys. Rev.* **73**, 1178 (1948).

[138] H. Jones, Infrared-microwave double-resonance techniques, *in* "Modern Aspects of Microwave Spectroscopy" (Ed. G. W. Chantry), Chapter 3, Academic Press London and New York (1979).

[139] L. Allen and J. H. Eberley, "Optical Resonance and Two-level Atoms", Wiley Interscience (1975).

[140] I. I. Rabi, *Phys. Rev.* **51**, 652 (1937).

[141] A. Javan, *Phys. Rev.* **107**, 1579 (1957).

[142] T. Shimizu and T. Oka, *Phys. Rev.* **A2**, 1177 (1970): T. Yajima, *J. Phys. Soc. Japan,* **16**, 1594 (1961); M. Takami, *Jap. J. Appl. Phys.* **15**, 1063, 1889 (1976); ibid. **17**, 125 (1977).

[143] A. Eyer and H. Jones, *J. Mol. Spectrosc.* **52**, 420 (1974).

[144] J. G. Baker, Microwave-microwave double resonance, *published in* "Modern Aspects of Microwave Spectroscopy" (Ed. G. W. Chantry), Academic Press, New York and London (1979).

[145] P. Glorieux, J. Legrand, B. Macke and J. Messelyn, *J. Quant. Spectrosc. Radiat. Transfer* **12**, 731 (1972).

[146] F. Bloch, *Phys. Rev.* **70**, 460 (1946).

[147] S. L. McCall and E. L. Hahn, *Phys. Rev. Lett.* **18**, 908 (1967); *Phys. Rev.* **183**, 457 (1969); ibid. **2A**, 861 (1970).

[148] R. G. Brewer, A. Z. Genack and S. B. Grossman, Coherent transients and pulse Fourier transform spectroscopy, *published in* "Laser Spectroscopy III" (Eds J. L. Hall and J. L. Carlsten), Springer Verlag Berlin, Heidelberg, New York (1977); R. G. Brewer, Coherent Optical Transients, *published in* "Coherence in Spectroscopy and Modern Physics" (Eds F. T. Arecchi, R. Bonifacio and M. O. Scully), Plenum Press, New York and London (1978).

[149] H. M. Gibbs and R. E. Slusher, *Appl. Phys. Lett.* **18**, 505 (1971).

[150] G. L. Lamb, Jr., *Phys. Lett.* **25A**, 181 (1967); *Rev. Mod. Phys.* **43**, 99 (1971); see also T. W. Barnard, *Phys. Rev.* **A7**, 373 (1973).

[151] F. H. Read, "Electromagnetic Radiation", John Wiley, Chichester, UK (1980).

[152] R. W. Ditchburn, "Light" (3rd Edition), Appendix 6D.1, Academic Press, London and New York (1976).

[153] F. Grum and R. J. Becherer, "Optical Radiation Measurements": Vol. 1 "Radiometry", Academic Press, New York and London (1979).

[154] T. Pearcey, "Tables of Fresnel Integrals to Six Decimal Places", Cambridge University Press (1956)

[155] H. E. Bennett and J. O. Porteus, *J. Opt. Soc. Am.* **51**, 123 (1961); J. O. Porteus, ibid. **53**, 1394, (1963).

[156] R. Kompfner, *Appl. Opt.* **11**, 2412 (1972).

[157] P. K. Yu and A. L. Cullen, *Proc. R. Soc. Lond.* **A380**, 49 (1982).

[158] J. A. Arnaud, "Beam and Fiber Optics", Academic Press, New York and London (1976).

[159] J. S. Toll, *Phys. Rev.* **104**, 1760 (1956); F. Stern, *Solid State Phys.* **15**, 299 (1963).

[160] E. C. Titchmarsh, "Introduction to the Theory of Fourier Integrals", Oxford University Press, Oxford (1937).

[161] J. Chamberlain, F. D. Findlay and H. A. Gebbie, *Nature (London)* **206**, 886 (1965).

[162] E. E. Bell, *Jap. J. Appl. Phys.* **4**, 412 (1965); *Infrared Phys.* **6**, 57 (1966); *Handbuch Phys.* **25**, 1 (1967); *J. Phys.* (Suppl. 3–4), **28**, C2-18, C2-25 (1967).

[163] E. E. Russell and E. E. Bell, *Infrared Phys.* **6**, 75 (1966).

[164] J. Gast and L. Genzel, *Opt. Commun.* **8**, 26 (1973).

[165] T. J. Parker and W. G. Chambers, *IEEE Trans. Microwave Theory Techn.* **MTT-22**, 1032 (1974); T. J. Parker, W. G. Chambers and J. F. Angress, *Infrared Phys.* **14**, 207 (1974); D. A. Ledsham, W. G. Chambers and T. J. Parker, ibid. **16**, 515 (1976); D. A. Ledsham, W. G. Chambers and T. J. Parker, ibid. **17**, 165 (1977).

[166] J. R. Birch, G. D. Price and J. Chamberlain, *Infrared Phys.* **16**, 311 (1976); J. R. Birch and D. K. Murray, *Infrared Phys.* **18**, 283 (1978); see also J. Chamberlain, M. N. Afsar, D. K. Murray, G. D. Price and M. S. Zafar, *IEEE Trans. Instrum. Meas.* **IM-23**, 483 (1974).

[167] P. R. Staal and J. E. Eldridge, *Infrared Phys.* **17**, 299 (1977).

[168] M. S. Zafar, J. B. Hasted and J. Chamberlain, *Nature Phys. Sci.* **243**, 106 (1973).

[169] J. Chamberlain, M. N. Afsar, J. B. Hasted, M. S. Zafar and G. J. Davies, *Nature Phys. Sci.* **255**, 319 (1975); M. N. Afsar and J. B. Hasted, *J. Opt. Soc. Am.* **67**, 902 (1977).

[170] T. S. Robinson and W. C. Price, *Proc. Phys. Soc.* **B65**, 910 (1952); T. S. Robinson and W. C. Price, ibid. **B66**, 969 (1953); see also D. M. Roessler, *Brit. J. Appl. Phys.* **16**, 1119 (1965); ibid. **17**, 1313 (1966).

[171] J. W. Fleming, D. Siapkis, J. Lewis and G. R. Wilkinson, "High Frequency Dielectric Measurement" (Eds J. Chamberlain and G. W. Chantry), pp. 122–126, IPC Press, Guildford, UK (1973).

[172] H. M. Gibbs, S. L. McCall, T. N. C. Venkatesan, A. C. Gossard, A. Passner and W. Wiegmann, *Appl. Phys. Lett.* **35**, 451 (1979).

[173] D. A. B. Miller, S. D. Smith and A. Johnston, *Appl. Phys. Lett.* **35**, 658 (1979).

[174] D. A. B. Miller, S. D. Smith and C. T. Seaton, *IEEE J. Quantum. Electron.* **QE-17**, 312 (1981); Optical Bistability (Eds C. M. Bowden, M. Ciftan and H. R. Robl), Plenum Press, New York and London (1981); see also A. Zardecki, *Phys. Rev.* **A22**, 1664 (1980); and also P. D. Drummond, K. J. McNeil and D. F. Walls, *Phys. Rev.* **A22**, 1672 (1980).

[175] D. A. B. Miller and S. D. Smith, *Opt. Commun.* **31**, 101 (1979).

[176] D. A. B. Miller, R. G. Harrison, A. M. Johnston, C. T. Seaton and S. D. Smith, *Opt. Commun.* **32**, 478 (1980).

[177] J. Chamberlain and H. A. Gebbie, *Nature (London)* **206**, 602 (1965); J. Chamberlain, E. G. C. Werner, H. A. Gebbie and W. Slough, *Trans. Faraday Soc.* **63**, 2605 (1967).

[178] E. V. Loewenstein and D. R. Smith, *Appl. Optics* **10**, 577 (1971); **12**, 398 (1973).

[179] P. J. Severin, *Appl. Opt.* **9**, 2381 (1970); P. F. Cox and A. F. Stalder, *J. Electrochem. Soc. Solid-State Sci. Technol.* **120**, 287 (1973).

[180] H. Kilp *J. Phys. E. (Scientific Instruments)* **10**, 985 (1977); H. Kilp, D. C. Barnes, F. W. J. Clutterbuck, M. N. Afsar and G. W. Chantry, *Infrared Phys.* **18**, 11 (1978).

[181] D. E. Williamson, *J. Opt. Soc. Am.* **42**, 712 (1952).

[182] H. Witte, *Infrared Phys.* **5**, 179 (1965); E. V. Loewenstein and G. Newell, *J. Opt. Soc. Am.* **59**, 407 (1969); D. H. Martin, *Infrared Physics*, **15**, 67 (1975).

[183] K. D. Moller and W. G. Rothschild, "Far Infrared Spectroscopy", pp. 68–72, Wiley Interscience, New York (1971).

[184] A. F. Harvey, "Microwave engineering", pp. 1040ff, Academic Press, New York and London (1963).

[185] E. A. J. Marcatili and R. A. Schmeltzer, *Bell System Tech. J.* **43**, 1783 (1964).

[186] H. Steffen, J. Steffen, J. F. Moser and F. K. Kneubuhl, *Phys. Lett.* **20**, 20 (1966); ibid. **21**, 425 (1966); P. Schwaller, H. Steffen, J. F. Moser and F. K. Kneubuhl, *Appl. Opt.* **6**, 827 (1967).

[187] H. Steffen and F. K. Kneubuhl, *IEEE J. Quant. Electron.* **QE-4**, 992 (1968).

[188] F. K. Kneubuhl and E. Affolter, Infrared and submillimetre-wave waveguides, *in* "Infrared and Millimetre Waves" (Ed. K. J. Button), Vol. 1, pp. 235–278, Academic Press, New York (1979).

[189] P. W. Smith, *Appl. Phys. Lett.* **19**, 132 (1971).

[190] T. J. Bridges, E. G. Burkhardt and P. W. Smith, *Appl. Phys. Lett.* **20**, 403 (1972); R. E. Jensen and M. S. Tobin, *Appl. Phys. Lett.* **20**, 508 (1972).

[191] G. Lockhard III and R. Yusek, *IEEE Electron Devices Meeting*, Washington DC, December 4–6 (1972).

[192] P. Belland, D. Veron and L. B. Whitbourn, *J. Phys. D. (Appl. Phys.)* **8**, 2113 (1975).

[193] J. J. Degnan and D. R. Hall, *IEEE J. Quant. Electron.* **QE-9**, 901 (1973); J. J. Degnan, *Appl. Phys.* **11**, 1 (1976); R. L. Abrams, Waveguide gas lasers, Chapter A2 of the "Laser Handbook" (Ed. M. L. Stitch), North Holland, Amsterdam (1979)

[194] E. Loh, *Phys. Rev.* **166**, 673 (1968).

[195] D. T. Hodges, F. B. Foote and R. E. Reel, *IEEE J. Quant. Electron.* **QE-13**, 491 (1977).

[196] F. K. Kneubuhl and E. Affolter, Distributed-feedback gas lasers, *in* "Infrared and Millimetre Waves" (Ed. K. J. Button), Vol. 6, Academic Press, New York and London (1982).

[197] F. K. Kneubuhl and E. Affolter, *J. Opt.* **11**, 449 (1980).

[198] C. V. Shank, J. E. Borkholm and H. Kogelnik, *Appl. Phys. Lett.* **18**, 395 (1971).

[199] P. Czerski and S. Baranski, "Biological Effects of Microwaves", Dowden, Hutchinson and Ross, Stroudsberg, PA (1976).

[200] P. J. B. Clarricoats, *Proc. IEE*, **108C**, 170 (1961); E. Snitzer, *J. Opt. Soc. Am.* **51**, 491 (1961); W. Schlosser and H. G. Unger *in* "Advances in Microwaves" (Ed. L. Young), Vol. 1, p. 319, Academic Press, New York and London (1966) N. S. Kapany and J. J. Burke, "Optical Waveguides", Academic Press, New York (1972); P. J. B. Clarricoats and K. B. Chan, *Proc. IEE*, **120**, 1371 (1973).

[201] D. Marcuse, "Theory of Dielectric Optical Waveguides", Academic Press, New York and London (1974).

[202] D. Marcuse, "Light Transmission Optics", Van Nostrand Reinhold, Princeton, New Jersey (1972).

[203] J. B. Keller and S. J. Rubinow, *Ann. Phys.* **9**, 24 (1960).

[204] E. G. Neumann and H. D. Rudolph, *IEEE J. Microwave Theory Tech.* **MTT-23**, 142 (1975).

[205] K. C. Kao and G. A. Hockman, *Proc. IEE* **113**, 1151 (1966); D. Gloge, *Rep. Prog. Phys.* **42**, 1777 (1979).

[206] E. A. J. Marcatili, *Bell Syst. Tech. J.* **48**, 2103 (1969); D. Gloge, *Appl. Opt.* **10**, 2252 (1971).

[207] Texas Instruments Inc. P.O. Box 225012, M/S 308, Dallas Texas 75265, USA.

[208] D. B. Keck and R. E. Love, Fiber optics for communications, *in* "Applied Optics and Optical Engineering" (Ed. R. Kingslake and B. J. Thompson), Vol. 6, Academic Press, London and New York (1980).

[209] J. E. Midwinter, "Optical Fibers for Transmission", John Wiley, (1979); see also C. P. Sandbank, Optical Fibre Communication Systems, John Wiley (1980), and Optical Fibre Communications, Vol. 4, Ed. M. J. Howes and D. V. Morgan, John Wiley, Chichester and New York (1980).

[210] S. E. Miller and A. G. Chynoweth, Optical Fiber Telecommunications, Academic Press, London and New York (1979).

[211] P. C. Schultz, Progress in Optical Waveguide processes and materials. Proceeding of the Optical Society of America Topical Meetings on Optical Fiber Communication, Washington DC, 6–8 March (1979); D. Charlton and P. C. Schultz, Electro Optical Systems Design, December 1980, pp. 23–29; M. R. Montierth, *J. Electron. Mat.* **6**, 271 (1977).

[212] J. LeSergent, Brit. Pat. Spec. 1554978, 31 Oct. (1979).

[213] T. Moriyama, O. Fukuda, K. Sanada and S. Tanaka, Fabrication of ultra-low OH$^-$ optical fibers with the VAD method. Proceedings of the Fifth European Conference on Optical Communications, York, England, 16–19 September (1980).

[214] K. J. Beals and C. R. Doug, Physics and Chemistry of Glasses, **21** (1980); R. Olshansky, *Rev. Mod. Phys.* **51**, 308 (1979).

[215] C. P. Sandbank, *Electl Commun.* **50**, 10 (1975).

[216] C. H. L. Goodman, *Solid-St. Electron Devices* **2**, 129 (1978).

[217] A. G. Steventon, R. E. Spillett, R. E. Hobbs, M. G. Burt, P. J. Fiddyment and J. V. Collins, *IEEE J. Quant. Electron.* **QE-17**, 602 (1981).

[218] D. R. Smith, R. C. Hooper and I. Garrett, *Opt. Quant. Electron.* **10**, 293 (1978).

[219] D. R. Smith, A. K. Chatterjee, M. A. Z. Rejman, D. Wake and B. R. White, *Electronics Lett.* **16**, 750 (1980).

[220] N. Susa, H. Nakagome, O. Mikami, H. Ando and H. Kanbe, *IEEE, J. Quant. Electron.* **QE-16**, 864 (1980).

[221] J. P. Gordon, Optics of general guiding media, *Bell Syst. Techn.* **15**, 321 (1966); R. Olshansky, *Appl. Opt.* **15**, 782 (1976).

[222] P. C. Hensel, *Electronics Lett.* **13**, 734 (1977); see also M. J. Adams, "An Introduction to Optical Waveguides", John Wiley, Chichester and New York (1981), Chapter 8.

[223] P. R. Cooper, J. S. Leach, A. B. Harding and M. A. Mathews, *Laser Technol.* **14**(2), 87, April (1982).

[224] British Telecom Research Laboratories, Optical Fibre Systems and Components, publicity leaflet, BPO Tel Consult, Room 202, Lutyens House, Finsbury Circus, London EC2M 7LY.

[225] M. Eve, P. C. Hensel, D. J. Malyon, B. P. Nelson, J. R. Stern, J. V. Wright and J. E. Midwinter, *Opt. Quant. Elec.* **10**, 253 (1978); J. E. Midwinter and J. R. Stern, *IEEE Trans. Commun.* **COM-26**, 1015 (1978).

[226] S. E. Miller, *IEEE J. Quant. Electron.* **QE-8**, 199 (1972).

[227] T. Tamir (Ed.) Integrated Optics, Topics in *Appl. Phys.* **7**, Springer Verlag, Berlin, Heidelberg, New York (1975); see also P. K. Tien and J. A. Giordmaine, Bell Laboratories Record, December (1980), p. 371, The Proceedings of the European Conference on Integrated Optics, IEE Conference Series (1981); and H. Kogelnik, *Proc. IEEE* **69**, 232 (1981).

[228] N. J. Harrick, "Internal Reflection Spectroscopy", John Wiley, New York (1967).

[229] J. Fahrenfort, *Spectrochim. Acta* **17**, 698 (1961).

[230] P. A. Wilks, A practical approach to internal reflection spectroscopy, *in* Laboratory Methods in Infrared Spectroscopy" 2nd edition (Eds R. G. J. Miller and B. C. Stace), Chapter 14, Heyden and Son, London (1972). H. A. Willis and V. J. I. Zichy, The examination of polymer surfaces by infrared spectroscopy, *in* "Polymer Surfaces" (Ed. D. T. Clark and W. J. Feast), Chapter 15, Wiley Interscience, Chichester and New York (1978).

[231] P. A. Wilks, *Appl. Spectrosc.* **22**, 782 (1968).

[232] B. L. Crawford, T. G. Goplen and D. Swansen, The measurement of optical constants in the infrared by attenuated total reflection, in "Advances in Infrared and Raman Spectroscopy" (Eds R. J. H. Clark and R. E. Hester), Vol. 4, Heyden and Son, London (1978).

[233] R. W. Ditchburn, Light, 3rd Edition, Vol. 1, Section 5.22, pp. 122–124, Academic Press, London and New York (1976).

[234] C. S. Evans, R. Hunneman and J. S. Seeley, Proceedings of the Conference on Infrared Techniques, University of Reading, Sept (1971), IERE Conference Proceedings, No. 22, p. 125; *J. Phys. D (Appl. Phys.)* **9**, 309 (1976).

[235] J. T. Houghton and S. D. Smith, "Infrared Physics", Oxford University Press (1966).

[236] J. S. Seeley, R. Hunneman and A. Whatley, *Infrared Phys.* **19**, 429 (1979).

[237] G. Mie, *Ann. d. Physik.* **25**(4), 377 (1908), but see Born and Wolf [ref. 60], pp. 633 onwards; for a modern application see W. J. Glantschnig and S. H. Chen, *Appl. Opt.* **20**, 2499 (1981).

[238] P. C. T. Stillwell, private communication, but see the article on Thermal Imaging by this author in *J. Phys. E. (Sci. Instrum.)* **14**, 1113 (1981).

[239] G. Duyckaerts, *Analyst* **84**, 201 (1959).

[240] A. E. Costley, K. H. Hursey, G. F. Neill and J. M. Ward, *J. Opt. Soc. Am.* **67**, 979 (1977).

[241] J. P. Auton, *Appl. Optics*, **6**, 1023 (1967).

[242] T. Larsen, *IRE Trans. Microwave Theory Tech.* **MTT-10**, 191 (1962); see also R. Petit, *Nouv. Rev. Optique* **6**, 129 (1975).

[243] J. A. Beunen, A. E. Costley, C. L. Mok, G. F. Neill, T. J. Parker and G. Tait, *J. Opt. Soc. Am.* **71**, 184 (1981).

[244] F. S. Ham and B. Segall, *Phys. Rev.* **124**, 1786 (1961); see also W. G. Chambers, C. L. Mok and T. J. Parker, *J. Phys. A. Math. Gen.* **13**, 1433 (1980).

[245] C. L. Mok, W. G. Chambers, T. J. Parker and A. E. Costley, *Infrared Phys.* **19**, 437 (1979).

[246] See for example M. N. Afsar, J. B. Hasted and J. Chamberlain, *Infrared Phys.* **16**, 301 (1976).

[247] D. H. Martin and E. Puplett, *Infrared Phys.* **10**, 105 (1970).

[248] P. A. R. Ade, A. E. Costley, C. T. Cunningham, C. L. Mok, G. F. Neill and T. J. Parker, *Infrared Phys.* **19**, 599 (1979); see also ref. 240.

[249] B. Walker, E. A. M. Baker and A. E. Costley, *J. Phys. E. (Sci. Instrum)* **14**, 832 (1981).

[250] R. Ulrich, K. F. Renk and L. Genzel, *IEEE Trans* **MTT-11**, 363 (1963).

[251] R. Ulrich, *Infrared Phys.* **7**, 37 (1967); ibid. p. 65; *Appl. Opt.* **7**, 1981 (1968).

[252] G. D. Holah and N. Morrison, *J. Opt. Soc. Am.* **67**, 971 (1977).

[253] M. Francon, "Optical Image Formation and Processing", Chapter 7, Academic Press, London and New York (1979); for specifically infrared applications see T. L. Williams, *Proc. Soc. Opt. Inst. Eng.* **46**, 305 (1974), and also C. J. Hutchinson, J. P. Jennings, C. Lewis and G. N. Turner, *J. Phys. E. (Sci. Instrum)*, **14**, 846 (1981).

[254] J. R. Birch, J. D. Dromey and E. A. Nicol, *Infrared Phys.* **21**, 17 (1981).

[255] R. L. Petritz, *Phys. Rev.* **104**, 1508 (1956); A. V. MacRae and H. Levinstein, *Phys. Rev.* **119**, 62 (1960); V. Radeka, Proc. ISPRA Nuclear Electronic Symposium, p. 1 (1969).

[256] F. N. Hooge, T. G. M. Kleinpenning and L. K. J. Vandamme, *Rep. Prog. Phys.* **44**, 479 (1981).

[257] J. Chamberlain, *Infrared Phys.* **11**, 25 (1971); J. Chamberlain and H. A. Gebbie, *Infrared Physics*, **11**, 56 (1971).

[258] D. P. Blair and P. H. Sydenham, *J. Phys. E. (Sci. Instrum)*, **8**, 621 (1975).

[259] P. C. G. Danby, *Electron. Engng*, January (1970).

[260] J. D. W. Abernethy, *Phys. Bull.* **24**, 591 (1973).

[261] R. W. Harris and T. J. Ledwige, "Introduction to Noise Analysis", Pion Press, Applied Physics Series No. 7 (1974).

[262] C. E. Shannon, Communication in the presence of noise, *Proc. IRE*, **37**, 10–21 (1949).

[263] G. Shorter, *Wireless World* 200–205 May (1975).

[264] B. Austin Barry, "Errors in Practical Measurement in Science, Engineering and Technology", Wiley Interscience, New York and Chichester (1978).

[265] F. Oberhettinger, Fourier transforms of distributions and their inverses; A collection of tables, "Probability and Mathematical Statistics Series", Academic Press, New York and London (1973).

[266] See for example "Handbook of Chemistry and Physics" published by the Chemical Rubber Publishing Company, Cleveland, Ohio.

[267] J. W. Goodman, Introduction to Fourier Optics, McGraw Hill New York (1968); H. J. Caulfield (Ed.), "Handbook of Holography", Academic Press, New York and London (1979); A Van der Lugt, *Proc. IEEE* **62**, 1300 (1974); J. W. Goodman, *Proc. IEEE*, **65**, 29 (1977).

[268] D. Marcuse, "Principles of Quantum Electronics", Academic Press, New York (1980).

[269] R. Loudon, "The Quantum Theory of Light", Clarendon Press, Oxford (1973).

[270] See L. C. Robinson, "Methods of Experimental Physics", Vol. 10, "Physical Principles of Far-Infrared Radiation" (Ed. L. Marton) Academic Press, New York and London (1973); 219ff, for a detailed description of this point.

[271] R. H. Hanbury-Brown, "The Intensity of Interferometer", Taylor and Francis, London (1974).

[272] E. R. Pike and E. Jakeman, Photon-statistics and photon-correlation spectroscopy, *in* Advances in Quantum Electronics, Vol. 2 (Ed. D. W. Goodwin), Academic Press, London and New York (1974).

[273] M. J. Colles and C. R. Pidgeon, Tunable lasers, *Rep. Prog. Phys.* **38**, 329, (1975).

[274] J. W. Fleming and J. Chamberlain, *Infrared Phys.* **14**, 277 (1974).

[275] J. W. Fleming, *IEEE Trans. Microwave Theory Tech.* **MTT-22**, 1023 (1974).

[276] H. A. Gebbie and R. Q. Twiss, *Rep. Prog. Phys.* **29**, 729 (1966).

[277] L. Genzel and A. Hadni, Private communications.

[278] P. Jaquinot, *J. Opt. Soc. Am.* **44**, 761 (1954): *Rep. Prog. Phys.* **23**, 267 (1960).

[279] G. W. Chantry and J. W. Fleming, *Infrared Phys.* **16**, 655 (1976).

[280] P. Fellgett, Thesis, University of Cambridge (1951); *J. Phys. Radium, Paris* **19**, 187, 236 (1958).

[281] F. Kahn, *Astrophys. J.* **129**, 518 (1959).

[282] E. R. Pike, Review of Reference (7), *Nature (London)* **283**, 700 (1980).

[283] P. Jaquinot, *Appl. Opt.* **8** (3), March (1969).

[284] J. Strong, Multiplex spectrometry *in* "Essays in Physics" (Eds G. K. T. Conn and G. N. Fowler), Vol. 5, Academic Press, New York and London (1973).

[285] J. A. Decker, *Appl. Opt.* **10**, 24 (1971).

[286] P. Hansen and J. Strong, *Appl. Opt.* **11**, 502 (1972).

[287] M. Harwit, P. G. Phillips, T. Fine and N. J. A. Sloane, *Appl. Opt.* **9**, 1149 (1970).

[288] P. R. Griffiths, Infrared fourier transform spectrometry: applications to analytical chemistry, *in* "Transform Techniques in Chemistry" (Ed. P. R. Griffiths), Chapter 6, Heyden, London and New York (1978).

[289] C. Corsi, G. Cappucio, A. D'Amico, G. Petrocco and G. Vitali, *Infrared Phys.* **16**, 37 (1976).

[290] H. W. Thompson, *Nature (London)* (1946).

[291] L. Genzel, *J. Molec. Spectrosc.* **4**, 241 (1960); H. Happ and L. Genzel, *Infrared Phys.* **1**, 39 (1961).

[292] R. C. Milward, *in* "Molecular Spectroscopy", p. 81, Institute of Petroleum (1969).

[293] J. Connes and P. Connes, *J. Opt. Soc. Am.* **56**, 896 (1966).

[294] Reference (7), pp. 197ff; see also J. R. Birch and C. E. Bulleid, *Infrared Phys.* **17**, 279 (1977).

[295] J. Chamberlain, Submillimetre-wave techniques, *in* "High Frequency Dielectric Measurement" (Eds J. Chamberlain and G. W. Chantry), IPC Press, Guildford, UK (1972).

[296] J. R. Birch and T. J. Parker, "Dispersive Fourier Transform Spectroscopy in Infrared and Millimetre Waves" (Ed. K. J. Button), Vol. 2, Academic Press, New York and London (1979).

[297] J. Chamberlain, J. E. Gibbs and H. A. Gebbie, *Infrared Phys.* **9**, 185 (1969).

[298] M. N. Afsar, J. Chamberlain, G. W. Chantry, *IEEE Trans. Instrum. Meas.* **IM-25**, 290 (1976).

[299] J. Chamberlain, *Infrared Phys.* **12**, 145 (1972).

[300] J. E. Allnutt and J. A. Staniforth, *J. Phys. E. (Sci. Instrum.)* **4**, 730 (1971); J. Chamberlain, J. Haigh and M. J. Hine, *Infrared Phys.* **11**, 75 (1971).

[301] J. Connes, *Rev. Opt. Theor. Instrum.* **440**, 45, 116, 171, 231 (1961); P. L. Richards *in* "Spectroscopic Techniques" (Ed. D. H. Martin), pp. 58ff, North Holland, Amsterdam (1967).

[302] The best known is that of M. L. Forman, W. H. Steel and G. Vanasse, *J. Opt. Soc. Am.* **56**, 59 (1966), but the subject is also discussed very informatively by L. Mertz, *Appl. Opt.* **2**, 1331 (1963); *Infrared Phys.* **7**, 17 (1967).

[303] J. W. Fleming, NPL Report DES 49, September (1978).

[304] J. R. Birch, *Infrared Phys.* **20**, 349 (1980), but see also R. P. Lowe, R. J. Niciejewski and D. N. Turnbull, *Infrared Phys.* **21**, 189 (1981).

[305] D. P. C. Thackeray, the infrared spectrometer, *in* Laboratory Methods in Infrared Spectroscopy" (Eds R. G. J. Miller and B. C. Stace), Chapter 1, Heyden, New York and London (1972).

[306] J. W. Fleming, *Infrared Phys.* **17**, 263 (1977).

[307] G. A. Vanasse and H. Sakai, "Prog. Opt.", Vol. 6, North Holland, Amsterdam (1967).

[308] A similar point is discussed by J. Butterworth, D. E. MacLaughlin and B. C. Moss, *J. Sci. Instrum.* **44**, 1029 (1967).

[309] T. S. Moss, Infrared Detectors, Pergamon Press (1976): this is a hardback version of *Infrared Phys.* **15** (1975); see also various papers in "Advanced Infrared Detectors and Systems" Proceedings of the International Conference, held at the Institution of Electrical Engineers, 29–30 Oct. (1981), IEE Conference Publication No. 204.

[310] N. E. Johnson, Spectral imaging with the Michelson interferometer, *published in* "Infrared Imaging Systems Technology", *Proc. Soc. Photo-Optical Instrum. Engineers (SPIE)*, **226**, 2 (1980).

[311] T. G. Blaney, Proceedings of the SPIE Conference on New Developments and Applications in Optical Radiation Measurement, NPL, Teddington UK, May 7–8 (1980), *SPIE* **234**, 22–26.

[312] Details of "black" paints suitable for infrared work have been given by J. L. Pipher and J. R. Houck, *Appl. Opt.* **10**, 567 (1971) and by S. Takahashi, *Infrared Phys.* **13**, 1 (1973).

[313] M. F. Kimmitt, A. C. Prior and V. Roberts, Far-infrared techniques, *in* Plasma Diagnostic Techniques (Eds R. H. Huddlestone and S. L. Leonard), Chapter 9, Academic Press, New York and London (1965).

[314] R. R. Selleck, E. W. McDonald and J. C. Wiltse, Digest 2nd International Conference on Submillimetre Waves and their Applications, San Juan Puerto Rico (1976), IEEE, 76 CH1152-8 MTT, pp. 49–50.

[315] E. H. Putley, Solid-State Devices for Infrared Detectors, *J. Sci. Instrum.* **43**, 857 (1965); Detectors, *in* "Spectroscopic Techniques" (Ed. D. H. Martin), North Holland, Amsterdam (1967); Modern infrared detectors, *Phys. Technol.* **4**, 202 (1973).

[316] T. G. Blaney, *J. Phys. E. (Sci. Instrum.)*, **11**, 856 (1978).

[317] D. E. Bode, Infrared detectors, *in* "Applied Optics and Optical Engineering" (Eds R. Kingslake and B. J. Thompson) Vol. 6, Chapter 8, Academic Press, New York and London (1980).

[318] T. G. Blaney, "Detection Techniques at Short Millimetre and Submillimetre

Wavelengths: An Overview in Infrared and Millimeter Waves", Vol. 3, "Submillimeter Techniques" (Ed. K. J. Button), Academic Press, New York and London (1980).

[319] M. Golay, *Rev. Sci. Instrum.* **18**, 347, 357 (1947).

[320] E. H. Putley, "Semiconductors and Semi-metals" (Ed. R. K. Williardson and A. C. Beer), Vol. 5, pp. 259–285, Academic Press, London (1970).

[321] A. Hadni, Pyroelectricity and pyroelectric detectors, *in* "Infrared and Millimetre Waves", Vol. 3, "Submillimetre Techniques" (Ed. K. J. Button), Academic Press, New York and London (1980); *J. Phys. E (Sci. Instrum.)* **14** (1981).

[322] A. Hadni, *IEEE Trans. Microwave Theory Tech.* **MTT-22**, 1016 (1974); E. L. Dereniak and F. G. Brown, *Infrared Phys.* **15**. 39 (1975).

[323] R. J. Mahler, R. J. Phelan and A. R. Cook, *Infrared Phys.* **12**, 57 (1972).

[324] R. J. Phelan, R. J. Mahler and A. R. Cook, *Appl. Phys. Lett.* **19**, 337 (1971); A. M. Glass, J. H. McFee and J. G. Bergman, *J. Appl. Phys.* **42**, 5219 (1971).

[325] P. J. Lock, *Appl. Phys. Lett.* **19**, 390 (1971).

[326] C. B. Roundy and R. L. Byer, *Appl. Phys. Lett.* **21**, 512 (1972).

[327] W. R. Blevin and J. Geist, *Appl. Opt.* **13**, 1171 (1974)

[328] M. R. Holter, S. Nudelman, G. H. Suits, W. L. Wolfe and G. J. Zissis, "Fundamentals of Infrared Technology", MacMillan, New York and London (1962).

[329] J. Clarke, P. L. Richards and N. H. Yeh, *Appl. Phys. Lett.* **39**, 664 (1977).

[330] H. D. Drew and A. J. Sievers, *Appl. Opt.* **8**, 2067 (1969).

[331] G. Chanin, J. P. Torre and L. Peccoud, *Infrared Phys.* **18**, 657 (1978).

[332] R. D. Britt and P. L. Richards, *Int. J. Infrared Millimetre Waves* **2**, 1083 (1981).

[333] The concept came originally from R. C. Jones, *Proc. Inst. Radio Eng.* **47**, 1481 (1959).

[334] A. H. Sommer, "Photo-emissive Materials", John Wiley, New York (1968).

[335] D. H. Martin and D. Bloor, *Cryogenics* **1**, 159 (1961).

[336] C. L. Bertin and K. Rose, *J. Appl. Phys.* **42**, 163 (1971); M. K. Maul and M. W. P. Strandberg, *J. Appl. Phys.* **40**, 2822 (1969); G. Gallinaro and R. Varone, *Cryogenics*, **15**, 292 (1975).

[337] T. G. Blaney, *Space Sci. Rev.* **17**, 691 (1975).

[338] A. E. Siegman, *Proc. IEEE*, **54**, 1350 (1966).

[339] T. G. Blaney and N. R. Cross, *J. Phys. E.* (Sci. Instrum.) **10**, 146 (1977).

[340] J. J. Gustincic, *Proc. Soc. Photo-Optical Instrum. Engineers (SPIE)*, **105**, 40 (1977).

[341] H. R. Fettermann, P. E. Tannenwald, B. J. Clifton, C. D. Parker, W. D. Fitzgerald and N. R. Erickson, to be published.

[342] D. H. Martin, *in* Infrared Detection Techniques for Space Research, Reidel (1972).

[343] G. T. Wrixon and W. M. Kelly, *Infrared Phys.* **18**, 413 (1978).

[344] D. H. Martin, Invited lectures to the Fifth International Conference on Infrared and Submm Waves, Wurzburg FRG (1980), and to the XXth Congress of URSI, Washington DC (1981).

[345] H. Krautle, E. Sauter and G. V. Schultz, *Infrared Phys.* **18**, 705 (1978).

[346] D. B. Rutledge, S. E. Schwarz and A. T. Adams, *Infrared Phys.* **18**, 713 (1978).

[347] D. B. Rutledge, S. E. Schwarz, T. L. Hwang, D. J. Angelakos, K. K. Mei and S. Yokota, *IEEE J. Quant. Electron.* **QE-16**, 508 (1980).

[348] B. J. Clifton, G. D. Alley, R. A. Murphy and I. H. Mroczkowski, *IEEE Trans. Electron. Devices*, **ED-28** 155 (1981).

[349] D. Buhl, G. Chin, G. A. Koepf, N. McAvoy, H. R. Fettermann, B. J. Clifton, D. D. Peck, P. E. Tannenwald, P. F. Goldsmith and N. R. Erickson, to be published.

[350] J. P. Sattler, T. L. Worchesky, K. J. Ritter and W. J. Lafferty, *Opt. Lett.* **5**, 21 (1980).

[351] J. R. Tucker, *IEEE Trans. Quant. Electron.* **QE-15**, 1234 (1979).

[352] W. G. Chambers, *J. Phys. A.* **14**, 138 (1981).

[353] M. McColl, A review of submillimeter-wave mixers, *Proc. Soc. Photo-Optical Engineers (SPIE)*, **105**, 24 (1977); P. E. Tannenwald, *Int. J. Infrared Millimetre Waves* **1** 159 (1980).

[354] C. D. Payne and B. E. Prewer, *Radio Electron. Eng.* **39**, 167 (1970); A. A. M. Saleh, "Theory of Resistive Mixers", MIT Press, Cambridge, Massachusetts (1971).

[355] J. W. Dees, *Microwave J.* **9**, 48 (1966); V. Daneu, D. Sokoloff, A. Sanchez and A. Javan, *Appl. Phys. Lett.* **15**, 398 (1969); S. I. Green, P. D. Coleman and J. R. Baird, The MOM electric tunnelling detector, *in* Submillimeter Waves, Proceedings of the Symposium, Polytechnic Press, Polytechnic Institute of Brooklyn, Volume XX, (1970); H. D. Riccius, *Appl. Phys.* **17**, 49 (1978); P. J. Epton, W. L. Wilson, F. T. Tittel and T. A. Rabson, *Infrared Phys.* **19**, 335 (1979).

[356] J. N. Crouch, *Infrared Phys.* **18**, 89 (1978); J. E. Muller and C. Hanke, *Infrared Phys.* **19**, 533 (1978).

[357] T. G. Phillips, *Astrophys. J. Lett.* **186**, L19 (1973); see also A. Arams, *Proc. IEEE.* **54**, 612 (1966).

[358] W. M. Kelly and G. T. Wrixon, Optimisation of Schottky barrier diodes for low conversion-loss operation at near mm wavelengths, *in* "Infrared and Millimeter Waves", Vol. 3, "Submillimeter Techniques" (Ed. K. J. Button), Academic Press, New York and London, (1980); B. J. Clifton, *IEEE Trans. Microwave Theory* Tech. **MTT-25**, 457 (1977); *Radio Electron. Engng* **49**, 333 (1979).

[359] P. L. Richards, *in* "Semiconductors and Semimetals", Vol. 12, "Infrared Detectors II" (Eds R. K. Willardson and A. C. Beer), p. 395, Academic Press, New York and London (1977); Y. Taur, J. H. Claasen and P. L. Richards, *IEEE Trans. Microwave Theory Tech.* **MTT-22**, 1005 (1974); J. Edrich, Proceedings of the Fourth International Conference on Infrared and Millimetre Waves, Conference Digest IEEE Cat. No. 79 CH 1384–7 MTT, p. 152 (1979).

[360] P. L. Richards, T. M. Shen, R. E. Harris and F. L. Lloyd, *Appl. Phys. Lett.* **34**, 345 (1979).

[361] J. P. Sattler, T. L. Worchesky, M. S. Tobin, K. J. Ritter and T. W. Daley, *Int. J. Infrared Millimetre Waves* **1**, 127 (1980); W. J. Lafferty, J. P. Sattler, T. L. Worchesky and K. J. Ritter, *J. Molec. Spectrosc.* **87**, 416, (1981).

[362] H. A. Gebbie, N. W. B. Stone, E. H. Putley and N. Shaw, *Nature (London)* **214**, 165 (1967).

[363] S. D. Personick, *Proc. IEEE*, **69**(2), 262 (1981).

[364] S. M. Sze, "Physics of Semiconductor Devices", John Wiley, New York (1969).

[365] M. McColl, D. T. Hodges and W. A. Garber, *IEEE Trans. Microwave Theory Tech.* **MTT-25**, 463 (1977).

[366] L. M. Matarrese and K. M. Evenson, *Appl. Phys. Lett.* **17**, 8 (1970).

[367] R. A. Murphy, C. O. Bozler, C. D. Parker, H. R. Fetterman, P. E. Tannenwald, B. J. Clifton, J. P. Donnelly and W. T. Lindley, *IEEE Trans. Microwave Theory Tech.* **MTT-25**, 494 (1977).

[368] T. G. Blaney, N. R. Cross and Th. de Graauw, to be published.

[369] B. D. Josephson, *Phys. Lett.* **1**, 251 (1962); *Rev. Mod. Phys.* **36**, 216 (1964); *Adv. Phys.* **56**, 14 (1965); see also J. R. Waldram, The Josephson effects in weakly-coupled superconductors. *Reps. Prog. Phys.* **39**, 751 (1976).

[370] J. Bardeen, L. N. Cooper and J. R. Schrieffer, *Phys. Rev.* **108**, 1175 (1957).

[371] P. L. Richards and M. Tinkham, *Phys. Rev.* **119**, 575 (1960); P. Wyder, *Infrared Phys.* **16**, 243 (1976).

[372] I. K. Harvey, J. C. Macfarlane and R. B. Frenkel, *Metrologia* **8**, 114, (1972); B. F. Field, T. F. Finnegan and J. Toots, *Metrologia*, **9**, 155 (1973); A. Hartland, T. J. Witt, D. Reymann and T. F. Finnegan, *IEEE Trans. Instrum. Meas.* **IM-27**, 470 (1978).

[373] T. G. Blaney, *Radio Electron. Engineer*, **42**, 303 (1972).

[374] C. C. Grimes, P. L. Richards and S. Shapiro, *J. Appl. Phys.* **39**, 3905 (1968); T. Poorter and H. Tolner, *Infrared Phys.* **19**, 317 (1979).

[375] P. L. Richards and S. A. Sterling, *Appl. Phys. Lett.* **14**, 394 (1969).

[376] D. G. McDonald, V. E. Kose, K. M. Evenson, J. S. Wells and J. D. Cupp, *Appl. Phys. Lett.* **15**, 121 (1969); D. G. McDonald, A. S. Risley, J. D. Cupp and K. M. Evenson, *Appl. Phys. Lett.* **18**, 162 (1971); T. G. Blaney, Josephson mixers at submillimetre wavelengths: Present experimental status and future developments, *in* "Future Trends in Superconductive Electronics" (Eds B. S. Deaver, C. M. Falco, J. H. Harris and S. A. Wolf), pp. 230–238, AIP Conference Proceedings, No. 44 (1978).

[377] T. G. Blaney, N. R. Cross and R. G. Jones, to be published, but see also J. Edrich, D. B. Sullivan and D. G. McDonald, *IEEE Trans. Microwave Theory Tech.* **MTT-25**, 476 (1977).

[378] J. R. Tucker, *Appl. Phys. Lett.* **36**, 477 (1980).

[379] F. L. Vernon, M. F. Millea, M. F. Bottjer, A. H. Silver, R. J. Pederson and M. McColl, *IEEE Trans. Microwave Theory Tech.* **MTT-25**, 286 (1977).

[380] J. H. Greiner, C. J. Kircher and I. Ames, *IBM J. Res. Devel.* **24**, 195 (1980); T. Gheewala, *Proc. IEEE*, **70**, 26 (1982).

[381] A. Matsuda, *J. Appl. Phys.* **51**, 4310 (1980).

[382] G. J. Dolan, T. G. Phillips and D. P. Woody, *Appl. Phys. Lett.* **34**, 347 (1979).

[383] P. L. Richards and T. M. Shen, *IEEE Trans. Electron Devices, ED-77*, **ED-77**, 1909 (1980).

[384] Work attributed to Heppner in private communication from P. L. Richards.

[385] G. J. Dolan, R. A. Linke, T. C. L. G. Sollner, D. P. Woody and T. G. Phillips, *IEEE Trans. Microwave Theory Tech.* **MTT-29**, 87 (1981); T. M. Shen and P. L. Richards, *IEEE Trans. Magn.* **MAG-17**, 677 (1981).

[386] D. W. Peterson, A. R. Kerr, M. J. Feldman, P. H. Siegel and S. K. Pan, to be published.

[387] T. M. Shen, *IEEE J. Quant. Electron.* **QE-17**, 1151 (1981).

[388] A. F. Gibson and M. F. Kimmitt, Photon-drag detection, *in* "Infrared and Millimeter Waves", Vol 3. "Submillimeter Techniques", (Ed. K. J. Button), Academic Press, New York and London (1980).

[389] A. A. Grinberg, *Sov. Phys. JEPT*, **31**, 531 (1970); K. Cameron, A. F. Gibson, J. Giles, C. B. Hatch, M. F. Kimmitt and S. Shafik, *J. Phys. C; Solid-State Phys.* **8**, 3137 (1975); A. F. Gibson and S. Montasser, *J. Phys. C; Solid-State Phys.* **8**, 3147 (1975).

[390] A. Rosencwaig, *Adv. Electronics Electron Phys.* **46**, 207 (1979); "Photoacoustics and Photoacoustic Spectroscopy", John Wiley, New York (1980); G. Busse, *J. Opt. (Paris)* **11**, 454 (1980); J. Badoz, D. Fournier and A. C. Boccara, Ibid. 399; see also "Optoacoustic Spectroscopy and Detection" (Ed. Y. H. Pao), Academic Press, New York and London (1977).

[391] G. Busse and K. F. Renk, *Infrared Phys.* **18**, 517 (1978).

[392] G. Busse and H. Schultz, *Infrared Phys.* **19**, 313 (1979).

[393] C. F. Dewey, R. D. Kamm and C. E. Hackett, *Appl. Phys. Lett.* **23**, 633 (1973).

[394] P. C. Claspy, Infrared optoacoustic spectroscopy and detection, *in* "Optoacoustic Spectroscopy and Detection" (Ed. Y. H. Pao), Chapter 6, Academic Press, New York and London (1977).

[395] D. H. Martin and E. Puplett, to be published.

[396] G. Busse and B. Bullemer, *Infrared Phys.* **18**, 631 (1978).

[397] D. W. Vidrine, *Appl. Spectrosc.* **34**, 314 (1980).

[398] K. F. Luft, *Z. Tech. Phys.* **24**, 97 (1943); *Angew. Chem.* **B19**, 2 (1947); *Compt. Rend.* **238**, 1651 (1954).

[399] M. L. Veingerov, *Dokl. Akad. Nauk. S.S.R.* **19**, 687 (1938); J. C. Waters and N. W. Hartz, *Instruments* **25**, 622 (1952).

[400] D. W. Hill and T. Powell, "Non-dispersive Infrared Gas Analysis in Science, Medicine and Industry", Plenum, New York (1968).

[401] C. O. Peterson, W. V. Dailey and W. G. Amthein, *Instrum. Technol.* **14**, 8, 45 (1967); D. J. Troy, *Control Eng.* **4**, 11, 116 (1957).

[402] T. Takahashi, M. O. Weaver and L. A. Prince, *J. Geophys. Res.* **81**, 3736 (1976).

[403] H. M. J. M. Dortmans, *J. Phys. E.* (Sci. Instrum.) **14**, 777 (1981); see also various articles in *Electronics and Power*, **27**(3), March (1981).

[404] B. W. Kernighan and P. J. Plauger, Software Tools.

[405] W. Neuhauser, M. Hohenstatt and P. E. Toschek, *in* Laser Spectroscopy IV (Eds H. Walther and K. W. Rothe), Springer Verlag, London, Heidelberg and New York (1979) pages 73–78.

[406] E. K. Plyler and L. R. Blaine, *J. Res. Nat. Bur. Stand.* **62**, 7 (1959).

[407] G. Guelachvili, *Appl. Opt.* **17**, 1322 (1978); L. Genzel and K. Sakai, *J. Opt. Soc. Am.* **67**, 871 (1977); see also K. Sakai, High resolving power fourier spectroscopy *in* "Spectrometric Techniques" (Ed. G. Vanasse), Vol. 1, Academic Press, New York and London (1977).

[408] See for example A. G. Maki and J. Sams, *J. Mol. Struct.* **26**, 107 (1975).

[409] M. Czerny, *Z. Phys.* **34**, 227 (1925); ibid. **45**, 476 (1927).

[410] H. Rubens and E. Aschkinass, *Astrophys J.* **8**, 176 (1898); H. Rubens and G. Hettner, *Ver. Deut. Phys. Ges.* **18**, 154 (1916); H. Rubens, *Berl. Ber.* 8 (1921).

[411] N. Bjerrum, *Nernst Festschrift*. 90 (1912).

[412] R. H. Partridge, *Infrared Phys.* **19**, 571 (1979).

[413] M. J. Bangham, A. Bonetti, R. H. Bradsell, B. Carli, J. E. Harries, F. Mencaraglia, D. G. Moss, S. Pollitt, E. Rossi and N. R. Swann, to be published.

[414] D. J. W. Kendall, H. L. Buijs and J. W. C. Johns, to be published.

[415] G. Guelachvili, *Appl. Opt.* **17**, 1322 (1978).

[416] J. Connes, H. De Louis, P. Connes, G. Guelachvili, J. P. Mailland and G. Michel, *Nouv. Revue d'Optique appliquee* **1**, 3 (1970).

[417] M. Cuisenier and J. Pinard, *J. Phys. (Suppl.)* **28**, 97 (1967); R. Beer and D. Marjamiemi, *Appl. Opt.* **5**, 1191 (1966); see also R. B. Sanderson and H. E. Scott, *Appl. Opt.* **10**, 1097 (1971).

[418] W. H. Steel, Interferometers for Fourier Spectroscopy, Aspen International Conference on Fourier Spectroscopy (1970) (Eds G. A. Vanasse, A. T. Stair Jr. and D. J. Baker), p. 43, AFCRL-71-0019, 5 Jan 1971, Spec. Rep. No 114.

[419] J. Connes and P. Connes, *J. Opt. Soc. Am.* **56**, 876 (1966); P. Connes, Astronomical fourier spectroscopy, *Ann. Rev. Astron. Astrophys.* **8**, 209 (1970).

[420] H. R. Schlossberg and P. L. Kelley, Infrared spectroscopy using tunable lasers, *in* Spectrometric Techniques (Ed. G. Vanasse) Vol II, chapter 4, Academic Press, London and New York (1981).

[421] J. M. Besson, W. Paul and A. R. Calawa, *Phys. Rev.* **173**, 699 (1968); A. S. Pine, C. J. Glassbrenner, and J. A. Kafalas, *IEEE J. Quant. Electron.* **QE-9**, 800 (1973).

[422] J. C. Hill, W. Lo and J. A. Sell, *Laser Focus*, **16**, 86 (1980); R. T. Ku, E. D. Hinkley and J. O. Sample, *Appl. Opt.* **14**, 854 (1975); E. D. Hinkley, *Opt. Quant. Electron.* **8**, 155 (1976); E. D. Hinkley, R. T. Ku, K. W. Nill and J. F. Butler, *Appl. Opt.* **15**, 1653 (1976); L. W. Chaney, D. G. Rickel, G. M. Russwurm and W. A. McCleany, *Appl. Opt.* **18**, 3004 (1979).

[423] J. P. Sattler and G. J. Simonis, *IEEE J. Quant. Electron.* **QE-13**, 461 (1977); T. L. Worchesky, K. J. Ritter, J. P. Sattler and W. A. Riesser, *Opt. Lett.* **2**, 70 (1978).

[424] J. P. Sattler, T. L. Worchesky and W. A. Riessler, *Infrared Phys.* **18**, 521 (1978).

[425] S. D. Smith, "Very High Resolution Spectroscopy" (Ed. R. A. Smith), Chapter 2, Academic Press, London and New York (1977).

[426] P. G. Buckley, J. H. Carpenter, A. McNeish, J. D. Muse, J. J. Turner and D. H. Whiffen, *J. Chem. Soc. (Faraday Trans. II)*, **74**, 129 (1978).

[427] C. K. N. Patel, *Appl. Phys. Lett.* **25**, 112 (1974).

[428] N. F. Ramsey, "Molecular Beams", Oxford University Press, New York and London (1956).

[429] Ch. J. Borde, M. Ouhayan, A. van Lerberghe, C. Salmon, S. Avrilliev, C. D. Cantrell and J. Borde, High resolution saturation spectroscopy with CO_2 lasers, application to the v_3 bands of SF_6 and OsO_4 *published in* "Laser Spectroscopy IV" (Eds H. Walther and K. W. Rothe), Springer-Verlag, Berlin, Heidelberg and New York (1979).

[430] T. Oka and T. Shimizu, *Appl. Phys. Lett.* **19**, 88 (1971).

[431] H. Jones and F. Kohler, *J. Molec. Spectrosc.* **58**, 125 (1975). H. Jones, ibid. **78**, 452 (1979).

[432] G. Grynberg and B. Cagnac, *Rep. Prog. Phys.* **40**, 791 (1977); see also S. M. Freund and T. Oka, *Phys. Rev.* **A13**, 2178 (1976); for an example see H. Jones, *Infrared Phys.* **18**, 449 (1978).

[433] T. W. Hansch, S. A. Lee, R. Wallenstein and C. Wieman, *Phys. Rev. Lett.* **34**, 307 (1975). C. Wieman and T. W. Hansch, *in* "Laser Spectroscopy III" (Eds J. L. Hall and J. L. Carlsten), Springer Series in Optical Sciences, Vol. 7, p. 39, Springer-Verlag, Berlin, Heidelberg and New York, (1977).

[434] A. Owyoung, High resolution coherent Raman spectroscopy of gases, *published in* "Laser Spectroscopy IV" (Eds H. Walther and K. W. Rothe), Springer-Verlag, Berlin and New York (1979).

[435] T. Oka, "Laser Spectroscopy" (Eds H. G. Brewer and A. Mooradian), pp. 413–431, Plenum, New York (1974).

[436] K. Shimoda and T. Shimizu, "Non-Linear Spectroscopy of Molecules", Pergamon, Oxford (1972).

[437] J. Lemaire, J. Thibault, F. Herlemont and J. Houriez, *Mol. Phys.* **27**, 611 (1974).

[438] V. P. Chebotayev, Multiple coherent interaction in optically separated fields, *in* "Laser Spectroscopy IV" (Eds H. Walther and K. W. Rothe), pp. 106-119, Springer-Verlag, Berlin, Heidelberg and New York (1979); see also the article by J. C. Berquist, R. L. Berger and D. J. Glaze in the same volume, pp. 120–129, and also J. C. Berquist, S. A. Lee and J. L. Hall, *Phys. Rev. Lett.* **38**, 159 (1972).

[439] S. A. Lee, J. Helmcke and J. L. Hall, High-resolution two-photon spectroscopy of Rb Rydberg levels, *in* "Laser Spectroscopy IV" (Eds H. Walther and K. W. Rothe), pp. 130–141, Springer-Verlag, Berlin, Heidelberg and New York (1979); see also article by J. L. Hall in *Science*, October (1978).

[440] J. L. Hall, Bureau Int. Poids Mesures, CCDM 78–9, (1978).

[441] P. Glorieux, J. Legrand, B. Macke and J. Messelyn, *J. Quant. Spectrosc. Radiat. Transfer* **12**, 731 (1972).

[442] C. Borde, *C. R. Acad. Sci. (Paris)* **282B**, 341 (1976); Ye V. Baklanov and B. Ya Dubetskii, *Sov. J. Quant. Electr.* **8**, 51 (1978).

[443] J. L. Hall, C. J. Borde and K. Uehara, *Phys. Rev. Lett.* **37**, 1339 (1976).

[444] R. E. Drullinger and D. J. Wineland, Laser cooling of ions bound in a Penning trap, *in* "Laser Spectroscopy IV" (Eds H Walther and K. W. Rothe), Springer-Verlag, Berlin, Heidelberg and New York (1979).

[445] D. J. Wineland, R. E. Drullinger and F. L. Walls, *Phys. Rev. Lett.* **40**, 1639 (1978); W. Neuhauser, M. Hohenstatt, P. Toschek and H. Dehmett, *Phys. Rev. Lett.* **41**, 233 (1978).

[446] A. Walsh, The application of atomic absorption spectroscopy to chemical analysis, *Spectrochim. Acta*, 7, 108 (1955); see also, R. J. Reynolds and K. Aldous, "Atomic Absorption Spectroscopy", Griffin, London (1970), and M. Slavin, "Atomic Absorption Spectroscopy", 2nd Edition, John Wiley, New York and Chichester (1978).

[447] G. F. Kirkbright and M. Sargent, "Atomic Absorption and Fluorescence Spectroscopy", Academic Press, London and New York (1974).

[448] T. W. Johnston, *Proc. IEEE*, **69**(2). 149 (1981).

[449] J. Connes, P. Connes, H. Delouis, G. Guelachvili, J. P. Maillard and G. Michel, *Nouv. Rev. Opt. Appl.* **1**, 3, (1970); for some preliminary work in assigning the spectrum of holmium, see J. Blaise, P. Camus, G. Guelachvii, J. Verges and J. F. Wyart, *C. r. Acad Sci. Paris*, **274**, 1302 (1972); ibid. **275**, 81 (1972).

[450] L. Pauling and E. B. Wilson, "Introduction to Quantum Mechanics", McGraw-Hill, New York (1935); J. C. Slater, "Quantum Theory of Molecules and Solids", Vols 1 and 2, McGraw-Hill, New York (1963).

[451] G. Herzberg, "Atomic Spectra and Atomic Structure", Dover Books, New York (1944); W. G. Richards and P. R. Scott, "Structure and Spectra of Atoms", Wiley, London (1976); P. W. Atkins, "Physical Chemistry", Oxford University Press, Oxford (1978): I. I. Sobelman, "Atomic Spectra and Radiative Transitions", Springer Series in Chemical Physics, Vol. 1, Springer-Verlag, Berlin and New York (1979).

[452] H. G. Kuhn, "Atomic Spectra", Longman, London (1962).

[453] H. G. Kuhn and G. W. Series, *Nature (London)* **162**, 373 (1948); *Proc. R. Soc.* **A202**, 127 (1950).

[454] For an account of the Dirac Theory, see J. McConnell, "Quantum Particle Dynamics", 2nd Edition, North-Holland, Amsterdam (1960).

[455] R. P. Feynman, "Quantum Electrodynamics", Benjamin, (1962); R. Delbourgo, How to deal with infinite integrals in Quantum Field Theory, *Rep. Prog. Phys.* **39**, 345 (1976).

[456] W. E. Lamb, Jr. and R. C. Retherford, *Phys. Rev.* **72**, 241 (1947).

[457] T. W. Hansch, S. A. Lee, R. Wallenstein and C. Wieman, *Phys. Rev. Lett.* **34**, 307 (1975); S. A. Lee, R. Wallenstein and T. W. Hansch, *Phys. Rev. Lett.* **35**, 1262 (1975).

[458] H. A. Bethe, *Phys. Rev.* **72**, 339 (1947); see also E. E. Salpeter, *Phys. Rev.* **89**, 92 (1953); for an account of modern studies of heavier single-electron (that is "hydrogenic") ions, see H. W. Kugel and D. E. Murnick, The Lamb-shift in hydrogenic ions, Rep. *Prog. Phys.* **40**, 297 (1977).

[459] T. A. Welton, *Phys. Rev.* **74**, 1157 (1948).

[460] C. E. Moore, Atomic Energy Levels (3 Vols), National Bureau of Standards, Circ. 467, Washington DC (1949), (1952), (1958); S. Bashkin and J. O. Stoner, "Atomic Energy-Level and Grotrian Diagrams", North-Holland, Amsterdam and New York, published in three volumes, Vol. 1, (1976), Vol. 2, (1978), Vol. 3, (1981); S. Fraga, J. Karwowski and K. M. S. Saxena, "Atomic Energy Levels", Elsevier, Amsterdam (1979).

[461] C. L. Pekeris, *Phys. Rev.* **112**, 1649 (1958).

[462] G. Herzberg, *Proc. R. Soc.* **A248**, 309 (1958).

[463] G. Racah, *Phys. Rev.* **61**, 186 (1942); **62**, 438, 523 (1942); **63**, 367 (1943); **76**, 1352 (1949).

[464] R. A. McFarlane, W. L. Faust, C. K. N. Patel and C. G. B. Garrett, *Proc. IEEE* **52**, 318 (1964); C. K. N. Patel, W. L. Faust, R. A. McFarlane and C. G. B. Garrett, *Appl. Phys. Lett.* **4**, 18 (1964): C. K. N. Patel, W. L. Faust, R. A. McFarlane and C. G. B. Garrett, *Proc. IEEE* **52**, 713 (1964).

[465] C. A. Coulson, "Valence", 2nd Edition, Oxford University Press, Oxford (1961); "The Shape and Structure of Molecules", Oxford University Press, Oxford (1973).

[466] G. Herzberg, "Infrared and Raman Spectroscopy", Van Nostrand, New York (1945).

[467] See for example G. W. Chantry, H. A. Gebbie, R. J. L. Popplewell and H. W. Thompson, *Proc. R. Soc.* **A304**, 45 (1968).

[468] J. K. G. Watson, *J. Molec. Spectrosc.* **40**, 536 (1971); K. Fox, *Phys. Rev. Lett.* **27**, 233 (1971).

[469] See for example I. Ozier and A. Rosenberg, *Can. J. Phys.* **51**, 1882 (1973).

[470] I. Ozier, *Phys. Rev. Lett.* **27**, 1329 (1971); R. F. Curl and T. Oka, *J. Chem. Phys.* **58**, 4908 (1973); C. W. Holt, M. C. L. Gerry and I. Ozier, *Phys. Rev. Lett.* **31**, 1033 (1973).

[471] J. W. Fleming and J. Chamberlain, *Infrared Phys.* **14**, 277 (1974); see also J. W. Fleming, MM and sub-mm interferometric spectrometry, in "Modern Aspects of Microwave Spectroscopy" (Ed. G. W. Chantry), Chapter 5, Academic Press, London and New York (1979).

[472] M. Tinkham and M. W. P. Strandberg, *Phys. Rev.* **97**, 937, 951 (1955).

[473] M. Lichtenstein, V. E. Derr and J. J. Gallagher, *J. Molec. Spectrosc.* **20**, 391 (1966); W. J. Burroughs, J. E. Harries and H. A. Gebbie, *Nature (London)* **222**, 658 (1969); P. Helminger, F. C. De Lucia and W. Gordy, *Bull. Am. Phys. Soc.* **16**, 531 (1971).

[474] A good account is given by Townes and Schawlow, reference [92].

[475] J. W. Fleming and M. J. Gibson, *J. Molec. Spectrosc.* **62**, 326 (1976).

[476] J. W. Fleming, *Spectrochim. Acta* **32A**, 787 (1976).

[477] J. W. Fleming and R. P. Wayne, *Chem. Phys. Lett.* **32**, 135 (1975); J. W. Fleming, *Spectrochim. Acta* **32A**, 787 (1976).

[478] D. Papousek, J. M. R. Stone and V. Sporko, *J. Molec. Spectrosc.* **48**, 17 (1973).

[479] The particular case of nitromethane is discussed by D. Papousek, K. Sarka, V. Sporko and B. Jordanov, *Collect. Czech. Chem. Commun.* **36**, 890 (1971), and by P. Bunker in Vibrational Spectra and Structure, (Ed. J. R. Durig), Vol. 3, Chapter 1, Dekker, New York (1975).

[480] H. A. Gebbie, G. Topping, R. Illsley and D. M. Dennison, *J. Molec. Spectrosc.* **11**, 229 (1963).

[481] P. R. Bunker, "Molecular Symmetry and Spectroscopy", Academic Press, New York and London (1979).

[482] J. O. Henningsen, "Spectroscopy of Molecules by Far Infrared Laser Emission in Infrared and Millimeter Waves", (Ed. K. J. Button), Vol. 5, Academic Press, New York (1980).

[483] Reference [482] but see also J. O. Henningsen and J. C. Petersen, *Infrared Phys.* **18**, 475 (1978).

[484] See for example S. M. Kirschner and J. K. G. Watson, *J. Molec. Spectrosc.* **47**, 347 (1973).

[485] G. Herzberg, "Spectra of Diatomic Molecules" (Molecular Spectra and Molecular Structure 1), 2nd Edition, Van Nostrand, New York (1950); K. P. Huber and G. Herzberg, "Constants of Diatomic Molecules" (Molecular Spectra and Molecular Structure 4), Van Nostrand-Reinhold, New York (1979).

[486] E. B. Wilson, J. C. Decius and P. C. Cross, "Molecular Vibrations: The Theory of Infrared and Raman Vibrational Spectra", McGraw-Hill, New York (1955).

[487] D. Steele, Absolute absorption intensities as measured in the gas-phase, *in* "Advances in Infrared and Raman Spectroscopy" (Eds R. J. H. Clark and R. E. Hester), Vol. 1, Heyden, London (1975).

[488] S. J. Cyvin, Mean amplitudes of vibration for organic molecules, *in* "Advances in Infrared and Raman Spectroscopy", (Eds R. J. H. Clark and R. E. Hester), Vol. 5, Heyden, London (1978).

[489] R. F. Curl, T. Oka and D. S. Smith, *J. Molec. Spectrosc.* **46**, 518 (1973); R. F. Curl and T. Oka, *J. Chem. Phys.* **58**, 4908 (1973); *J. Molec. Spectrosc.* **48**, 165 (1973).

[490] For frequency measurements on the stabilised laser, see for example B. G. Whitford and D. S. Smith, *Opt. Commun.* **20**, 280 (1977); D. J. E. Knight, G. J. Edwards, P. R. Pearce and N. R. Cross, *Nature (London)* **285**, 388 (1980).

[491] K. M. Evenson, J. S. Wells, F. R. Petersen, B. L. Danielson, G. W. Day, R. L. Barger and J. L. Hall, *Phys. Rev. Lett.* **29**, 1346 (1972).

[492] M. Born and K. Huang, "Dynamical Theory of Crystal Lattices", Oxford University Press, New York (1954).

[493] S. S. Mitra and P. J. Gielisse, "Progress in Infrared Spectroscopy", (Ed. H. A. Szymanski), Vol. 2, 47, Plenum Press, New York and London (1964). A good modern account of the same ground is given by I. Nakagowa and Y. Morioka in "Vibrational Spectra and Structure" (Ed. J. R. Durig), Vol. 9, Chapter 1, Elsevier, Amsterdam and New York (1980).

[494] J. Howard and T. C. Waddington, Molecular spectroscopy with neutrons, *in* "Advances in Infrared and Raman Spectroscopy" (Eds R. J. H. Clark and R. E. Hester), Vol. 7. Heyden & Son, London, Philadelphia and Rheine (1980).

[495] K. D. Moller and W. G. Rothschild, "Far Infrared Spectroscopy", pp. 412–471, Wiley Interscience, Chichester and New York. (1971).

[496] A. Hadni, The interaction of infrared radiation with crystals, *in* "The Infrared Spectra of Minerals" (Ed. V. C. Farmer), The Mineralogical Society, London (1974).

[497] R. H. Lyddane, R. G. Sachs and E. Teller, *Phys. Rev.* **59**, 673 (1941).

[498] D. W. Berreman, *Phys. Rev.* **130**, 2193 (1963).

[499] See for example J. A. B. Beairsto and J. E. Eldridge, *Can. J. Phys.* **51**, 2550 (1973).

[500] A. D. B. Woods, W. Cochran and B. N. Brockhouse, *Phys. Rev.* **119**, 980 (1960).

[501] W. C. Price and G. R. Wilkinson, Final Report, US Army Contract, DA-91-591-EUC-1308-01-4201-60 (1960); see also, G. O. Jones *et al.*, *Proc. R. Soc.* **261**, 10 (1961) and L. Genzel, H. Happ and R. Weber, *Z. Phys.* **154**, 13 (1959).

[502] See for example, M. Lax and E. Burstein, *Phys. Rev.* **93**, 674 (1954); J. R. Hardy and S. D. Smith, *Phil. Mag.* **6**, 1163 (1961), F. A. Johnson, *Prog. Semicond.* **9**, 181 (1965).

[503] T. J. Parker, J. R. Birch and C. L. Mok, *Solid-State Commun.* **36**, 581 (1980).

[504] A. S. Barker, *in* "Far Infrared Properties of Solids" (Eds S. S. Mitra and S. Nudelman), Plenum Press, New York and London (1970).

[505] R. A. Cowley, *Phys. Rev.* **134A**, 981 (1964).

[506] W. Cochran, *Adv. Phys.* **9**, 387 (1960).

[507] W. G. Nilsen and J. G. Skinner, *J. Chem. Phys.* **48**, 2240 (1968).

[508] P. A. Fleury and J. M. Worlock, *Phys. Rev. Lett.* **18**, 665 (1968); *Phys. Rev.* **174**, 613 (1968); P. A. Fleury, J. F. Scott and J. M. Worlock, *Phys. Rev. Lett.* **21**, 16 (1968).

[509] E. Anastassakis, S. Iwasa and E. Burstein, *Phys. Rev. Lett.* **17**, 1051 (1966).

[510] G. Dolling, Lattice dynamics of molecular solids, *in* "Molecular Dynamics and Structure of Solids" (Eds R. S. Carter and J. J. Rush), pp. 289–314, NBS Special Publication No. 301 (1969).

[511] See for example B. Wincke, A. Hadni and X. Gerbaux, *J. Phys.* **31**, 893 (1970).

[512] See for example A. Anderson and H. A. Gebbie, *Spectrochim. Acta*, **21**, 883 (1965).

[513] A. Anderson, H. A. Gebbie and S. H. Walmsley, *Molec. Phys.* **7**, 401 (1964).

[514] J. W. Fleming, P. A. Turner and G. W. Chantry, *Molec. Phys.* **19**, 853 (1970).

[515] G. W. Chantry, H. A. Gebbie and H. N. Mirza, *Spectrochim. Acta* **23A**, 2749 (1967).

[516] G. W. Chantry, A. Anderson and H. A. Gebbie, *Spectrochim. Acta* **20**, 1465 (1964).

[517] P. Debye, *Ann. Phys.* **39**, 789 (1912).

[518] J. N. Plendl, New spectral and atomistic relations in the physics and chemistry of solids, in "Optical Properties of Solids" (Ed. S. S. Mitra), Plenum Press, New York (1969).

[519] A. Sommerfeld, *Naturwissenschaften* **15**, 825 (1927); *Z. Phys.* **47**, 1, 43 (1928).

[520] F. Bloch, *Z. Phys.* **52**, 555 (1928); ibid. **59**, 208 (1929).

[521] R. Peierls, *Ann. Phys. Lpz.* **4**, 121 (1930); *Z. Phys.* **80**, 763 (1933); ibid. **81**, 186 (1933).

[522] N. F. Mott, *Proc. R. Soc.* **A153**, 699 (1936).

[523] See for example A. H. Wilson, "The Theory of Metals", Cambridge University Press, Cambridge (1953); S. Raines, "The Wave-Mechanics of Electrons in Metals", North-Holland, Amsterdam (1961).

[524] R. Kronig and W. G. Penney, *Proc. R. Soc.* (*London*) **A130**, 499 (1931). For a modern treatment of the concept of deep traps, see H. Scher and E. W. Montroll, *Phys. Rev. B*, **12**, 2455 (1975).

[525] R. A. Stradling, *Infrared Phys.* **18**, 435 (1978).

[526] J. H. Reuszer and P. Fisher, *Phys. Rev.* **140A**, 245 (1965).

[527] A. S. Ramdas and S. Rodriguez, *Rep. Prog. Phys.* **44**, 1297 (1981).

[528] C. Jagannath, Z. W. Grabowski and A. K. Ramdas, *Solid-State Commun.* **29**, 355 (1979); C. Jagannath and A. K. Ramdas, *J. Phys. Soc. Japan* **49**, Suppl. A201 (1980); C. Jagannath, *Phys. Rev.* **B23**, 2082, 4426 (1981).

[529] M. N. Afsar, K. J. Button, A. Y. Cho and H. Morkoc, *Int. J. Infrared Millimetre Waves*, **2**, 1113, (1981).

[530] Ya Pokrovskii, *Phys. Status. Solidi* **a11**, 385 (1972).

[531] E. Otsuka, H. Ohyama, H. Nakata and Y. Okada, *J. Opt. Soc. Am.* **67**, 931 (1977).

[532] T. Ohyama, T. Sanada and E. Otsuka, *Phys. Rev. Lett.* **33**, 647 (1974); T. Kawabata, T. K. Muro and S. Narita, *Solid-State Commun.* **23**, 267 (1977).

[533] C. C. Bradley, J. R. Birch and J. R. Stockton, Proceedings of the Conference on Infrared Techniques, Reading UK, 21–23 September (1971), pp. 187–196, IERE Conference Proceedings No. 22; J. R. Birch, *Infrared Phys.* **12**, 29 (1972); J. R. Birch and C. C. Bradley, ibid. **13**, 99 (1973).

[534] A. Kamgar, P. Kneschaurek, G. Dorda and J. F. Koch, *Phys. Rev. Lett.* **32**, 1251 (1974); D. C. Tsui and E. Gornik, *Appl. Phys. Lett.* **32**, 365 (1978).

[535] A. Anderson, G. W. Chantry, H. A. Gebbie, D. H. Whiffen and A. J. Wright, *Spectrochim. Acta* **20**, 1875 (1964).

[536] G. Birnbaum, *Molec. Phys.* **25**, 241 (1973); J. A. Cugley and A. D. E. Pullin, *Chem. Phys. Lett.* **17**, 406 (1972).

[537] P. Debye, "Polar Molecules", Chemical Catalogue Co., New York (1929).

[538] L. D. Landau and E. M. Lifschitz, "Statistical Physics", Section 3, Pergamon Press, London and Paris (1958).

[539] See for example A. Rahman, *Phys. Rev.* **A136**, 405 (1964); P. S. Y. Cheung and J. G. Powles, *Molec. Phys.* **30**, 921 (1975); K. Singer, J. V. L. Singer and A. J. Taylor, *Molec. Phys.* **37**, 1239 (1979).

[540] For a modern examples of the use of the Langevin equation, see J. H.

Calderwood, W. T. Coffey, A. Morita and S. Walker, *Proc. R. Soc.*, **A352**, 275 (1976); G. W. Ford, J. T. Lewis and J. McConnell, *Phys. Rev. A*, **19**, 907 (1979), and G. J. Davies and M. Evans, *J. Chem. Soc. Faraday Trans II*, **72**, 1194 (1976).

[541] J. S. Rowlinson and M. W. Evans, *Ann. Rep. Chem. Soc. A* 5 (1975); see also B. Quentrec and P. Bezot, *Molec. Phys.* **27**, 879 (1974).

[542] F. Hufnagel, *Z. Naturforsch.* **25a**, 1143 (1970).

[543] R. Delker and G. Klages, *Z. Naturforsch.* **36a**, 611 (1981).

[544] See the long series of papers by C. P. Smyth and his students published in *J. Chem. Phys., J. Am. Chem. Soc.* and *J. Phys. Chem.* between the early fifties and the mid sixties. As a typical example, one might quote R. W. Rampolla, R. C. Miller and C. P. Smyth, *J. Chem. Phys.* **30**, 566 (1959). Much of the earlier work is summarised in C. P. Smyth, "Dielectric Behaviour and Structure", McGraw-Hill, New York (1955).

[545] K. S. Cole and R. H. Cole, *J. Chem. Phys.* **9**, 341 (1941); R. H. Cole, *J. Chem. Phys.* **23**, 493 (1955).

[546] D. W. Davidson and R. H. Cole, *J. Chem. Phys.* **18**, 1417 (1950); ibid. **19**, 1484 (1951).

[547] S. Havriliak and S. Negami, *J. Polymer Sci.* **C14**, 99 (1966); *Polymer* **8**, 161 (1967).

[548] C. J. F. Bottcher and P. Bordewijk, "Theory of Electric Polarisation" Vol. 2, "Dielectrics in Time-dependent Fields", Elsevier, Amsterdam (1978).

[549] A. K. Jonscher, *Nature (London)* **267**, 673 (1977); K. L. Ngai, A. K. Jonscher and C. T. White, *ibid.* **277**, 185 (1979).

[550] A. K. Jonscher, L. A. Dissado and R. M. Hill, *Phys. Stat. Solidi*, **B102**, 351 (1980).

[551] R. A. Sack, *Proc. Phys. Soc.* **B70**, 402, 414 (1957)

[552] H. S. Wall, "Analytic Theory of Continued Fractions", Van Nostrand, New York (1948).

[553] Y. Rocard, *J. Phys. Radium* **4**, 247 (1933).

[554] J. G. Powles, *Trans. Faraday Soc.* **44**, 802 (1948).

[555] J. McConnell, "Rotational Brownian Motion and Dielectric Theory", Academic Press, London and New York (1980).

[556] G. W. Ford, J. T. Lewis and J. McConnell, *Proc. R. Ir. Acad. Sec. A* **76**, 117 (1976).

[557] for example see J. McConnell, *Proc. R. Ir. Acad. Sec. A* **78**, 87 (1978).

[558] W. A. Steele, *J. Chem. Phys.* **43**, 2598 (1965); G. Williams, *Chem. Rev.* **72**, 55 (1972); *Chem. Soc. Rev.* **7**, 89 (1978); M. W. Evans, "Dielectric and Related Molecular Processes" (Ed. Mansel Davies), Vol. 3, Chemical Society London (1977).

[559] B. K. P. Scaife, European Molecular Liquids Group, Conference on Analytical and Computational Studies of Basic Problems in Molecular Liquids, Dublin, April (1982).

[560] K. M. Case, *Transp. Theor. Stat. Phys.* **2**, 129 (1972).

[561] R. Kubo, *J. Phys. Soc. Japan* **12**, 570 (1957); "Lectures in Theoretical Physics", Vol. 1, Chapter 4, Interscience, New York (1958); *Rep. Prog. Phys.* **29**, 255 (1966).

[562] G. Birnbaum and E. R. Cohen, *J. Chem. Phys.* **53**, 2885 (1970).

[563] J. T. Lewis, European Molecular Liquids Group, Conference on Analytical and Computational Studies of Basic Problems in Molecular Liquids, Dublin, April (1982).

[564] G. J. Evans and M. W. Evans, *J. Chem. Soc. Faraday Trans.* 2, **72**, 1169 (1976): G. J. Davies and M. W. Evans, ibid. 1194.

[565] H. Mori, *Prog. Theor. Phys.* **33**, 423 (1965): see also the article by P. Schofield *in* "Statistical Mechanics" (Ed. K. Singer), Vol. 2, Specialised Periodical Reports, The Chemical Society (1975).

[566] An early, but still very pertinent, contribution to this topic can be found in B. J.

Berne and G. Harp, Adv. Chem. Phys., **17**, 63, (1970): G. Harp and B. J. Berne, *Phys. Rev., A2*, 975, (1970).

[567] J. Ph. Poley, *J. Appl. Sci.* **B4**, 337 (1955).

[568] G. W. Chantry, and H. A. Gebbie, *Nature (London)* **208**, 378 (1965).

[569] G. W. Chantry, *IEEE Trans.* **MTT-25**, 496 (1977).

[570] N. E. Hill, *J. Phys. C. (Solid-State Phys.)*, **10**, 459 (1977); ibid. **11**, 815 (1978).

[571] G. W. Chantry, J. R. Birch, J. H. Calderwood and J. McConnell, *Int. J. Infrared Millimetre Waves*, to be published.

[572] H. Kilp, D. C. Barnes, F. W. J. Clutterbuck, M. N. Afsar and G. W. Chantry, *Infrared Phys.* **18**, 11 (1978).

[573] C. C. Bradley, H. A. Gebbie, A. C. Gilby, V. V. Kechin and J. H. King, *Nature (London)* **211**, 839 (1966).

[574] G. W. Chantry, J. W. Fleming, E. A. Nicol, H. A. Willis and M. E. A. Cudby, *Infrared Phys.* **12**, 101 (1972); G. W. Chantry, J. W. Fleming, R. J. Cook, D. G. Moss, E. A. Nicol, H. A. Willis and M. E. A. Cudby, *Infrared Phys.* **13**, 157 (1973).

[575] V. Hoffmann, W. Frank and W. Zeil, *Kolloid-Z., Z. Polymers* **241**, 1044 (1970).

[576] E. A. Nicol, B. E. Read and G. W. Chantry, *Br. Polym. J.* **5**, 379 (1973).

[577] see for example, D. D. Klug,. D. E. Kranbuehl and W. E. Vaughan, *J. Chem. Phys.* **50**, 3904 (1969).

[578] R. G. Gordon, *J. Chem. Phys.* **42**, 3658 (1965); ibid. **43**, 1307 (1965); ibid. **44**, 1830 (1966); *Adv. Magn. Resonance* **3**, 1 (1968).

[579] M. W. Evans and G. J. Davies, *J. Chem. Soc. Faraday Trans.* 2, **72**, 1206 (1976).

[580] E. Fatuzzo and P. R. Mason, *Proc. Phys. Soc. (London)* **90**, 741 (1967); see also U. M. Titulaer and J. M. Deutch, *J. Chem. Phys.* **60**, 1502 (1974), and R. Finsy and R. Van Loon, *J. Chem. Phys.* **63**, 4831 (1975).

[581] J. L. Rivail, *J. Chim. Phys.* **66**, 981 (1969).

[582] T. W. Nee and R. Zwanzig, *J. Chem. Phys.* **52**, 6353 (1970); see also W. E. Vaughan, Dielectric relaxation, *in Ann. Rev. Phys. Chem.* **30**, 103 (1979).

[583] G. Bossis, *Physica* **110A**, 408 (1982).

[584] G. Bossis, B. Quentrec and C. Brot, *Molec. Phys.* **39**, 1233 (1980); G. Bossis and C. Brot, *Molec. Phys.* **43**, 1095 (1981).

[585] A. Bellemans, European Molecular Liquids Group, Meeting at Dublin April (1982); see also M. W. Evans, *Adv. Molec. Rel. Int. Proc.* **10**, 203 (1977).

[586] J. G. Powles, W. A. B. Evans, E. McGrath, K. E. Gubbins and S. Murad, *Molec. Phys.* **38**, 893 (1979); J. G. Powles, E. McGrath and K. E. Gubbins, ibid. **40**, 179 (1980); S. Murad, K. E. Gubbins and J. G. Powles, ibid. **40**, 253 (1980).

[587] see M. W. Evans and J. Yarwood, *Adv. Molec. Rel. Int. Proc.* **21**, 1 (1981).

[588] R. Zwanzig, *Ann. Rev. Phys. Chem.* **16**, 67 (1965).

[589] R. E. D. McClung, *J. Chem. Phys.* **57**, 5478 (1972).

[590] C. J. Reid and M. W. Evans, *Molec. Phys.* **40**, 1357 (1980).

[591] S. Bratos and R. M. Pick, "Vibrational Spectroscopy of Molecular Liquids and Solids", Plenum Press, New York (1980).

[592] M. Takami and K. Shimoda, *Jap. J. Appl. Phys.* **10**, 658 (1971).

[593] K. M. Evenson, H. E. Radford and J. M. Moran, *Appl. Phys. Lett.* **18**: F. D. Wayne and H. E. Radford, *Molec. Phys.* **32**, 1407 (1976).

[594] A. Kaldor, W. B. Olson and A. G. Maki, *Science* **176**, 508 (1972); P. A. Bonczyck, *Chem. Phys. Lett.* **18**, 147 (1973); H. J. Zeiger, F. A. Blum and K. Nill, *J. Chem. Phys.* **59**, 3968 (1973); K. Hakuta and H. Uehara, *J. Molec. Spectrosc.* **58**, 316 (1975); R. M. Dale, J. W. C. Johns, A. R. W. McKellar and M. Riggin, *J. Molec. Spectrosc.* **67**, 440 (1977).

[595] S. M. Freund, J. T. Hougen and W. J. Lafferty, *Can. J. Phys.* **53**, 1929 (1975).

[596] S. V. Broude, Y. M. Gershenzon, S. D. Il'in and S. A. Kolesnikov, *Doklady Phys. Chem.* (English translation), **223**, 705 (1975).

[597] J. M. Brown, J. Buttenshaw, A. Carrington and C. R. Parent, *Molec. Phys.* **33**, 589 (1977).

[598] M. Dagenais, J. W. C. Johns and A. R. W. McKellar, *Can. J. Phys.* **54**, 1438 (1976); J. W. C. Johns, A. R. W. McKellar and M. Riggin, *J. Chem. Phys.* **66**, 3962 (1977).

[599] J. W. C. Johns, A. R. W. McKellar and M. Riggin, *J. Chem. Phys.* **67**, 2427 (1977).

[600] N. Muira, G. Kido and S. Chikazumi, *Solid-State Commun.* **18**, 885 (1976).

[601] D. J. E. Ingram, "Free-Radicals as studied by Electron-Spin-Resonance", Butterworth, London (1958); P. B. Ayscough, "Electron-Spin-Resonance in Chemistry", Methuen, London (1967); W. T. Dixon, "Theory and Interpretation of Magnetic Resonance Spectra", Plenum Press, London and New York (1972); A. Carrington, "Microwave Spectroscopy of Free-Radicals", Academic Press, London and New York (1974); for a review of applications see A. Carrington, *Quart. Rev.* **17**, 67 (1963).

[602] J. W. Emsley, J. Feeney and L. H. Sutcliffe, "High Resolution Nuclear Magnetic Resonance Spectroscopy", Vols 1 and 2, Pergamon Press, Oxford (1966).

[603] R. A. Stradling, Proceedings of International Conference, Physics of Semiconductors, p. 1187, Edinburgh, Institute of Physics, Conference Series, Number 43 (1979).

[604] E. D. Palik and J. R. Stevenson, *Phys. Rev.* **130**, 1344 (1963); Y. Couder, *Phys. Rev. Lett.* **22**, 890 (1969).

[605] K. J. Button, H. A. Gebbie and B. Lax, *IEEE J. Quant. Electron.* **QE-2**, 202 (1966).

[606] C. C. Bradley, K. J. Button B. Lax and L. G. Rubin, *IEEE J. Quant. Electron.* **QE-4**, 733 (1968); K. J. Button, B. Lax and C. C. Bradley, *Phys. Rev. Lett.* **21**, 350 (1968); ibid. **23**, 14 (1969): J. Waldman, D. M. Larsen, P. E. Tannenwald, C. C. Bradley, D. R. Cohn and B. Lax, *Phys. Rev. Lett.* **23**, 1033 (1969).

[607] C. C. Bradley, P. E. Simmonds, J. R. Stockton and R. A. Stradling, *Solid-State Commun.* **12**, 413 (1973); J. Leotin, J. C. Ousset, R. Barbaste, S. Askenazy, M. S. Skolnick, R. A. Stradling and G. Poiblaud, *Solid-State Commun.* **16**, 353 (1975).

[608] R. J. Nicholas, R. A. Stradling, J. C. Ramage, J. C. Portal and S. Askenazy, to be published.

[609] U. Strom, H. D. Drew and J. F. Koch, *Phys. Rev. Lett.* **26**, 1110 (1971).

[610] Reference [235] (J. T. Houghton and S. D. Smith), pp. 262 to 266.

[611] R. G. Wheeler, F. M. Reames and E. J. Wachtel, *J. Appl. Phys.* **39**, 915 (1968).

[612] P. L. Richards, *J. Appl. Phys.* **34**, 1237 (1963); **35**, 850 (1964).

[613] J. C. Wright and H. W. Moos, *Phys. Lett.* **29A**, 495 (1969).

[614] T. G. Blocker, M. A. Kinch and F. G. West, *Phys. Rev. Lett.* **22**, 853 (1969).

[615] P. L. Richards, *Bull. Am. Phys. Soc.* **10**, 33 (1965).

[616] P. L. Richards, reference [375], but see also M. Tinkham, *Phys. Rev.* **124**, 311 (1961); A. J. Sievers and M. Tinkham, ibid. 321.

[617] A. J. Sievers and M. Tinkham, *J. Appl. Phys.* **34**, 1235 (1963).

[618] R. H. Hughes and E. B. Wilson, *Phys. Rev.* **71**, 562 (1947); A. H. Sharbaugh, *Rev. Sci. Instrum.* **21**, 120 (1950); M. L. Stitch, A. Honig and C. H. Townes, *Rev. Sci. Instrum.* **25**, 759 (1954); C. O. Britt, *Rev. Sci. Instrum.* **38**, 1496 (1967).

[619] M. H. Alexander, *Phys. Rev.* **178**, 34 (1969).

[620] I. M. Mills, *in* "Molecular Spectroscopy Modern Research" (Ed. K. Narahari Rao and C. W. Mathews), pp. 115–140, Academic Press, New York and London (1972).

[621] G. Duxbury and R. G. Jones, *Molec. Phys.* **20**, 721 (1971).

[622] K. M. Evenson, J. S. Wells and H. E. Radford, *Phys. Rev. Lett.* **25**, 199 (1970).

[623] K. M. Evenson, H. E. Radford and J. Moran, *Appl. Phys. Lett.* **18**, 426 (1971).

[624] H. E. Radford, K. M. Evenson and C. J. Howard, *J. Chem. Phys.* **60**, 3178 (1974).

[625] H. E. Radford and M. M. Litvak, *Chem. Phys. Lett.* **34**, 561 (1975); F. D. Wayne and H. E. Radford, *Molec. Phys.* **32**, 1407 (1976).

[626] P. B. Davies, D. K. Russell, B. A. Thrush and F. D. Wayne, *J. Chem. Phys.* **62**, 3739 (1975).

[627] P. B. Davies, D. K. Russell and B. A. Thrush, *Chem. Phys. Lett.* **36**, 280 (1975).

[628] P. B. Davies, D. K. Russell and B. A. Thrush, *Chem. Phys. Lett.* **37**, 43 (1976).

[629] J. M. Cook, K. M. Evenson, C. J. Howard and R. F. Curl, *J. Chem. Phys.* **64**, 1381 (1976).

[630] H. E. Radford and D. K. Russell, *J. Chem. Phys.* **66**, 2222 (1977).

[631] B. M. Landsberg, A. J. Merer and T. Oka, *J. Molec. Spectrosc.* **67**, 459 (1977).

[632] J. W. C. Johns and A. R. W. McKellar, *J. Chem. Phys.* **66**, 1217 (1977).

[633] A nice example will be found in the paper by W. A. Kreiner, T. Oka and A. G. Robiette, *J. Chem. Phys.* **68**, 3236 (1978).

[634] J. H. Taylor, C. S. Rupert and J. Strong, *J. Opt. Soc. Am.* **41**, 626 (1951).

[635] W. Brugel, *Z. Phys.* **127**, 400 (1950); S. Silvermann, *J. Opt. Soc. Am.* **38**, 989 (1948).

[636] C. S. Willett, "Introduction to Gas Lasers: Population Inversion Mechanisms", Chapter 4, Pergamon Press, Oxford and New York (1974).

[637] W. W. Duley, "CO_2 Lasers: Effects and Applications", Academic Press, New York and London (1976).

[638] L. O. Hocker, D. R. Sokoloff, V. Daneu, A. Szoke and A. Javan, *Appl. Phys. Lett.* **12**, 401 (1968); T. Y. Chang, N. Van Tran and C. K. N. Patel, **13**, 401 (1968).

[639] T. J. Bridges and T. Y. Chang, *Phys. Rev. Lett.* **22**, 811 (1969).

[640] J. C. Siddoway, *J. Appl. Phys.* **39**, 4854 (1968).

[641] C. Freed, A. H. M. Ross and R. G. O'Donnell, *J. Molec. Spectrosc.* **49**, 439 (1974).

[642] K. M. Evenson, J. S. Wells and L. M. Matarrese, *Appl. Phys. Lett.* **16**, 251 (1970).

[643] T. G. Blaney, C. C. Bradley, G. J. Edwards, B. W. Jolliffe, D. J. E. Knight, W. R. C. Rowley, K. C. Shotton and P. T. Woods, *Proc. R. Soc. Lond.* **A355**, 61 (1977).

[644] For example B. G. Whitford, *Opt. Commun.* **31**, 363 (1979).

[645] A. L. S. Smith and P. G. Browne, *J. Phys. D: Appl. Phys.* **7**, 1652 (1974).

[646] R. L. Johnson and J. D. O'Keefe, *Appl. Opt.* **11**, 2926 (1972).

[647] J. A. Dobrowolski and F. Ho, *Appl. Opt.* **21**, 288 (1982)

[648] W. Wild and K. Wilner, *Appl. Opt.* **18**, 3880 (1979); R. B. James, E. Schweig, D. L. Smith and T. C. McGill, *Appl. Phys. Lett.* **40**, 231 (1982).

[649] A. J. Glass and A. H. Guenther, NBS Special Reports No. 414 (1974) and 435 (1975).

[650] M. J. Soileau and V. Wang, *Appl. Opt.* **13**, 1286 (1974).

[651] T. A. Wiggins and R. S. Reid, *Appl. Opt.* **21**, 1675 (1982).

[652] W. R. C. Rowley, B. W. Jolliffe, K. C. Shotton, A. J. Wallard and P. T. Woods, *Opt. Quant. Electron.* **8**, 1 (1976).

[653] Edinburgh Instruments Ltd., Commercial Literature.

[654] G. W. Flynn, L. O. Hocker, A. Javan, M. A. Kovacs and C. K. Rhodes, *IEEE J. Quant. Electron.* **QE-2**, 378 (1966)

[655] O. R. Wood and S. E. Schwarz, *Appl. Phys. Lett.* **11**, 88 (1967); T. Y. Chang, C. H. Wang and P. K. Cheo, *Appl. Phys. Lett.* **15**, 157 (1969); A. Szoke, V. Daneu, J. Goldhar and N. A. Kurmit, *Appl. Phys. Lett.* **15**, 376 (1969).

[656] See the article, "CO_2 Short-pulse Technology" by T. E. Stratton in the Institute of Physics Conference Series, Vol. 29, devoted to the Conference on High-Power Gas Lasers (1975).

[657] O. R. Wood, *Proc. IEEE* **62**, 355 (1974).

[658] See for example H. J. Seguin, K. Manes and J. Tulip, *Rev. Sci. Instrum.* **43**, 1134 (1972).

[659] A. J. Beaulieu, *Appl. Phys. Lett.* **16**, 504 (1970); K. A. Laurie and M. Hale, *IEEE J. Quant. Electron.* **QE-6**, 530 (1970).

[660] W. Rogowski, *Arch. f. Elekt.* **12**, 1 (1923); W. Rogowski and H. Rengier, ibid. **16**, 73 (1926).

[661] P. R. Pearson and H. M. Lamberton, *IEEE J. Quant. Electron.* **QE-8**, 145 (1972).

[662] H. M. Lamberton and P. R. Pearson, *Electron. Lett.* **7**, 141 (1971).

[663] B. Norris and A. L. S. Smith, *J. Phys. E. (Sci. Instrum.)* **10**, 551 (1977).

[664] C. A. Fenstermacher, M. J. Nutter, J. P. Rink and K. Boyer, *Bull. Am. Phys. Soc.* **16**, 42 (1971); C. A. Fenstermacher, M. J. Nutter, W. T. Leland and K. Boyer, *Appl. Phys. Lett.* **20**, 56 (1972).

[665] J. D. Daugherty, E. R. Pugh and D. H. Douglas-Hamilton, *Bull. Am. Phys. Soc.* **17**, 399 (1972); J. D. Daugherty, E. R. Pugh, D. A. House and D. H. Douglas-Hamilton, *Phys. Today* **25**, 18 (1972).

[666] See for example W. B. Tiffany, R. Targ and J. D. Foster, *Appl. Phys. Lett.* **15**, 91 (1969); see also R. T. Brown, *IEEE J. Quant. Electron.* **QE-9**, 1120 (1973).

[667] For a discussion of high-power continuous-wave CO_2 lasers see A. J. Maria, *Proc. IEEE* **61**, 731 (1973).

[668] K. J. Siemsen and J. Reid, *Opt. Commun.* **20**, 284 (1977).

[669] K. J. Siemsen and B. G. Whitford, *Opt. Commun.* **22**, 11 (1977).

[670] R. L. Abrams and W. B. Bridges, *IEEE J. Quant. Electron.* **QE-9**, 940 (1973).

[671] D. Crocker, C. D. Clack and R. J. Butcher, *J. Phys. E. (Sci. Instrum.)*, **14**, 121 (1981).

[672] R. L. Abrams, *IEEE J. Quant. Electron.* **QE-8**, 838 (1972); J. J. Degnan, *Appl. Phys.* **11**, 1 (1976).

[673] J. J. Degnan and D. R. Hall, *IEEE J. Quant. Electron.* **QE-9**, 901 (1973).

[674] D. R. Hall, Waveguide lasers, *in* "Laser Advances and Applications" (Ed. B. A. Wherrett), Wiley-Interscience, Chichester and New York (1980).

[675] G. M. Carter and S. Marcus, *Appl. Phys. Lett.* **35**, 129 (1979).

[676] M. B. Klein and R. L. Abrams, *IEEE J. Quant. Electron.* **QE-11**, 609 (1975).

[677] P. W. Smith, P. J. Maloney and O. R. Wood, *Appl. Phys. Lett.* **23**, 524 (1973).

[678] S. Lovold and G. Wang, *Appl. Phys. Lett.* **40**, 13 (1982).

[679] F. O'Neill and W. T. Whitney, *Appl. Phys. Lett.* **26**, 454 (1975).

[680] N. W. Harris, F. O'Neill and W. T. Whitney, *Opt. Commun.* **16**, 57 (1976); F. O'Neill and W. T. Whitney, *Appl. Phys. Lett.* **28**, 539 (1976).

[681] L. B. Kreuzer, N. D. Kenyon and C. K. N. Patel, *Science* **177**, 347 (1972); R. R. Patty, G. M. Russwurm, W. A. McClenny and D. R. Morgan, *Appl. Opt.* **13**, 2850 (1974).

[682] R. L. Byer, *Opt. Quant. Electron.* **7**, 147 (1975).

[683] N. G. Basov and A. N. Oraevskii, *Sov. Phys. J.E.T.P.* **17**, 1171 (1963).

[684] J. D. Anderson, "Gas Dynamic Lasers—an Introduction", Academic Press, New York and London (1976); S. A. Losev, "Gas Dynamic Lasers", Springer series in Chemical Physics, Vol. 12, Springer-Verlag, Berlin and New York (1982).

[685] J. D. Anderson, Gas dynamic lasers—a state of the art survey, *NEREM Record* **14**, 180 (1972); see also for example J. Tulip and H. Sequin, *Appl. Phys. Lett.* **19**, 263 (1971).

[686] See for example E. T. Gerry, *IEEE Spectrum* **7**, 51 (1970).

[687] E. L. Klosterman and A. L. Hoffman, "Recent Developments in Shock Tube Research", p. 156, Stanford University Press, Stanford, California (1973).

[688] F. Legay and N. Legay–Sommaire *C.r. hebd. Séanc., Acad. Sci.* **A259**, 99, (1964); C. K. N. Patel and R. J. Kerl, *Appl. Phys. Lett.* **5**, 81 (1964).

[689] E. L. Klosterman and S. R. Byron, to be published; M. M. Mann, D. K. Rice and R. G. Eguchi, *IEEE J. Quant. Electron.* **QE-10**, 682 (1974); M. J. W. Boness and R. E. Center, *J. Appl. Phys.* **48**, 2705 (1977); a good account of high-power CO

laser research is given by R. E. Center, *in* "Laser Handbook" (Ed. M. L. Stitch), Vol. 3, Chapter A3, North-Holland, Amsterdam (1979).

[690] N. Legay–Sommaire, L. Henry and F. Legay, *C. r. hebd. séanc. Acad. Sci.* **A260**, 3339 (1965); C. K. N. Patel, *Appl. Phys. Lett.* **7**, 246 (1965).

[691] R. M. Osgood and W. C. Eppers, *Appl. Phys. Lett.* **13**, 409 (1968); R. M. Osgood, E. R. Nichols and W. C. Eppers, *Appl. Phys. Lett.* **15**, 69 (1969); R. M. Osgood, W. C. Eppers and E. R. Nichols, *IEEE J. Quant. Electron.* **QE-6**, 145 (1970).

[692] M. A. Pollock, *Appl. Phys. Lett.* **8**, 237 (1966); D. W. Gregg and S. J. Thomas, *J. Appl. Phys.* **39**, 4399 (1968); G. Hancock and I. W. M. Smith, *Trans. Faraday Soc.* **67**, 2586 (1971); C. Wittig, J. C. Hassler and P. D. Coleman, *J. Chem. Phys.* **55**, 5523 (1971); W. Q. Jeffers and H. Y. Ageno, *Appl. Phys. Lett.* **27**, 227 (1975); C. J. Ultee, Chemical and gas dynamic lasers, *in* "Laser Handbook" (Ed. M. L. Stitch), Vol. 3, North-Holland, Amsterdam (1979).

[693] R. L. McKenzie, *Appl. Phys. Lett.* **17**, 462 (1970); W. S. Watt, *Appl. Phys. Lett.* **18**, 487 (1971); E. L. Klosterman, Mathematical Sciences, Northwest Inc., Final Technical Report 75–123–1, Contract N00014–24–C–0258 March (1975).

[694] J. D. Dougherty, E. R. Pugh and D. H. Douglas-Hamilton, *Bull. Am. Phys. Soc.* **17**, 3 (1972); C. H. Fenstermacher, M. J. Nutter, W. T. Leland and K. Boyer, *Bull. Am. Phys. Soc.* **17**, 399 (1972); M. M. Mann, D. K. Rice and R. G. Eguchi, *IEEE J. Quant. Electron.* **QE-10**, 682 (1974); M. J. W. Boness and R. E. Center, *J. Appl. Phys.* **48**, 2705 (1977).

[695] M. L. Bhaumik, W. B. Lacina and M. M. Mann, *IEEE J. Quant. Electron.* **QE-6**, 575 (1970); **QE-8**, 150 (1972); M. L. Bhaumik, *Appl. Phys. Lett.* **17**, 188 (1970); **20**, 342 (1972).

[696] See for example L. R. Boedeker, J. A. Shirley and B. R. Bronfin, *Appl. Phys. Lett.* **21**, 247 (1972).

[697] R. C. Millikan and D. R. White, *J. Chem. Phys.* **39**, 3209 (1963); R. Joeckle and M. Peyron, *J. Chim. Phys.* **67**, 1175 (1970); C. E. Caledonia and R. E. Center, *J. Chem. Phys.* **55**, 552 (1971).

[698] C. E. Treanor, J. W. Rich and R. G. Rehm, *J. Chem. Phys.* **48**, 1728 (1968).

[699] See for example R. E. Center and G. E. Caledonia, *Appl. Opt.* **10**, 1795 (1971).

[700] J. C. Polanyi, Chemical lasers, *Appl. Opt. Suppl.* **2**, 109 (1965).

[701] W. B. Roh and K. N. Rao, *J. Molec. Spectrosc.* **49**, 317 (1974).

[702] H. M. Mould, W. C. Price and G. R. Wilkinson, *Spectrochim. Acta* **16**, 479 (1960).

[703] R. T. Menzies, *Appl. Opt.* **10**, 1532 (1971).

[704] G. W. Chantry and G. Duxbury, Molecular lasing systems, *in* "Methods of Experimental Physics", 2nd Edition (Ed. Dudley Williams), Vol. 3, Part A, Academic Press, New York and London (1974); F. K. Kneubuhl and Ch. Sturzenegger, Electrically excited sub-mm wave lasers, *in* "Infrared and mm-Waves", Vol. 3, "Submillimeter Techniques" (Ed. K. J. Button), Academic Press, New York and London (1980).

[705] J. Chamberlain, E. G. C. Werner, H. A. Gebbie and W. Slough, *Trans. Faraday Soc.* **63**, 2605 (1967); see also J. Chamberlain *et al.*, *Infrared Phys.* **11**, 25 (1971).

[706] See for example C. C. Bradley and D. J. E. Knight, *Phys. Lett.* **32A**, 59 (1970).

[707] See for example H. A. Gebbie, *New Scientist* **25**, 426 (1965).

[708] J. Chamberlain H. A. Gebbie, A. George and J. D. E. Beynon, *J. Plasma Phys.* **3**, 75 (1969); G. J. Parkinson, A. E. Dangor and J. Chamberlain, *Appl. Phys. Lett.* **13**, 233 (1968).

[709] See for example A. H. Brittain, A. P. Cox, G. Duxbury, T. G. Hersey and R. G. Jones, *Molec. Phys.* **24**, 843 (1972); K. M. Evenson, H. E. Radford and M. M. Moran, *Appl. Phys. Lett.* **18**, 426 (1971); R. F. Curl, K. M. Evenson and J. S. Wells, *J. Chem. Phys.* **56**, 5143 (1972).

[710] A. G. Maki, *Appl. Phys. Lett.* **12**, 122 (1968); L. O. Hocker and A. Javan, *Appl. Phys. Lett.* **12**, 124 (1968).

[711] A. Crocker, H. A. Gebbie, M. F. Kimmitt and L. E. S. Mathias, *Nature (London)* **201**, 250 (1964); W. J. Witteman and R. Bleekrode, *Phys. Lett.* **13**, 126 (1964); W. W. Muller and G. T. Flesher, *Appl. Phys. Lett.* **8**, 217 (1966).

[712] See for example P. Belland, D. Veron and L. B. Whitbourn, *Appl. Opt.* **15**, 3047 (1976).

[713] See for example V. Daneu, L. O. Hocker, A. Javan, D. Ramachandra Rao, A. Szoke and F. Zernike, *Phys. Lett.* **29A**, 319 (1969).

[714] K. Uehara, T. Shimizu and K. Shimoda, *IEEE J. Quant. Electron.*, **QE-4**, 728 (1968).

[715] L. E. S. Mathias, A. Crocker and M. S. Wills, *Phys. Lett.* **14**, 33 (1965); D. P. Akitt and C. F. Wittig, *J. Appl. Phys.* **40**, 902 (1969).

[716] D. R. Lide, *Phys. Lett.* **24A**, 599 (1967).

[717] D. J. E. Knight, "Laser Handbook", to be published.

[718] A. Voss, B. W. Davis, W. J. Firth and C. R. Pidgeon, to be published.

[719] A. Javan, *Phys. Rev.* **107**, 1579 (1957); T. Y. Chang, Optical Pumping in Gases, "Topics in Applied Physics" (Ed. Y. T. Shen), Vol. 16, p. 215, Springer-Verlag, Berlin (1977); see also G. A. Koepf and K. Smith, *IEEE J. Quant. Electron.* **QE-14**, 333 (1978), who give further references.

[720] D. T. Hodges, Advances in optically pumped far infrared lasers, *in* "Submillimetre Waves and Their Applications" (Ed. G. W. Chantry), Pergamon Press, Oxford (1978).

[721] H. Jones, *J. Molec. Spectrosc.* **53**, 118 (1978).

[722] L. D. Fesenko and S. F. Dyubko, *Sov. J. Quant. Electron.* **6**, 839 (1976); see also G. A. Koepf and N. McAvoy, *IEEE J. Quant. Electron.* **QE-13**, 418 (1977).

[723] D. T. Hodges and T. S. Hartwick, *Appl. Phys. Lett.* **23**, 252 (1973); M. Yamanaka, *J. Opt. Soc. Am.* **67**, 952 (1977).

[724] G. Dodel, G. Magyar and D. Veron, *Infrared Phys.* **18**, 529 (1978).

[725] See for example R. J. Temkin, D. R. Cohn, Z. Dirozdowicz and F. Brown, *Opt. Commun.* **14**, 314 (1975); Z. Dirozdowicz, R. J. Temkin, K. J. Button and D. R. Cohn, *Appl. Phys. Lett.* **29**, 328 (1975); A. Serret and N. C. Lukmann, *Appl. Phys. Lett.* **28**, 659 (1976); D. E. Evans, L. E. Sharp, W. A. Peebles and G. Taylor, *IEEE J. Quant. Electron.* **QE-13**, 54 (1977); Z. Dirozdowicz, P. Woskoboinikow, K. Isobe, D. R. Cohn, R. J. Temkin, K. J. Button and J. Waldman, *IEEE J. Quant. Electron.* **QE-13**, 413 (1977).

[726] E. J. Danielewicz, T. K. Plant and T. A. DeTemple, *Opt. Commun.* **13**, 366 (1975); E. J. Danielewicz and P. D. Coleman, *Appl. Opt.* **15**, 761 (1976); M. R. Schubert, M. S. Durschlag and T. A. DeTemple, *IEEE. J. Quant. Electron.* **QE-13**, 455 (1977); D. T. Hodges, F. B. Foote and R. D. Reel, *Appl. Phys. Lett.* **29**, 662 (1976).

[727] D. P. Hutchinson, S. P. I. E. Seminar on Far Infrared/submillimeter Wave Technology and Applications, *SPIE* **105**, 80 (1977).

[728] F. Shimizu, *J. Chem. Phys.* **52**, 3572 (1970); **53**, 1149 (1970).

[729] R. A. Wood, B. W. Davis, A. Vass and C. R. Pidgeon, *Opt. Lett.* **5**, 153 (1980); B. W. Davis, A. Vass, C. R. Pidgeon and G. R. Allan, *Opt. Commun.* **37**, 303 (1981).

[730] T. Y. Chang, T. J. Bridges and E. G. Burkhardt, *Appl. Phys. Lett.* **17**, 357 (1970).

[731] T. Y. Chang and J. D. McGee, *Appl. Phys. Lett.* **28**, 526 (1976); **29**, 725 (1976).

[732] T. Oka, *J. Chem. Phys.* **48**, 4919 (1968).

[733] P. K. Gupta and R. G. Harrison, *IEEE J. Quant. Electron.* **QE-17**, 2238 (1981).

[734] B. W. Davis, A. Vass, C. R. Pidgeon and G. R. Allan, *Opt. Commun.* **37**, 303 (1981).

[735] C. R. Pidgeon, W. J. Firth, R. A. Wood, A. Vass and B. W. Davis, *Int. J. Infrared Millimetre Waves* **2**, 207 (1981).

[736] C. K. N. Patel, T. Y. Chang and V. T. Nguyen, *Appl. Phys. Lett.* **28**, 603 (1976).

[737] B. Walker, G. W. Chantry and D. G. Moss, *Opt. Commun.* **23**, 8 (1977).

[738] E. D. Shaw and C. K. N. Patel, *Opt. Commun.* **27**, 419 (1978); A. Z. Girasiuk, *Appl. Phys.* **21**, 173 (1980).

[739] R. G. Harrison, P. K. Gupta, A. K. Kar and M. R. Taghizadeh, *Opt. Commun.* **34**, 445 (1980); P. K. Gupta, A. K. Kar, M. R. Taghizadeh, *Opt. Commun.* **34**, 445 (1980); P. K. Gupta, A. K. Kar, M. R. Taghizadeh and M. G. Harrison, *Appl. Phys. Lett.* **39**, 32 (1981).

[740] C. R. Pidgeon, W. J. Forth, R. A. Wood, A. Voss and B. W. Davis, *Int. J. Infrared Millimetre Waves* **2**, 207 (1981).

[741] S. J. Petuchowski, A. T. Rosenberger and T. A. DeTemple, *IEEE J. Quant. Electron.* **QE-13**, 476 (1977); J. D. Wiggins, Z. Dirozdowicz and R. J. Temkin, *ibid.* **QE-14**, 23 (1978).

[742] T. K. Plant, L. A. Newman, E. J. Danielewicz, T. A. De Temple and P. D. Coleman, *IEEE Trans. Microwave Theor. Techniques* **MTT-22**, 988 (1974); G. Taylor, G. F. D. Levy, D. A. Huckridge and D. E. Evans, *Infrared Phys.* **18**, 509 (1978).

[743] R. L. Abrams, *Appl. Phys. Lett.* **25**, 304 (1974); M. F. Lyszyk, F. Herlemont and J. Lemaire, *J. Phys. E*, (Sci. Instrum.), **10**, 1110 (1977); A. Leberghe, S. Avrillier and C. J. Borde, *IEEE J. Quant. Electron.* **QE-14**, 481 (1978); D. Crocker and R. J. Butcher, *Infrared Phys.* **21**, 85 (1981); M. S. Tobin, J. P. Sattler and T. W. Daley, *IEEE J. Quant. Electron.* **QE-18**, 79 (1982).

[744] M. Redon, C. Gastaud and M. Fourrier, *Int. J. Infrared Millimetre Waves* **1**, 95 (1980); *Infrared Phys.* **20**, 93 (1980); C. Gastaud, A. Sentz, M. Redon and M. Fourrier, *IEEE J. Quant. Electron.* **QE-16**, 1285 (1980); *Opt. Lett.* **6**, 449 (1981).

[745] F. C. Vanden Heuvel, W. L. Meerts and H. Dymanus, to be published.

[746] G. A. Koepf, H. R. Fetterman and N. McAvoy, *Int. J. Infrared Millimeter Waves*, **1**, 597 (1980).

[747] K. L. Kompa, Topics in Current Chemistry No. 37, "Chemical Lasers", Springer-Verlag, Berton, Heidelberg and New York (1973).

[748] N. Jonathan, C. M. Melliar-Smith, S. Okuda, D. H. Slater and D. H. Timlin, *Molec. Phys.* **22**, 561 (1971).

[749] A. N. Chester, "Chemical Lasers", *Inst. Phys. Conf. Ser.* No. 29, p. 162 (1976).

[750] See for example K. L. Kompa and J. Wanner, *Chem. Phys. Lett.* **12**, 560 (1972).

[751] R. P. Sorokin and J. R. Lankard, *J. Chem. Phys.* **51**, 2929 (1969); *ibid.* **54**, 2184 (1971).

[752] O. R. Wood and T. Y. Chang, *Appl. Phys. Lett.* **20**, 77 (1972).

[753] T. F. Deutsch, *Appl. Phys. Lett.* **10**, 234 (1967).

[754] For example R. K. Pearson, J. O. Cowles, G. I. Hermann, D. W. Gregg and J. R. Creighton, *IEEE J. Quant. Electron.* **QE-9**, 879 (1973).

[755] J. Goldhar, R. M. Osgood and A. Javan, *Appl. Phys. Lett.* **18**, 167 (1971).

[756] T. F. Deutsch, *Appl. Phys. Lett.* **11**, 18 (1967); D. Pakitt and J. T. Yardley, *IEEE J. Quant. Electron.* **QE-6**, 113 (1970); N. Skribanovitz, I. P. Herman, R. M. Osgood, M. S. Feld and A. Javan, *Appl. Phys. Lett.* **20**, 428 (1972).

[757] D. E. Mann, B. A. Thrush, D. R. Lide, J. J. Ball and N. Acquista, *J. Chem. Phys.* **34**, 420 (1961).

[758] See for example I. Burak, Y. Notev, A. M. Bonnard, A. Szoke, *Chem. Phys. Lett.* **13**, 322 (1972).

[759] T. F. Deutsch, *IEEE J. Quant. Electron.* **QE-7**, 174 (1971).

[760] C. R. Giuliano and L. D. Hess, *J. Appl. Phys.* **40**, 2428 (1969).

[761] J. D. Campbell and J. V. V. Kasper, *Chem. Phys. Lett.* **10**, 436 (1971).

[762] M. A. Pollack, *Appl. Phys. Lett.* **9**, 94 (1966); C. R. Giuliano and L. D. Hess, *J. Appl. Phys.* **38**, 4451 (1967).

[763] J. V. V. Kasper and G. C. Pimentel, *Appl. Phys. Lett.* **5**, 231 (1964); J. V. V. Kasper, J. H. Parker and G. C. Pimentel, *J. Chem. Phys.* **43**, 1827 (1965).

[764] G. Brederlow, R. Brodmann, K. Eidmann, H. Krause, M. Nippus, R. Petsch, R. Volk, S. Witkowski and K. J. Witte, Projektgruppe fur Laserforschung, D-8046 Garching, West Germany, Laboratory Report, PLF5, July (1979); G. Brederlow, K. J. Witte, E. Fill, K. Hohla and R. Volk, *IEEE J. Quant. Electron.* **QE-12**, 152 (1976).

[765] D. R. Gray, H. J. Baker and T. A. King, *J. Phys. D: Appl. Phys.* **10**, 169 (1977).

[766] H. G. Basov, V. V. Gromov, E. L. Koshelev, E. P. Markon and A. N. Oraevskii, *JETP Lett.* (English translation), **10**, 2 (1969).

[767] R. W. F. Gross, *J. Chem. Phys.* **50**, 1889 (1969); T. O. Pockler, M. Shandov and R. E. Walker, *Appl. Phys. Lett.* **20**, 497 (1972); G. K. Vasil'ev, E. F. Makarov, V. G. Papin and V. L. Tal'rose, *Soviet Phys. JETP* (English translation), **34**, 51 (1972), R. S. Chang, R. A. McFarlane and G. J. Wolya, *J. Chem. Phys.* **56**, 667 (1972); R. R. Stephens and T. A. Cool, *J. Chem. Phys.* **56**, 5863 (1972); J. K. Hancock and W. H. Green, *J. Chem. Phys.* **56**, 2474 (1972).

[768] D. J. Spencer, T. A. Jacobs, H. Mirels and R. W. F. Gros, *Int. J. Chem. Kinet.* **1**, 493 (1969); T. A. Cool, R. R. Stephens, and T. J. Falk, *Int. J. Chem. Kinet.* **1**, 495 (1969).

[769] H. Mirels and D. J. Spencer, *IEEE J. Quant. Electron* **QE-7**, 501 (1971).

[770] See for example "Dye Lasers", 2nd Revised Edition (Ed. F. P. Schafer), Vol. 1, Springer-Verlag, Topics in Applied Physics, Berlin, Heidelberg and New York (1977); Visible region dyes are assessed by T. F. Johnston, R. H. Brady and W. Proffitt, *Appl. Opt.* **21**, 2307 (1982).

[771] H. W. Kogelnik E. P. Ippen, A. Dienes and C. V. Shank, *IEEE, J. Quant. Electron.* **QE-8**, 374 (1972).

[772] W. Kranitzky, B. Kopainsky, W. Kaiser, K. H. Drexage and G. A. Reynolds, *Opt. Commun.* **36**, 149 (1981).

[773] L. F. Mollenauer and D. H. Olson, *J. Appl. Phys.* **46**, 3109 (1975).

[774] F. Seitz, *Rev. Mod. Phys.* **18**, 384 (1946); *ibid.* **26**, 7 (1954).

[775] C. Kittel, "Introduction to Solid-State Physics", 2nd Edition, John Wiley, New York (1956); J. H. Schulman and W. Dale Compton, "Colour-Centres in Solids", Pergamon Press, Oxford (1962).

[776] W. B. Fowler, *in* Physics of Colour Centers (Ed. W. B. Fowler), Chapter 2, Academic Press, New York and London (1968).

[777] G. Litfin, R. Beigang and H. Welling, *Appl. Phys. Lett.* **31**, 381 (1977).

[778] J. V. V. Kasper, C. R. Pollock, R. F. Curl and F. K. Tittel, *Appl. Opt.* **21**, 236 (1982).

[779] G. Litfin, C. R. Pollock, J. V. V. Kasper, R. F. Curl and F. K. Tittel, *IEEE J. Quant. Electron.* **QE-16**, 1154 (1980).

[780] W. Koechner, "Solid-State Laser Engineering", Springer Series in Optical Sciences, Vol. 1, Springer-Verlag, Berlin and New York (1976).

[781] D. C. Brown, "High-Peak-Power Nd: Glass Laser Systems", Springer Series in Optical Sciences, Vol. 25, Springer-Verlag, Berlin and New York (1981).

[782] D. C. Brown, in Chapter 8 of Reference [781].

[783] K. H. Drexage and U. T. Muller-Westerhoff, *IEEE J. Quant. Electron.* **QE-8**, 759 (1972).

[784] L. A. Riseberg and M. J. Weber, "Progress in Optics", Vol. XIV, North-Holland, Amsterdam (1976).

[785] Some details are given by Brown in Reference [781].

[786] B. R. Judd, *Phys. Rev.* **127**, 750 (1962); G. S. Ofeld, *J. Chem. Phys.* **37**, 511 (1962).

[787] M. J. Weber, M. Bass, T. E. Varitimos and D. P. Bua, *IEEE J. Quant. Electron.* **QE-9**, 1086 (1973).

[788] L. F. Johnson and H. J. Guggenheim, *J. Appl. Phys.* **38**, 4837 (1967).

[789] L. F. Johnson, R. E. Dietz and H. J. Guggenheim, *Phys. Rev. Lett.* **11**, 318 (1963).

[790] T. B. Reed, R. E. Fahey and P. F. Moulton, *J. Crystal Growth* **42**, 569 (1977); P. F. Moulton, A. Mooradian and T. B. Reed, *Opt. Lett.* **3**, 164 (1978); P. F. Moulton and A. Mooradian, *in* "Laser Spectroscopy" (Eds H. Walther and K. W. Rothe), Vol. 4, p. 584, Springer-Verlag, Berlin and New York (1979); *Appl. Phys. Lett.* **35**, 838 (1979).

[791] H. C. Casey and M. B. Panish, "Heterostructure Lasers", Part A, Fundamental Principles, Academic Press, New York and London (1978); G. H. B. Thompson, "Physics of Semiconductor Laser Devices", John Wiley, Chichester and New York (1980).

[792] C. H. L. Goodman, private communication.

[793] R. J. Nelson, R. B. Wilson, P. D. Wright, P. A. Barnes and N. K. Dutta, *IEEE J. Quant. Electron.* **QE-17**, 202 (1981); P. D. Wright, R. J. Nelson and R. B. Wilson, *ibid.* **QE-18**, 249 (1982).

[794] A. D. Ceruzzi, T. E. Stockton and J. B. McNeeley, Electro-Optical Systems Design, March (1982), pp. 29–35.

[795] P. L. Kelley, article in "Optoacoustic Spectroscopy and Detection" (Ed. Y. H. Pao), p. 113, Academic Press, New York and London (1977).

[796] H. R. Schlossberg and P. L. Kelley, Infrared spectroscopy using tunable lasers, *in* 'Spectrometric Techniques" (Ed. G. Vanasse), Vol. 2, Academic Press, New York and London (1981).

[797] E. D. Hinkley, R. T. Ku, K. W. Nill and J. F. Butler, *Appl. Opt.* **15**, 1653 (1976); commercial suppliers are Laser Analytics in the USA and Telefunken in Germany.

[798] See for example R. S. Eng, P. L. Kelley, A. Mooradian, A. R. Calawa and T. C. Harman, *Chem. Phys. Lett.* **19**, 524 (1973).

[799] C. K. N. Patel and E. D. Shaw, *Phys. Rev. B*, **3**, 1279 (1971); S. D. Smith, R. B. Dennis and R. G. Harrison, *Prog. Quant. Electron* **5**, Part 4 (1977).

[800] J. P. Sattler, B. A. Weber and J. Nemaritch, *Appl. Phys. Lett.* **25**, 491 (1974); *ibid.* **27**, 93 (1975); P. W. Kruse, *Appl. Phys. Lett.* **28**, 90 (1976).

[801] C. K. N. Patel and E. D. Shaw, *Phys. Rev. Lett.* **24**, 451 (1970).

[802] C. K. N. Patel, E. D. Shaw and R. J. Kerl, *Phys. Rev. Lett.* **25**, 8 (1970).

[803] A. Mooradian, S. R. J. Brueck and F. A. Blum, *Appl. Phys. Lett.* **17**, 481 (1970); S. R. J. Brueck and A. Mooradian, *Phys. Rev. Lett.* **28**, 161 (1972).

[804] See for example R. A. Wood, R. B. Dennis and J. W. Smith, *Opt. Commun.* **4**, 383 (1972); R. L. Allwood, R. B. Dennis, W. J. Firth, S. D. Smith, B. S. Wherrett and R. A. Wood, *Radio Electron. Eng.* **42**, 243 (1972); R. A. Wood, A. McNeish, N. L. Brignall and C. R. Pidgeon, *Opt. Commun.* **8**, 248 (1973).

[805] B. Walker, C. C. Bradley, D. G. Moss and G. W. Chantry, *Opt. Commun.* **13**, 235 (1975).

[806] H. A. MacKenzie, S. D. Smith and R. B. Dennis, *Opt. Commun.* **15**, 151 (1975); T. Scragg and S. D. Smith, *Opt. Commun.* **15**, 166 (1975); S. R. J. Brueck and A. Mooradian, *IEEE J. Quant. Electron.* **QE-10**, 634 (1974); B. Walker, G. W. Chantry, D. G. Moss and C. C. Bradley, *J. Phys. D: Appl. Phys.* **9**, 1501 (1976).

[807] B. Walker, D. G. Moss and G. W. Chantry, *Opt. Commun.* **22**, 8 (1977); B. Walker, G. W. Chantry and D. G. Moss, *Indian J. Pure Appl. Phys.*, Raman Centenary Issue, **16**, 417 (1978); S. R. J. Brueck and A. Mooradian, *IEEE J. Quant. Electron.* **QE-12**, 218 (1976).

[808] T. Scragg and S. D. Smith, *Opt. Commun.* **15**, 166 (1975).

[809] This point is discussed by, amongst others, W. J. Firth, B. S. Wherrett and D. Weaire, *IEEE J. Quant. Electron.* **QE-12**, 218 (1976); A. McNeish, R. L. Allwood, D. H. Whiffen and B. S. Wherrett, *Opt. Quant. Electron.* **10**, 495 (1978); A. McNeish and D. H. Whiffen, *Opt. Quant. Electron.* **12**, 303 (1980).

[810] A. D. Buckingham, *J. Phys. Chem.* **86**, 1175 (1982).

[811] The relation between the symmetry of matter tensors and the operation of time reversal is discussed by L. D. Barron and E. Norby Svendsen in the article Antisymmetric light scattering and time reversal, *in* "Advances in Infrared and Raman Spectroscopy" (Eds R. J. H. Clark and R. E. Hester), Vol. 8, Chapter 6, Heyden, London (1981).

[812] A very readable introduction to non-linear optics will be found in the article Optics at Bell Laboratories—Lasers in science by J. A. Giordmaine, *Appl. Opt.* **11**, 2435 (1972).

[813] N. Bloembergen, "Non-linear Optics", Benjamin, New York (1965); Non-linear optics and spectroscopy, Nobel Laureate Lecture, *Science*, **216**, 1057 (1982).

[814] D. A. Kleinman, *Phys. Rev.* **126**, 1977 (1962).

[815] R. L. Byer, Parametric oscillators and nonlinear materials, *in* "Nonlinear Optics" (Eds P. G. Harper and B. S. Wherrett), Chapter 2, Academic Press, London and New York (1977).

[816] P. A. Franken and J. F. Ward, *Rev. Mod. Phys.* **35**, 23 (1963).

[817] A. Yariv, "Quantum Electronics", 2nd Edition, John Wiley, New York (1975).

[818] D. A. Kleinman, A. Ashkin and G. D. Boyd, *Phys. Rev.* **145**, 338 (1966).

[819] J. E. Geusic, H. J. Levinstein, S. Singh, R. G. Smith and L. G. Van Uitert, *Appl. Phys. Lett.* **12**, 306 (1968).

[820] C. G. B. Garrett, *IEEE J. Quant. Electron.* **QE-4**, 70 (1968).

[821] A. C. Albrecht, *J. Chem. Phys.* **34**, 1476 (1961).

[822] R. C. Miller, *Appl. Phys. Lett.* **5**, 17 (1964).

[823] N. Menyuk, K. Dwight and J. W. Pierce, *Appl. Phys. Lett.* **21**, 159 (1972).

[824] D. C. Hanna, B. Luther-Davies, H. N. Rutt, R. C. Smith and C. R. Stanely, *IEEE J. Quant. Electron.* **QE-8**, 317 (1972).

[825] See for example H. Kildal and J. C. Mikkelson, *Opt. Commun.* **10**, 306 (1974).

[826] G. H. Sherman and P. D. Coleman, *IEEE J. Quant. Electron.* **QE-9**, 403 (1973).

[827] See for example R. G. Mellish, R. B. Dennis and R. L. Allwood, *Opt. Commun.* **4**, 249 (1971).

[828] S. R. J. Brueck and H. Kildal, *Opt. Lett.* **2**, 33 (1978).

[829] G. D. Boyd, H. M. Kasper, J. H. McFee and F. G. Storz, *IEEE J. Quant. Electron.* **QE-8**, 900 (1972); H. Kildal and J. C. Mikkelsen, *Opt. Commun.* **9**, 315 (1973).

[830] D. S. Chemla, P. J. Kupeck, D. S. Robertson and R. C. Smith, *Opt. Commun.* **3**, 29 (1971).

[831] R. L. Byer, M. M. Choy, R. L. Herbst, D. S. Chemla and R. S. Feigelson, *Appl. Phys. Lett.* **24**, 65 (1974).

[832] F. Zernike and P. R. Berman, *Phys. Rev. Lett.* **15**, 999 (1965).

[833] T. Yajima and K. Inoue, *Phys. Lett.* **26A**, 281 (1969); *IEEE J. Quant. Electron.* **QE-5**, 140 (1968).

[834] T. Y. Chang, N. Van Tran and C. K. N. Patel, *Appl. Phys. Lett.* **13**, 357 (1968).

[835] N. Van Tran and C. K. N. Patel, *Phys. Rev. Lett.* **22**, 463 (1969).

[836] C. F. Dewey and L. O. Hocker, *Appl. Phys. Lett.* **18**, 58 (1971).

[837] A. S. Pine, *J. Opt. Soc. Am.* **64**, 1683 (1974).

[838] A. S. Pine and A. G. Robiette, *J. Molec. Spectrosc.* **80**, 388 (1980).

[839] D. C. Hanna, R. C. Smith and C. R. Stanley, *Opt. Commun.* **4**, 300 (1971); C. D. Decker and F. K. Tittel, *Opt. Commun.* **8**, 244 (1973).

[840] G. C. Bhar, D. C. Hanna, B. Luther-Davies and R. C. Smith, *Opt. Commun.* **6**, 323

(1972); D. C. Hanna, V. V. Rampal and R. C. Smith, *Opt. Commun.* **8**, 151 (1973).

[841] T. J. Bridges, V. T. Nguyen, E. G. Burkhardt and C. K. N. Patel, *Appl. Phys. Lett.* **27**, 600 (1975).

[842] R. G. Smith, Optical parametric oscillators, *in* "The Laser Handbook" (Eds F. T. Arecchi and E. O. Schulz-Dubois), Vol. 1, Chapter C8, North-Holland, Amsterdam (1972).

[843] H. L. Dai, A. H. Kung and C. B. Moore, *Phys. Rev. Lett.* **43**, 761 (1979).

[844] L. S. Goldberg, *Appl. Phys. Lett.* **17**, 489 (1970); A. I. Izrailenko, A. I. Kovrigin and P. V. Nikles, *JETP Lett.* **12**, 331 (1970); A. J. Campillo and C. L. Tang, *Appl. Phys. Lett.* **12**, 376 (1971).

[845] D. C. Hanna, B. Luther–Davies and R. C. Smith, *Appl. Phys. Lett.* **22**, 440 (1973).

[846] R. L. Herbst and R. L. Byer, *Appl. Phys. Lett.* **21**, 189 (1972); A. A. Davydov, L. A. Kulevskii, A. M. Prokhorov, A. D. Savel'ev, V. V. Smirnov and A. V. Shirkov, *Opt. Commun.* **9**, 234 (1973).

[847] R. G. Wenzel and G. P. Arnold, *Appl. Opt.* **15**, 1322 (1976).

[848] V. J. Corcoran, J. M. Martin and W. T. Smith, *Appl. Phys. Lett.* **22**, 517 (1973).

[849] G. M. Carter, Proceedings SPIE Technical Symposium, East, April (1980).

[850] A. Fenner Milton, *Appl. Opt.* **11**, 2311 (1972).

[851] V. Evtuhov, B. H. Soffer and D. Y. Tseng, *Appl. Opt.* **11**, 2998 (1972); K. F. Hulme and J. Warner, *Appl. Opt.* **11**, 2957 (1972).

[852] J. E. Midwinter and J. Warner, *J. Appl. Phys.* **38**, 519 (1967); J. Warner, *Appl. Phys. Lett.* **12**, 222 (1968); J. Falk and J. M. Yarborough, *Appl. Phys. Lett.* **19**, 68 (1971); A. C. Walker and A. J. Alcock, *Rev. Sci. Instrum.* **47**, 915 (1976); P. A. Jaanimagi, M. C. Richardson and N. R. Isenor, *Opt. Lett.* **4**, 45 (1979).

[853] H. J. Hartmann and A. Laubereau, *Appl. Opt.* **20**, 4259 (1981).

[854] J. Brandmuller and H. Moser, Einfuhrung in die Raman Spektroscopie, Dietrich Steinkopf Verlag (1962).

[855] C. E. Hathaway, Raman Instrumentation and Techniques, *in* "The Raman Effect" (Ed. A. Anderson), Vol. 1, Chapter 4, Marcel Dekker, New York (1971).

[856] J. Howard and T. C. Waddington, Raman microprobe analysis, *in* "Advances in Infrared and Raman Spectroscopy" (Eds R. J. H. Clark and R. E. Hester), Vol. 7, Heyden, London, (1980).

[857] E. J. Woodbury and W. K. Ng, *Proc. Inst. Radio Eng.* **50**, 2367 (1962); G. Eckhardt, R. W. Hellwarth, F. J. McClung, S. E. Schwarz, D. Weiner and E. J. Woodbury, *Phys. Rev. Lett.* **9**, 455 (1962).

[858] P. Lallemand, The stimulated Raman effect, *in* "The Raman Effect" (Ed. A. Anderson), Vol. 1, Chapter 5, Marcel Dekker, New York (1971).

[859] G. W. Chantry, The polarisability theory of the Raman effect, *in* "The Raman Effect" (Ed. A. Anderson), Vol. 1, Chapter 2, Marcel Dekker, New York (1971).

[860] J. Behringer and J. Brandmuller, *Z. Elektrochem.* **60**, 643 (1956); J. Behringer, *Z. Elektrochem.* **62**, 906 (1958); D. G. Rea, *J. Molec. Spectrosc.* **4**, 499 (1960).

[861] R. E. Slusher, C. K. N. Patel and P. A. Fleury, *Phys. Rev. Lett.* **18**, 77 (1967).

[862] Y. Kato and H. Takuma, *J. Chem. Phys.* **54**, 5398 (1971).

[863] G. W. Chantry and R. A. Plane, *J. Chem. Phys.* **33**, 634 (1960).

[864] See for example R. W. Hellwarth, *Phys. Rev.* **130**, 1850 (1963); A. D. Buckingham, *J. Chem. Phys.* **43**, 25 (1965).

[865] J. Gelbswachs, R. H. Pantell, H. E. Puthoff and J. M. Yarborough, *Appl. Phys. Lett.* **14**, 259 (1969).

[866] H. E. Puthoff, R. H. Pantell, B. G. Huth and M. A. Chacon, *J. Appl. Phys.* **39**, 2144 (1968).

[867] J. M. Yarborough, S. S. Sussman, H. E. Puthoff, R. H. Pantell and B. C. Johnson, *Appl. Phys. Lett.* **15**, 102 (1969).

[868] N. Lee, R. L. Aggarwal and B. Lax, *Opt. Commun.* **19**, 401 (1976); see also R. G. Harrison, S. R. Butcher and R. A. Wood, *Opt. Laser Technol.*, June (1979), p. 133.

[869] P. P. Sorokin, J. J. Wynne and J. R. Lankard, *Appl. Phys. Lett.* **22**, 342 (1973).

[870] R. Byer and W. R. Trutna, *Opt. Lett.* **3**, 144 (1978).

[871] J. R. Birch, R. J. Cook, A. F. Harding, R. G. Jones and G. D. Price, *J. Phys. D: Appl. Phys.* **8**, 1353 (1975).

[872] D. E. McCarthy, *Appl. Opt.* **2**, 591, 596 (1963); *ibid.* **4**, 317, 507 (1965); *ibid.* **7**, 1997 (1968).

[873] For a recent study of laser damage to metallic mirrors see S. J. Thomas, R. F. Harrison and J. F. Figuerra, *Appl. Phys. Lett.* **40**, 200 (1982).

[874] E. H. Putley, *Phys. Stat. Solidi* **6**, 571 (1964); *Appl. Opt.* **4**, 649 (1969).

[875] M. A. Kinch and B. V. Rollin, *Br. J. Appl. Phys.* **14**, 672 (1963).

[876] QMC Industrial Instruments Ltd. 229, Mile-End Rd. London E1 4AA.

[877] An equivalent smoothing procedure is described by F. J. J. Clarke in IEE Conference Proceedings No. 103, p. 136 (1973).

[878] See for example R. C. Milward in "Molecular Spectroscopy", p. 81, Institute of Petroleum, London (1969).

[879] R. C. Milward, private communication.

[880] R. C. Milward, Technical Bulletin No. 1, Polytec GMBH, D7501, Reichenbach-Karlsruhe, West Germany.

[881] G. W. Chantry, *J. Molec. Struct.* **45**, 307 (1978).

[882] For example H. Sakai, AFCRL Technical Report 74-0571, p. 224 (1974).

[883] J. E. Chamberlain, G. W. Chantry, F. D. Findlay, H. A. Gebbie, J. E. Gibbs, N. W. B. Stone and A. J. Wright, *Infrared Phys.* **6**, 195 (1966).

[884] R. S. Sternberg and J. F. James, *J. Sci. Instrum.* **41**, 225 (1964).

[885] M. A. Spivack and G. Ferrante, *J. Electrochem. Soc.* **116**, 4 (1969); M. A. Spivack, *Rev. Sci. Instrum.* **41**, 13 (1970); *ibid.* **43**, 7 (1972).

[886] For an example of near-infrared measurements in the area of crop management see P. J. H. Sharpe and H. N. Barber, *Appl. Opt.* **11**, 2902 (1972).

[887] F. Keilmann and K. F. Renk, *Appl. Phys. Lett.* **18**, 452 (1971); for a different approach based on similar ideas see U. Martens and F. K. Kneubuhl, *Appl. Opt.* **13**, 1455 (1974); U. Martens, P. Jeannet and F. K. Kneubuhl, *ibid.* **14**, 1177 (1975).

[888] W. Clark, Photography by Infrared, 2nd Edition, Wiley, New York (1946).

[889] A. Blackburn, M. V. Blackman, D. E. Charlton, W. A. E. Dunn, M. D. Jenner, K. J. Oliver and J. T. M. Wotherspoon, *Infrared Phys.* **22**, 57 (1982).

[890] G. Dodel and W. Kunz, *J. Opt. Soc. Am.* **67**, 975 (1977).

[891] R. Walton, S. Smith, B. Harper and M. Wreathall, *IEEE Trans. Electron. Devices.* **ED-21**, 462 (1974).

[892] R. Kurczewski, *Proc. SPIE* **62**, 207 (1975).

[893] A. Hadni, *J. Phys. E (Sci. Instrum.)*, **14**, 1233 (1981).

[894] D. S. Mountain and J. M. B. Webber, Proceedings of the Fourth European Electro-Optics Conference, Utrecht (1978), SPIE, Vol. 164, p. 189.

[895] K. F. Hulme and J. Warner, *Appl. Opt.* **11**, 2956 (1972).

[896] J. U. White, *J. Opt. Soc. Am.* **32**, 285 (1942); E. R. Stephens, *Infrared Phys.* **1**, 187 (1961).

[897] J. R. Gott, NPL, unpublished work.

[898] H. A. Gebbie, *Phys. Rev.* **107**, 1174 (1957).

[899] J. E. Harries, *J. Opt. Soc. Am.* **67**, 880 (1977).

[900] J. E. Harries, D. G. Moss, N. R. W. Swann, G. F. Neill and P. Gildwarg, *Nature (London)* **259**, 300 (1976).

[901] J. W. Fleming and M. J. Gibson, *J. Molec. Spectrosc.* **62**, 326 (1976).

[902] J. W. Fleming and G. F. Neill, *J. Molec. Spectrosc.* **59**, 493 (1976).

[903] J. W. Fleming, *Infrared Phys.* **18**, 791 (1978).

[904] P. G. Abell, P. Ellis, J. T. Houghton, G. Peckham, C. D. Rodgers, S. D. Smith and E. J. Williamson, *Proc. R. Soc.* **A320**, 35 (1970); F. W. Taylor, J. T.

Houghton, G. Peskett, C. D. Rodgers and E. J. Williamson, *Appl. Opt.* **11**, 135 (1972).

[905] S. D. Smith and C. R. Pidgeon, *Mem. Soc. R. Sci. Liege* **9**, 336 (1964).

[906] J. E. Harries and John Chamberlain, *Appl. Opt.* **15**, 2667 (1976).

[907] For a recent example using aircraft see J. W. Waters, J. J. Gustincic, R. K. Kakar, H. K. Roscoe, P. N. Swanson, T. G. Phillips, T. De Graauw, A. R. Kerr and R. J. Mattauch, *J. Geophys. Res.* **84**, 7034 (1979).

[908] C. K. N. Patel, private communication.

[909] For a readable introduction to millimetre and sub-millimetre astronomy see the article by M. Rowan-Robinson in *New Scientist* 78, 11 July (1974).

[910] G. Neugebauer, D. E. Martz and R. B. Leighton, *Astrophys. J.* **142**, 399 (1965).

[911] J. Rast and F. K. Kneubuhl, *Astron. Astrophys.* **68**, 229 (1978).

[912] For good surveys of the present situation in laser-induced chemical reactions see the following volumes in the Springer series in Chemical Physics: Vol. 3, "Advances in Laser Chemistry", (Ed. A. H. Zewail) (1978); Vol. 6, "Laser-Induced Processes in Molecules" (Eds K. L. Kompa and S. D. Smith) (1979); Vol. 10, "Lasers and Chemical Change" by A. Ben-Shaul, Y. Haas, K. L. Kompa and R. D. Levine, Springer, Berlin and New York (1981).

[913] See for example the paper by Ruffler and Gurs, *Opt. Laser Technol.* December (1972) p. 265.

[914] For example, M. A. Saifi and U. Paek, *Ceramic Bull.* **52**, 838 (1973).

[915] I. Kaplan, *Opt. Laser Technol.* **14**, 41 (1982).

[916] K. Dittrich, Second CIRP Conference, Zurich (1980).

[917] A. E. Costley, Cyclotron radiation from magnetically confined plasmas, Fourth European Physical Society General Conference, Trends in Physics, Institute of Physics Conference Series (1979), Chapter 5, p. 351.

[918] H. A. Gebbie, C. F. Osborne and N. W. B. Stone, *New Scientist*, 17 Oct. (1968), p. 29; H. A. Gebbie, C. F. Osborne, N. W. B. Stone and B. K. Taylor, *The Engineer*, 25 Oct. (1968).

[919] K. D. Froome, *Surv. Rev.* **21**, 28 (1971); *Sci. Prog.* **59**, 199 (1971).

[920] R. L. Barger and J. L. Hall, *Appl. Phys. Lett.* **22**, 196 (1973).

[921] K. M. Baird and G. R. Hanes, Stabilisation of wavelengths from gas lasers, *Rep. Prog. Phys.* **37**, 927 (1974).

[922] P. Cerez and S. J. Bennett, *Appl. Opt.* **18**, 1079 (1979).

[923] F. Spieweck, Conference Proceedings Laser 77 Optoelectronics pp. 130–136 IPC Science and Technology Press (1977).

[924] F. Spieweck, CPEM Conference Braunschweig (1980); *IEEE Trans. Instrum. Meas.* **IM-27**, 398 (1978).

[925] C. C. Bradley and D. J. E. Knight, *Phys. Lett.* **32A**, 59 (1970).

[926] T. G. Blaney, D. J. E. Knight and E. Murray-Lloyd, to be published.

[927] G. Duxbury and W. J. Burroughs, *J. Phys. Soc. B*, **3**, 98 (1970).

[928] L. Frenkel, T. Sullivan, M. A. Pollack and T. J. Bridges, *Appl. Phys. Lett.* **11**, 344 (1967).

[929] L. O. Hocker, J. G. Small and A. Javan, *Phys. Lett.* **29A**, 321 (1969).

[930] K. M. Evenson, J. S. Wells, L. M. Matarrese and L. B. Elwell, *Appl. Phys. Lett.* **16**, 159 (1970).

[931] V. Daneu, L. O. Hocker, A. Javan, D. Ramachandra Rao and A. Szoke, *Phys. Lett.* **29A**, 319 (1969).

[932] D. J. E. Knight, G. J. Edwards, P. R. Pearce and N. R. Cross, *Nature (London)* **285**, 88 (1980).

[933] K. M. Evenson, private communication.

[934] G. J. Edwards, private communication.

[935] J. W. Cooley and J. W. Tukey, *Math. Comput.* **19**, 297 (1965)

[936] J. Connes *in* "Aspen International Conference on Fourier Spectroscopy 1970" (Eds G. A. Vanasse, A. T. Stair Jr and D. J. Baker) AFCRL-71-0019, Special Report No. 114, p. 83 (1971).

[937] R. J. Bell, "Introductory Fourier Transform Spectroscopy", Academic Press, New York and London (1972)

[938] J. W. Fleming, National Physical Laboratory Report No. Mat. App. 27, February (1973).

General Bibliography

History of infrared research

Warren N. Arnquist, A survey of early infrared developments, *Proc. IRE*, **47**, 1420–1430 (1959).

E. Scott Barr, The infrared pioneers, I Sir William Herschel, *Infrared Phys.* **1**, 1 (1961); II Macedonio Melloni, ibid. **2**, 67 (1962); III Samuel Pierpont Langley, ibid. **3**, 195 (1963).

E. Scott Barr, Historical Survey of the Early Development of the Infrared Spectral Region, *Am. J. Phys.* **28**, 42–54 (1960); ibid, **18**, 76 (1965).

E. D. Palik, History of far infrared research I. The Rubens era *J. Opt. Soc. Am.* **67**, 857–865 (1977).

N. Ginsburg, History of far infrared research II. The grating era, 1925–1960, *J. Opt. Soc. Am.* **67**, 865–871 (1977).

L. Genzel and K. Sakai, Interferometry from 1950 to the present, *J. Opt. Soc. Am.* **67**, 871–879 (1977).

G. A. Vanasse, Infrared spectrometry, *Appl. Opt.* **21**, 189 (1982).

E. H. Putley, History of infrared detection, Part 1 *Infrared Phys.* **22**, 127, (1982): Part 2, The first thermographs, ibid. **22**, 189 (1982).

Literature surveys

E. D. Palik, *J. Opt. Soc. Am.* **50**, 1329 (1960).

S. Passman, Current research papers in infrared physics, *Infrared Phys.* **2**, 205 (1962); ibid. **4**, 73 (1964).

D. Bloor, A bibliography on far infrared spectroscopy, *Infrared Phys.* **10**, 55 (1970).

B. Ellis and A. K. Walton, A bibliography on optical modulators, *Infrared Phys.* **11**, 85 (1971).

E. D. Palik, A far infrared bibliography, Appendix VII, pp. 679–759 of "Far Infrared Spectroscopy" by K. D. Moller and W. G. Rothschild, Wiley Interscience, New York (1971).

T. L. Hsu, A Nd: YAG laser bibliography, *Appl. Opt.* **11**, 1287 (1972).

C. N. R. Rao, S. K. Dikshit, S. A. Kudchadker, D. S. Gupta, V. A. Narayan and J. J. Comeford, "Bibliography of Infrared Spectroscopy through 1960", US Department of Commerce, National Bureau of Standards, Special Publication No. 428, Parts 1, 2 and 3 (1976).

J. R. Birch and T. J. Parker, A bibliography on dispersive Fourier transform spectrometry, *Infrared Phys.* **19**, 201–215 (1979).

G. Simonis, Index to the Literature dealing with the Near-mm wave properties of materials, *Int. J. Infrared Millimetre Waves* **3**, 439 (1982).

Standard works

J. A. Stratton, "Electromagnetic Theory", McGraw-Hill, New York (1941).

G. Herzberg, "Infrared and Raman Spectra", Van Nostrand, New York (1945).

N. F. Mott and R. W. Gurney, "Electronic Processes in Ionic Crystals", 2nd Edition, Oxford University Press, Oxford (1948).

W. Gordy, N. V. Smith and R. F. Tramburalo, "Microwave Spectroscopy", Dover Publications, London (1953).

M. Born and K. Huang, "Dynamical Theory of Crystal Lattices", Oxford University Press, New York (1954).

C. H. Townes and A. L. Schawlow, "Microwave Spectroscopy", McGraw Hill, New York (1955).

E. B. Wilson, J. C. Decius and P. C. Cross, "Molecular Vibrations: The Theory of Infrared and Raman Vibrational Spectra", McGraw-Hill, New York (1955).

L. J. Bellamy, "The Infrared Spectra of Complex Molecules", Methuen, London (1958).

M. Born and E. Wolf, "Principles of Optics", Pergamon Press Oxford (1959).

H. L. Hackforth, "Infrared Radiation", McGraw-Hill, New York (1960).

P. W. Kruse *et al.* "Elements of Infrared Technology", John Wiley, New York (1962).

W. Summer, "Ultraviolet and Infrared Engineering", Pitman, London (1962).

W. Brugel "An Introduction to Infrared Spectroscopy", Methuen, London, John Wiley, New York (1962).

M. R. Holter, S. Nudelman, G. H. Suits, W. L. Wolfe and G. J. Zissis, "Fundamentals of Infrared Technology", Macmillan, New York and London (1962).

H. G. Kuhn, "Atomic Spectra", Longmans, London (1962).

J. A. Jamieson, R. H. McFee, G. N. Plass, R. H. Grube and R. G. Richards, "Infrared Physics and Engineering", McGraw-Hill, New York (1963).

W. J. Potts, "Chemical Infrared Spectroscopy", Vol. 1, John Wiley, New York (1963).

H. C. Allen and P. C. Cross, "Molecular Vib-Rotors: The Theory and

Interpretation of High-Resolution Infrared Spectra", John Wiley, New York (1963).

N. Bloembergen, "Non-Linear Optics", Benjamin, New York (1965).

J. T. Houghton and S. D. Smith, "Infrared Physics", Oxford University Press, Oxford (1966).

A. E. Martin, "Infrared Instrumentation and Techniques", Elsevier, Amsterdam (1966).

I. Simon, "Infrared Radiation", Van Nostrand, New York (1966).

A. Hadni, "Essentials of Modern Physics Applied to the Study of the Infrared", International Series of Monographs in Infrared Science and Technology, Vol. 2, Pergamon Press, Oxford (1967).

D. H. Martin (Ed.), "Spectroscopic Techniques for Far Infrared and mm-Waves", North Holland, Amsterdam (1967).

N. J. Harrick, "Internal Reflection Spectroscopy", John Wiley, London and New York (1967).

R. A. Smith, F. E. Jones and R. P. Chasmar, "The Detection and Measurement of Infrared Radiation", Revised 2nd Edition, Oxford University Press, Oxford (1968).

A. Vasko, 'Infrared Radiation", Translation of the original Czech Edition of (1963), Iliffe Books, London (1968).

M. A. Bramson, "Infrared Radiation: A Handbook of Applications", Plenum Press, New York (1968).

R. D. Hudson, "Infrared System Engineering", John Wiley, New York (1969).

J. F. James and R. S. Sternberg, "Design of Optical Spectrometers", Chapman and Hall, London (1969).

F. A. Benson (Ed.), "Millimetre and Sub-Millimetre Waves", Iliffe, London (1969).

D. Lorrain and D. Corson, Electromagnetic Fields and Waves, 2nd Edition, Freeman, San Francisco (1970).

J. E. Stewart, "Infrared Spectroscopy", Marcel Dekker, New York (1970).

M. F. Kimmitt, "Far Infrared Techniques", Pion Press, London (1970).

A. F. Harvey, "Coherent Light", Wiley-Interscience, London and New York (1970).

A. Finch, P. N. Gates, K. Radcliffe, F. N. Dickson and F. F. Bentley, "Chemical Applications of Far Infrared Spectroscopy", Academic Press, London and New York (1970).

A. K. Willardson and A. C. Beer, "Semiconductors and Semimetals", Vol. 5, "Infrared Detectors 1", Academic Press, New York and London (1970).

G. W. Chantry, "Submillimetre Spectroscopy", Academic Press, London and New York, (1971).

K. D. Moller and W. G. Rothschild, "Far Infrared Spectroscopy", John Wiley, New York (1971).

R. J. Bell, "Introductory Fourier Transform Spectroscopy", Academic Press, New York and London (1972).

R. G. J. Miller and B. C. Stace (Eds), "Laboratory Methods in Infrared Spectroscopy", Heyden, London and New York (1972).

G. Turrell, "Infrared and Raman Spectra of Crystals", Academic Press, London and New York (1972).

L. Mandel and E. Wolf, "Coherence and Quantum Optics", Plenum Press, New York and London (1973).

L. C. Robinson, "Physical Principles of Far-Infrared Radiation", Vol. 10 of "Methods of Experimental Physics" (Ed. L. Marton), Academic Press New York and London, (1973).

F. Bitter and H. A. Medicus, "Fields and Particles, An Introduction to Electromagnetic Wave Phenomena and Quantum Physics", Elsevier, Amsterdam (1973).

L. M. Sverdlov, M. A. Kovner and E. P. Krainov, "Vibrational Spectra of Polyatomic Molecules", John Wiley, New York (1974).

A. Yariv, "Quantum Electronics", 2nd Edition, John Wiley, New York (1975).

L. Lewin, "The Theory of Waveguides", Newnes-Butterworth, London (1975).

W. W. Duley, "CO_2 Lasers, Effects and Applications", Academic Press, New York and London (1976).

A. Yariv, Introduction to Optical Electronics, 2nd Edition, Holt, Rinehart and Winston, New York (1976).

A. Blanchard, "Phase-Locked Loops: Applications to Coherent Receiver Design", John Wiley, New York (1976).

A. K. Willardson and A. C. Beer, "Semiconductors and Semimetals", Vol. 11, "Infrared Detectors 2", Academic Press, New York and London (1977).

W. L. Wolfe and G. J. Zissis, "The Infrared Handbook", Infrared Information and Analysis (IRIA) Center, Environmental Research Institute of the University of Michigan (1978).

J. L. Steinfeld, "Laser and Coherence Spectroscopy", Plenum Press, New York (1978).

R. H. Kingston, "Detection of Optical and Infrared Radiation", Springer-Verlag, Berlin (1978).

J. Chamberlain, "The Principles of Interferometric Spectroscopy" (completed, revised and edited by G. W. Chantry and N. W. B. Stone) John Wiley, Chichester (1979).

G. W. Chantry (Ed.), "Modern Aspects of Microwave Spectroscopy", Academic Press, London and New York (1979).

F. M. Gardner, "Phase-Lock Techniques", 2nd Edition, John Wiley, New York (1979).

F. H. Read, "Electromagnetic Radiation", John Wiley, Chichester (1980).

C. M. Bowden, M. Ciftan and H. R. Robl (Eds), Optical Bistability, Plenum Press, New York and London (1981).

M. J. Weber (Ed.), CRC Handbook Series of Laser Science and Technology Vol. 1, Lasers and Masers; Vol. 2, Gas Lasers, CRC Press Inc. Boca Raton Florida (1981).

W. Demtruder, "Laser Spectroscopy: Basic Concepts and Instrumentation", Springer-Verlag, Berlin (1981).

P. M. Narendra (Ed.), "Infrared Technology for Target Detection and Classification", Vol. 302, SPIE, (1981).

I. J. Spiro (Ed.), "Modern Utilisation of Infrared Technology VII", Vol. 304, SPIE (1982).

Compilations of infrared standards

"Tables of Wavenumbers for the Calibration of Infrared Spectrometers", IUPAC, Butterworths, London (1961).

K. N. Rao et al.,. "Wavelength Standards in the Infrared", Academic Press, New York and London (1966).

J. S. Wells, F. R. Peterson and A. G. Maki, Appl. Opt. 18, 3567 (1979).

D. J. E. Knight, Ordered list of far-infrared laser lines (continuous, $\lambda > 12\,\mu$m), NPL Report Qu-45, (1st Revision), February (1981).

F. M. Zweibaum and H. Kaplan (Eds), "Contemporary Infrared Standards and Calibration", Vol. 308, SPIE (1982).

Subject Index